Remote Sensing
for Sustainability

Taylor & Francis Series in

Remote Sensing Applications

Series Editor

Qihao Weng

Indiana State University
Terre Haute, Indiana, U.S.A.

Remote Sensing for Sustainability

Edited by **Qihao Weng**

CRC Press
Taylor & Francis Group
Boca Raton London New York

CRC Press is an imprint of the
Taylor & Francis Group, an **informa** business

CRC Press
Taylor & Francis Group
6000 Broken Sound Parkway NW, Suite 300
Boca Raton, FL 33487-2742

First issued in paperback 2019

ISBN-13: 978-0-4987-0071-9 (hbk)
ISBN-13: 978-0-367-87140-6 (pbk)

Library of Congress Cataloging-in-Publication Data

Names: Weng, Qihao, editor.
Title: Remote sensing for sustainability / [edited by] Qihao Weng.
Description: Boca Raton : Taylor & Francis, a CRC title, part of the Taylor & Francis imprint, a member of the Taylor & Francis Group, the academic division of T&F Informa, plc, [2017] | Includes bibliographical references and index.
Identifiers: LCCN 2016023093 | ISBN 9781498700719 (hardback : acid-free paper)
Subjects: LCSH: Environmental monitoring--Remote sensing. | Sustainable development--Remote sensing. | Sustainable urban development--Remote sensing. | Sustainable agriculture--Remote sensing. | Natural resources--Remote sensing. | Renewable energy sources--Remote sensing.
Classification: LCC GE45.R44 R47 2017 | DDC 363.7/063028--dc23
LC record available at https://lccn.loc.gov/2016023093

Visit the Taylor & Francis Web site at
http://www.taylorandfrancis.com

and the CRC Press Web site at
http://www.crcpress.com

Contents

SECTION I   Remote Sensing for Sustainable Cities

SECTION II   Remote Sensing for Sustainable Natural Resources

SECTION III   Remote Sensing for Sustainable Environmental Systems

SECTION IV   Remote Sensing for Sustainable Energy

SECTION IV • Remote Sensing for Sustainable Energy

Foreword

Dr. Qihao Weng has authored another outstanding remote sensing book entitled *Remote Sensing for Sustainability* (CRC Press, 2016). Well known for his expertise in remote sensing and GIS analysis of urban and environmental systems, this professor from Indiana State University, director of the Center for Urban and Environmental Change, series editor for the Taylor & Francis Series in *Remote Sensing Applications*, and co-editor-in-chief of the *ISPRS Journal of Photogrammetry and Remote Sensing* also is involved in the Group on Earth Observations or GEO. Established in 2005, GEO is a partnership of governments and organizations with a vision toward a future when global decisions benefiting humankind are informed by Earth observations and information derived from coordinated and sustained remote sensing (http://www.earthobservations.org). Professor Weng's work as the task lead and coordinator for GEO SB-04, Global Urban Observation and Information Task, led him to the idea to write a book on the use of remote sensing for multiple and sustainable societal benefits, including biodiversity, disaster resilience, energy and mineral resources management, food security, sustainable agriculture, infrastructure and transportation management, public health surveillance, sustainable urban development, and water resources management. In other words, his book covers remote sensing of all the requirements to sustain life on Earth and ensure environments that promote human health, productivity, and security.

In this book, *Remote Sensing for Sustainability*, Weng recognizes that grand and global challenges require international collaboration and cooperation in order to optimize remote sensing data acquisition, storage, access, and dissemination. He understands the value of sharing methods and techniques of image processing and data analysis for producing accurate and actionable knowledge from images, point clouds, sensor networks, ground measurements, and modeled predictions. Experts from around the world were gathered to report on specialties of remote sensing for sustainable cities, natural resources, environmental systems, and energy. Chapters within these major topic areas cover critical issues of urban planning and growth, endangered species conservation, water and air quality, forest damage by diseases and insects, energy supplies, and sustainable wind and solar energy potential—all using geospatial data, techniques, and analyses. As Barbara Ryan, GEO Secretariat Executive Director, notes in the Preface of this book, there is a critical need to understand the character of urban environments where approximately half of the human global population live in only about 3% of the Earth surface. Rapidly expanding, dense, sprawling, and vulnerable to human and natural impacts, urban systems must be understood in order to make timely and informed decisions aimed toward minimizing deforestation, air and water pollution, energy consumption, food deserts, marginal housing, and limited green spaces. This book makes a giant step forward

in advancing global cooperation in remote sensing efforts and sharing knowledge
desperately needed for Earth sustainability in a changing climate.

Marguerite Madden
Center for Geospatial Research Department of Geography
The University of Georgia
Athens, GA

Preface

TOWARD A SUSTAINABLE EARTH THROUGH REMOTE SENSING

SUSTAINABILITY AND EARTH OBSERVATION

In recent years, we have witnessed an emerging trend worldwide to pursue green technologies and low-carbon economies and lifestyles. However, the sustainability issue is not new. At the 1992 United Nations Conference on Environment and Development (UNCED) in Rio, UNCED Principle Three characterized sustainable development as "the right to development must be fulfilled so as to equitably meet developmental and environmental needs of present and future generations," while Principle Four stated that "in order to achieve sustainable development, environmental protection shall constitute an integral part of the development process and cannot be isolated from it." These principles were reaffirmed at the 2002 Johannesburg World Summit on Sustainable Development and have since produced a profound implication for use and stewardship of natural resources, ecology, and environment (Weng and Yang 2003). In practice, sustainable development is a multifaceted concept and has been viewed from many perspectives, depending on one's personal experience, viewpoint, and discipline (Weng and Yang 2003). As we consider sustainable development from an ecological perspective, we must have an appreciation of ecosystem integrity, which implies the existence of the system structure and function, maintenance of system components, interactions among them, and the resultant dynamic of the ecosystem (Campbell and Heck 1997). Shaller (1990) suggested that "sustainable agriculture over the long-term enhances environmental quality and the resource base upon which agriculture depends, provides for basic human food and fiber needs, is economically viable, and enhances the quality of life for farmers and society as a whole." This definition illustrates well three pillars of sustainable development: ecological, economic, and social objectives (Weng and Yang 2003). The ecological objective seeks to preserve the integrity of the ecosystem, while the economic objective attempts to maximize human welfare within the existing capital stock and technologies and use economic units (i.e., money or perceived value) as a measurement standard (Campbell and Heck 1997). The social objective stresses the needs and desires of people and uses standards of well-being and social empowerment. The ecological perspective for sustainable development balances the ecological, economic, and societal values, and falls at the intersection of the spheres that represent the three components (Weng and Yang 2003). Incoordination among the three components will likely result in failure to achieve sustainability (Zonneveld and Forman 1990).

The spatial and temporal scales are key elements in assessing ecological and environmental sustainability, because they module the objectives of what we wish to sustain, over what time scale and in which geographical scope (Weng 2014). With more than one-half of the world population living in cities, the 21st century has become the first *urban century*. Cities are human-central ecosystems and are the most complex of all human settlements. While the economic and societal values are stressed, the ecological value is often ignored, and therefore, the ecological perspective of sustainable development asserts the importance of the coordination and balance among the three objectives for sustainability of cities (Weng and Yang 2003). Because of the nature of cities as dynamic, complex settlements and human-central ecosystems, the development of evaluation methods for sustainable cities should reflect these characteristics (Weng and Yang 2003). The US Climate Change Science Program (CCSP 2008) defines one of its five goals to be "understanding the sensitivity and adaptability of different natural and managed ecosystems and human systems to climate and related global changes." Although vulnerabilities of settlements to impacts of climate change vary regionally, they generally include some or many of the following impact concerns: health, water and infrastructures, severe weather events, energy requirements, urban metabolism, sea level rise, economic competitiveness, opportunities and risks, and social and political structures. CCSP (2008) further recommends that research on climate change effects on human settlements in the United States be given a much higher priority in order to provide better metropolitan-area scale decision-making.

Earth observation technology, in conjunction with in situ data collection, has been used to observe, monitor, measure, and model many of the components that comprise natural and human ecosystems cycles (Weng 2012a). Driven by the societal needs and improvement in sensor technology and image processing techniques, we have witnessed a great increase in research and development, technology transfer, and engineering activities worldwide since the turn of the 21st century. Commercial satellites acquire imagery at spatial resolutions previously only possible to aerial platforms, but these satellites have advantages over aerial imageries including their capacity for synoptic coverage, inherently digital format, short revisit time, and capability to produce stereo image pairs conveniently for high-accuracy 3D mapping thanks to their flexible pointing mechanism (Weng 2012b). Hyperspectral imaging affords the potential for detailed identification of materials and better estimates of their abundance in the Earth's surface, enabling the use of remote sensing data collection to replace data collection that was formerly limited to laboratory testing or expensive field surveys (Weng 2012b). While LiDAR technology provides high-accuracy height and other geometric information for urban structures and vegetation, radar technology has been re-invented since the 1990s mainly because of the increase of spaceborne radar programs (Weng 2012b). These technologies are not isolated at all. In fact, their integrated uses with more established aerial photography and multispectral remote sensing techniques have been the main stream of current remote sensing research and applications (Weng 2012b).

With these recent advances, techniques of and data sets from remote sensing and Earth observation have become an essential tool for understanding the Earth, monitoring the world's natural resources and environments, managing exposures to natural and man-made risks and disasters, and helping the sustainability and productivity of natural and human ecosystems (Weng 2012b). This book aims at introducing to the current state of remote sensing knowledge needed for sustainable development and management with selected studies. These studies either explore the methods and techniques of remote sensing for application to various aspects of sustainable development or provide important insights into sustainability science from the perspective of remote sensing technology. Therefore, this book would be of great value for students, professors, and researchers in both remote sensing and sustainability science. It can help narrow the gap between the two disciplines. This book may also serve as a textbook for undergraduate and graduate students or as an important supplement for those majoring in sustainability, remote sensing, geography, geosciences, planning, environmental science and engineering, civic engineering, resources science, land use, energy, and geographic information system (GIS). On college campuses, we have recently witnessed an ever-increasing number of classes and programs on sustainability or related subjects. Remote sensing is emerging as an essential geospatial tool in sustainability; this book would meet the need of those classes and programs. In addition, this book may be used as a reference book for sustainability officers, practitioners, and professionals alike in the government, commercial, and industrial sectors. Since its contents cover numerous applications of remote sensing to sustainability, this book, indeed, provides a useful toolbox.

INTERNATIONAL COOPERATION AND COLLABORATION ON EARTH OBSERVATION FOR A SUSTAINABLE EARTH

The 2002 World Summit on Sustainable Development in Johannesburg highlighted the urgent need for coordinated observations relating to the state of the Earth. The First Earth Observation Summit in Washington, DC, in 2003 adopted a declaration to establish the ad hoc intergovernmental Group on Earth Observations (ad hoc GEO) to draft a 10-Year Implementation Plan. Since 2003, GEO has been working to strengthen the cooperation and coordination among global observing systems and research programs for integrated global observations. The GEO process has outlined a framework document calling for Global Earth Observation System of Systems (GEOSS) and has defined nine areas of societal benefits (Group on Earth Observation 2008), including the following:

- Reducing loss of life and property from natural and human-induced disasters,
- Understanding environmental factors affecting human health and well-being,
- Improving the management of energy resources,
- Understanding, assessing, predicting, mitigating, and adapting to climate variability and change,
- Improving water resource management through better understanding of the water cycle,
- Improving weather information, forecasting and warning,
- Improving the management and protection of terrestrial, coastal and marine ecosystems,
- Supporting sustainable agriculture and combating desertification, and
- Understanding, monitoring and conserving biodiversity.

On September 25, 2015, the United Nations adopted a set of sustainable development goals (SDGs), each of which has specific targets to be achieved over the next

15 years (United Nations Development Programme 2015). These goals represent the United Nation's responses to numerous societal challenges and the efforts to build a sustainable Earth. Through large-scale, repetitive acquisition of the Earth surface image data, remote sensing can provide essential information and knowledge to supplement statistical analyses in the assessment of indicators toward the attainment of the SDGs. Because Earth observation offers an indispensable tool to measure and monitor the progress toward SDGs, in the recently developed "GEO Strategic Plan 2016–2025: Implementing GEOSS," GEO is determined to develop a concerted direction with the SDGs (Group on Earth Observations 2015). The GEO Global Urban Observation and Information Initiative (GI-17) has set the following goals for the period 2012–2015: (1) improving the coordination of urban observations, monitoring, forecasting, and assessment initiatives worldwide; (2) supporting the development of a global urban observation and analysis system; (3) producing up-to-date information on the status and development of the urban system—from a local to a global scale; (4) filling existing gaps in the integration of global urban observation with data that characterize urban ecosystems, environment, air quality and carbon emission, indicators of population density, environmental quality, quality of life, and the patterns of human environmental and infectious diseases; and (5) developing innovative techniques in support of effective and sustainable urban development. These goals will be implemented by developing and expanding selected activities and programs that the GEO Global Urban Observation (SB-04) task team has been working (Weng et al. 2014). This book intends to contribute to the GEO's Strategic Plan by addressing and exemplifying a number of societal benefit areas of remote sensing data sets, methods, and techniques for sustainable development.

Synopsis of the Book

This book consists of four sections. Section I deals with remote sensing for sustainable cities; Section II discusses remote sensing techniques and methods for forest resources; Section III presents remote sensing studies for sustainable environmental systems, with topics ranging from air, water, to land; and Section IV includes various contributions in remote sensing of sustainable energy systems.

Section I includes five chapters dealing with theories and methods as well as practical applications of sustainable development for cities using remote sensing. Chapter 1 provides key concepts and principles for assessing sustainability of cities. It then discusses the typical parameters derivable from remotely sensing imagery that can be used to define indicators for sustainable cities. Chapter 2 introduces a selection of applications and data products that provide support for day-to-day decision-making activities in urban and regional planning. Chapters 3 through 5 present various studies in which remotely sensed data, methods, and techniques are used for studying cities or urban clusters. In Chapter 3, Zhang and Weng monitor urban growth process in the Pearl River Delta, Guangdong Province, China, by using time series Landsat imagery from 1987 to 2014. This chapter demonstrates the effectiveness of time series data mining for assessing urban growth pattern over a long period and the usefulness of generated data set and information for exploring the relationship between urban growth and environmental sustainability. In Chapter 4, the damage of subsidence on urban structures in the St. Louis Metropolitan area, Missouri, is explored by using Synthetic Aperture Radar images in the period from 1992 to 2011. Their results show hot spots of ongoing and potential land collapses in the region, which are valuable not only to individual homeowners but also to city planners, insurance companies, and regional policymakers charged with assessing risks of abandoned coal mines. In the last chapter in the section, Chapter 5, Xiao and Weng examine urban growth in the North Carolina urban crescent from the Research Triangle Park (Raleigh–Durham–Chapel Hill) to Greensboro, and Guiyang City to Anshun City in Guizhou Province, China, from the 1980s to the 2010s, when accelerated industrialization and urbanization occurred. They compare and contrast the spatial patterns, paths, and driving forces of urbanization in the two regions and countries of different socioeconomic development stages.

Section II focuses on remote sensing methods and techniques for sustainable natural resources. In Chapter 6, Tong et al. present a few case studies on the application of remote sensing in grassland management for a mixed-grass prairie ecosystem in North America; on the basis of the case studies, they further discuss challenges and opportunities for remote sensing in grassland management. Chapter 7 assesses the relationship between species extinction and biodiversity loss using the example of palila (*Loxioides bailleui*), an endangered bird species on the island of Hawaii. To understand its population trend, tree species in its habitat were identified analyzed with high spatial resolution satellite imagery at both pixel and object levels. Chapter 8 takes this direction further along biodiversity and conservation by examining how evolved remote sensing techniques can be employed to investigate forest damages by diseases and insects. The last chapter in Section II, Chapter 9, focuses on water resources. Matsushita and his colleagues first survey major satellite sensors for water quality studies, followed by a discussion of representative algorithms for water area delineation, atmospheric correction, and water quality parameter estimation. This chapter ends with a proposed framework for water quality assessment using remote sensing.

What remote sensing methods and techniques can do for the sustainability of environmental systems and how they do it are addressed in Section III. In Chapter 10, Hu explores the use of satellite remote sensing data to expand ground network of PM2.5 observation by using aerosol optical depth (AOD). Various AOD products and methods that are widely used in PM2.5 concentration estimation were assessed. This discussion was followed by a case study in the Atlanta metropolitan area, Georgia, to estimate ground-level PM2.5 concentration from MODIS AOD. Chapter 11 continues the discussion on public health but from the perspective of heat hazards. Jiang and Weng develop methods to analyze daily and hourly variations of land surface temperature (LST), which were derived from remotely sensed thermal infrared image data, and discuss the impact of evapotranspiration on LST over a variety of urban surfaces. Both Chapters 12 and 13 study ecosystems in semi-arid and arid regions but with distinct approaches of remote sensing. The former investigates the tendency of desertification along the Mediterranean to arid transition zones in central Israel, and the latter assesses soil moisture condition in

the Umer Kot region of Pakistan. In Chapter 12, Shoshany develops three indicators using remote sensing data, that is, Green Vegetation Cover, life-forms' subpixel compositions, and spatial recovery versus erosion potential, and by assessing
these indicators, he suggests that no significant shift occurred in the transition zone
between 1996 and 2011, but that there was high vulnerability to future degradation. Soil moisture holds the key to drought and agricultural monitoring in semi-
arid and arid regions. Chapter 13 investigates the potential of near-infrared and red
reflectance space (i.e., the RSSMM method) and temperature vegetation dryness
index (TVDI) for the assessment of soil moisture. Results of the remote sensing
methods were compared with in situ soil moisture measurements at different land
surface depths of 0–15, 15–30, and 30–45 cm. RSSMM was found satisfactory in
determining soil moisture, but TVDI provided a more reliable estimate of moisture
condition.

The last section of this book, Section IV, examines the issues of energy use and
sustainable energy sources using remote sensing technology. The development of
renewable energies is one of the key challenges in the 21st century. In this context, it
is of central importance to focus on the review of surface potentials, the determination of suitable sites, the consideration of user interests, and the detection of trends
and impacts on the landscape. To meet these requirements and tasks, timely, spatial,
and thematic high-quality geospatial data are indispensable. In Chapter 14, Esch
et al. exemplify applications of remote sensing and related technologies and geoinformation products for land management in Germany, which is geared toward the
development of potentials of renewable energies. Chapter 15 assesses the capability
of DMSP-OLS nighttime light imagery for analyzing the decadal trends of energy
consumption (EC) in China, from 2000 to 2012. Here, Xie and Weng demonstrate
a moderate to rapid growth of EC for coastal and capital cities, but a slow growth
for the majority of central, northeastern, and western cities. They further find the
total and urban EC at the prefectural level to be regionally clustered, which may
have an implication in future Chinese energy policy and the spatial distribution of
EC. The last two chapters examine renewable energy sources. While Chapter 16
focuses on wind energy, Chapter 17 evaluates solar energy potential. Wind speed
and wind flow are strongly influenced by land surface properties. Three different
remote sensing–based parameters can help characterize wind resources: (1) land
cover and land use, (2) digital elevation models, and (3) phenological information.
In Chapter 16, Esch et al. discuss how Earth Observation (EO) data can be used
to support wind resource modeling, especially the possibilities brought about by
the Copernicus Sentinel satellites. They conclude that by using EO-based information on the surface (e.g., roughness) and in situ wind measurements, realistic
wind fields for sufficiently large areas can be derived by considering shadowing
effects and wind shear as well. Chapter 17 provides two case studies to demonstrate
the applicability of remote sensing techniques on sustainable energy development
in Indianapolis, Indiana. The first case study demonstrates a method to estimate
the solar energy potential of building roofs, and the second case study examines
the correlation between energy use and building morphological attributes such as
ground area, total floor area, height, surface area, compactness, aspect ratio, and
orientation.

REFERENCES

Group on Earth Observation, 2008. GEO Information Kit. http://www.earthobservations.org
    /documents/geo_information_kit.pdf.
Group on Earth Observations, 2015. The GEO 2016–2025 Strategic Plan: Implementing
    GEOSS, https://www.earthobservations.org/geoss_wp.php, last accessed March 5,
    2016.
United Nations Development Programme, 2015. Sustainable Development Goals (SDGs),
    http://www.undp.org/content/undp/en/home/sdgoverview/post-2015-development
    -agenda.html, last accessed March 5, 2016.
United Nations Environment Programme, 1992. Rio Declaration on Environment and
    Development. United Nations publication, Sales No. E.73.II.A.14 and corrigendum,
    Chap. I. http://www.unep.org/documents.multilingual/default.asp?documentid=78&art
    icleid=1163.
U.S. Climate Change Science Program (CCSP), 2008. Analyses of the Effects of Global
    Change on Human Health, Welfare, and Human Systems. A Report by the U.S. Climate
    Change Science Program and the Subcommittee on Global Change Research. [Gamble,
    J.L. (ed.), K.L. Ebi, F.G. Sussman, T.J. Wilbanks]. U.S. Environmental Protection
    Agency, Washington, DC.
Weng, Q. 2012a. Remote sensing of impervious surfaces in the urban areas: Requirements,
    methods, and trends. *Remote Sensing of Environment,* 117(2), 34–49.
Weng, Q. 2012b. *An Introduction to Contemporary Remote Sensing.* New York: McGraw-Hill
    Professional, 320 pp.
Weng, Q. 2014. *Scale Issues in Remote Sensing.* Hoboken, NJ: John Wiley & Sons, Inc.,
    352 pp.
Weng, Q., Esch, T., Gamba, P., Quattrochi, D.A. and Xian, G. 2014. Global urban observation
    and information: GEO's effort to address the impacts of human settlements. In Weng,
    Q. (ed.). *Global Urban Monitoring and Assessment through Earth Observation.* Boca
    Raton, FL: CRC Press/Taylor & Francis, Chapter 2, pp. 15–34.

Acknowledgments

I thank all the contributors for making this endeavor possible. Furthermore, I offer my deepest appreciation to all the reviewers, who have taken precious time from their busy schedules to review the chapters submitted to this book. Finally, I am indebted to my family for their enduring love and support. It is my hope that the publication of this book will provide inspiration to students, researchers, and practitioners to conduct more in-depth studies on remote sensing for sustainability. It becomes apparent that the realization of many societal benefits of Earth Observation technology and sustaining our Earth requires international collaboration and cooperation.

The reviewers of the chapters for this book are listed below in alphabetical order: Wanda De Keersmaecker, Chengbin Deng, Pinliang Dong, Paolo Gamba, Feng Gao, Anatoly Gitelson, Peter Hofmann, Husi Letu, Yangfan Li, Lu Liang, Dengsheng Lu, Ryo Michishita, Zina Mitraka, Hiroyuki Miyazaki, Daniel Odermatt, Dale A. Quattrochi, Conghe Song, Hannes Taubenboeck, Cuizhen Wang, Guangxing Wang, Charles Wong, Changshan Wu, George Xian, Klemen Zaksek, Hui Zeng, Caiyun Zhang, Qingling Zhang, Xiaoyang Zhang, Wei Zhuang, and Bin Zou.

Qihao Weng

Editor

Dr. Qihao Weng is the director of the Center for Urban and Environmental Change and a professor of geography at Indiana State University. In addition, he serves as editor-in-chief of *ISPRS Journal of Photogrammetry and Remote Sensing* and is the series editor for Taylor & Francis Series in *Remote Sensing Applications*. He received his PhD degree in geography from the University of Georgia in 1999. In the same year, he joined the University of Alabama as an assistant professor. Since 2001, he has been a member of the faculty in the Department of Earth and Environmental Systems at Indiana State University, where he has taught courses on remote sensing, digital image processing, remote sensing–GIS integration, GIS, and environmental modeling. He has mentored 15 doctoral and 13 master students.

Dr. Weng's research focuses on remote sensing and GIS analysis of urban ecological and environmental systems, land-use and land-cover change, environmental modeling, urbanization impacts and sustainability, and human–environment interactions. He is the author of more than 180 peer-reviewed journal articles and other publications and 9 books. Dr. Weng has worked extensively with optical and thermal infrared remote sensing data, and more recently with LiDAR data, primarily for urban heat island study, land-cover and impervious surface mapping, urban growth detection, image analysis algorithms, and the integration with socioeconomic characteristics, with financial support from US funding agencies that include NSF, NASA, USGS, USAID, NOAA, the National Geographic Society, and Indiana Department of Natural Resources. From the American Association of Geographers (AAG), Dr. Weng has received the Outstanding Contributions Award in Remote Sensing in 2011 and the Willard and Ruby S. Miller Award in 2015 for his outstanding contributions to geography. In May 2008, he received a prestigious NASA senior fellowship. Dr. Weng is also the recipient of the Robert E. Altenhofen Memorial Scholarship Award by the American Society for Photogrammetry and Remote Sensing (1999), the Best Student-Authored Paper Award by the International Geographic Information Foundation (1998), and the 2010 Erdas Award for Best Scientific Paper in Remote Sensing by ASPRS (first place). At Indiana State, he received the Theodore Dreiser Distinguished Research Award in 2006 and was selected as a Lilly Foundation Faculty Fellow in 2005. Dr. Weng has given more than 90 invited talks (including keynote addresses, public speeches, colloquia, and seminars) and has presented more than 100 papers at professional conferences (including co-presenting).

Dr. Weng is currently the coordinator for GEO's GI-17, Global Urban Observation and Information Task (2016–2019). He has been the organizer and program committee chair of the biennial IEEE/ISPRS/GEO-sponsored International Workshop on Earth Observation and Remote Sensing Applications conference series since 2008. He previously held the following positions: national director of the American Society for Photogrammetry and Remote Sensing (2007–2010), chair of AAG China Geography Specialty Group (2010–2011), secretary of the ISPRS Working Group VIII/1 (2004–2008), and panel member of US DOE's Cool Roofs Roadmap and Strategy in 2010.

Contributors

Iftikhar Ali
Vienna University of Technology
(Tu Wien)
Department of Geodesy and
Geoinformation (GEO)
Vienna, Austria

Gang Chen
Department of Geography and Earth
Sciences
and
Laboratory for Remote Sensing and
Environmental Change (LRSEC)
University of North Carolina at
Charlotte
Charlotte, North Carolina

Qi Chen
Department of Geography
University of Hawaii at Mānoa
Honolulu, Hawaii

Janik Deutscher
Joanneum Research
Institute for Information and
Communication Technologies
Graz, Austria

Paul Elsner
Birkbeck
University of London
Department of Geography, Environment
and Development Studies
London, United Kingdom

Thomas Esch
German Aerospace Center (DLR)
German Remote Sensing Data Center
(DFD)
Weßling, Germany

Takehiko Fukushima
Faculty of Life and Environmental
Sciences
University of Tsukuba
Tsukuba, Ibaraki, Japan

Abduwasit Ghulam
Center for Sustainability
Saint Louis University
St. Louis, Missouri

Mark Grzovic
Center for Sustainability
and
Department of Earth and Atmospheric
Sciences
Saint Louis University
St. Louis, Missouri

Xulin Guo
University of Saskatchewan
Saskatoon, Saskatchewan, Canada

Charlotte Bay Hasager
Technical University of Denmark
Department of Wind Energy
Risø Campus
Roskilde, Denmark

Yuhong He
University of Toronto Mississauga
Mississauga, Ontario, Canada

Manuela Hirschmugl
Joanneum Research
Institute for Information and
Communication Technologies
Graz, Austria

Xuefei Hu
Department of Environmental Health
Rollins School of Public Health
Emory University, Atlanta, Georgia

Lalu Muhamad Jaelani
Geomatics Engineering Department
Institut Teknologi Sepuluh Nopember
Surabaya, Indonesia

Yitong Jiang
Center for Urban and Environmental
    Change
Department of Earth and
    Environmental Systems
Indiana State University
Terre Haute, Indiana

Bing Lu
University of Toronto Mississauga
Mississauga, Ontario, Canada

Maitiniyazi Maimaitijiang
Center for Sustainability
Saint Louis University
St. Louis, Missouri

and

College of Management
Xinjiang Agricultural University
Urumqi, Xinjiang, China

Mattia Marconcini
German Aerospace Center (DLR)
German Remote Sensing Data Center
    (DFD)
Weßling, Germany

Bunkei Matsushita
Faculty of Life and Environmental
    Sciences
University of Tsukuba
Tsukuba, Ibaraki, Japan

Ross K. Meentemeyer
Center for Geospatial Analytics
and
Department of Forestry and
    Environmental Resources
North Carolina State University
Raleigh, North Carolina

Annekatrin Metz
German Aerospace Center (DLR)
German Remote Sensing Data Center
    (DFD)
Weßling, Germany

Siraj Munir
Pakistan Space and Upper Atmosphere
    Research Commission
Islamabad, Pakistan

and

Department of Remote Sensing and GIS
Institute of Space Technology
Karachi, Pakistan

Claudia Notarnicola
EURAC Research
Institute for Applied Remote Sensing
Bolzano, Italy

Salman Qureshi
School of Architecture
Birmingham City University
Birmingham, United Kingdom

and

Department of Geography
Humboldt University of Berlin
Berlin, Germany

Said Rahman
Pakistan Space and Upper Atmosphere
   Research Commission
Regional Office
Peshawar, Pakistan

and

Department of Remote Sensing and GIS
Institute of Space Technology
Karachi, Pakistan

Achim Roth
German Aerospace Center (DLR)
German Remote Sensing Data Center
   (DFD)
Weßling, Germany

Mamat Sawut
Center for Sustainability
Saint Louis University
St. Louis, Missouri

and

College of Resources and
   Environmental Sciences
Xinjiang University
Urumqi, Xinjiang, China

Fajar Setiawan
Graduate School of Life and
   Environmental Sciences
University of Tsukuba
Tsukuba, Ibaraki, Japan

Maxim Shoshany
Mapping and GeoInformation
   Engineering
Faculty of Civil and Environmental
   Engineering
Technion, Israel Institute of Technology
Haifa, Israel

Alexander Tong
University of Toronto Mississauga
Mississauga, Ontario, Canada

John C. Trinder
School of Civil and Environmental
   Engineering
The University of New South Wales
Sydney, New South Wales, Australia

Markus Tum
German Aerospace Center (DLR)
German Remote Sensing Data Center
   (DFD)
Weßling, Germany

Mudassar Umar
Pakistan Space and Upper Atmosphere
   Research Commission
and
Department of Remote
   Sensing Commission and GSIS
Institute of Space Technology
Karachi, Pakistan

Qihao Weng
Center for Urban and Environmental
   Change
Department of Earth and
   Environmental Systems
Indiana State University
Terre Haute, Indiana

Honglin Xiao
History and Geography
Elon University
Elon, North Carolina

Yanhua Xie
Center for Urban and Environmental
 Change
Department of Earth and
 Environmental Systems
Indiana State University
Terre Haute, Indiana

Wei Yang
Center for Environmental
 Remote Sensing
Chiba University
Chiba, Japan

Julian Zeidler
German Aerospace Center (DLR)
German Remote Sensing Data Center
 (DFD)
Weßling, Germany

Lei Zhang
Center for Urban and Environmental
 Change
Department of Earth and
 Environmental Systems
Indiana State University
Terre Haute, Indiana

and

The State Key Laboratory of
 Information Engineering in
 Surveying, Mapping and Remote
 Sensing
Wuhan University
Wuhan, People's Republic of China

Yuanfan Zheng
Center for Urban and
 Environmental Change
Department of Earth and
 Environmental Systems
Indiana State University
Terre Haute, Indiana

Section I

Remote Sensing
for Sustainable Cities

1 Extraction of Parameters from Remote Sensing Data for Environmental Indices for Urban Sustainability

John C. Trinder

CONTENTS

1.1 INTRODUCTION

It has been recognized over the past few decades that actions by humans have modified and altered the energy and mass exchanges that occur between the atmosphere, oceans, and biota, and researchers now understand that the changes being wrought on the planet could be beyond the resilience of natural systems to absorb. The consequence of these changes can also be a loss or a severe decline in the ecosystem services on which we rely, thus affecting our quality of life. McGlade (2007) referred to even more serious consequences of a *green backlash*, where dramatic shifts in the structure and behavior of ecosystems can occur without warning.

Sustainable development has been proposed as a means of ensuring that human impacts are within the capacity of the Earth's environment* to cope with changes. While there have been many definitions presented, *sustainability* refers to the adoption of practices in relation to environmental use and management that provide a satisfactory standard of living for today's population and that do not impair the capacity of the environment to provide for and support the needs of future generations. Alternatively, sustainable development is that which meets the needs of society today without foreclosing the needs or options of the future (Blanco et al. 2001; Mahi 2001). The concept of sustainability in respect of the use of the environment's resources includes the notion that the outputs derived, whether they are from land, water, or air, can be produced continuously over time, and that a balance can be achieved between the rate of economic growth, the use of resources, and environmental quality, thus minimizing the risk of long-term environmental degradation. Sustainable practice is one which is sensitive to ecological constraints and seeks to minimize the undesirable effects of exploitation and use, and which might negatively affect the longer-term viability of a resource. It is also one in which the full economic and environmental replacement costs associated with the use of a resource should be met. Turner (1993) described strong versus weak sustainability, which is based on "the economic concept of capital, defined as a stock of resources with the capacity to give rise to the flow of goods and services." Strong sustainability requires "the stock of natural capital to be maintained above critical levels" (Turner 1993). Weak sustainability presumes that the "total capital stock does not decline" and all types of capital are substitutable. Karlsson et al. (2007) indicated that where current policies and actions are heading in regard to human welfare, on the scale of weak to strong sustainability, can be mapped out provided they are formulated in monetary terms.

Brandon and Lombardi (2011) referred to *community capital* that contributes to humans' well-being, which includes built and financial, human and social, and natural capital, each being measured in different ways. When they are out of balance, chaos or disaster can occur.

Kates (2000) reviewed the relationship between population and consumption in terms of the formula $I = P \times C$, where I is environmental degradation or resource depletion, P is the number of people or households, and C is the transformation of energy, materials, and information. This simple formula shows that as population increases, resource depletion also increases. In addition, the value of C must be controlled according to an optimal level of transformation of natural capital. Therefore, as populations increase, or a certain section of the population wishes to increase its use of resources to the same level as other groups in a society, since natural capital available for transformation effectively remains constant, there will be fewer natural resources available per capita and a redistribution of resources will be required.

---

* In the context of this chapter, *environment* refers to the surroundings or conditions in which a person, animal, or plant lives or operates, while *ecology* is the branch of biology that deals with the relations of organisms to one another and to their physical surroundings.

The maintenance of sustainability therefore becomes even more difficult and yet more critical.

Sustainable development cannot be divorced from issues of equity, welfare, lifestyle, and the expectation of improved standards of living in most countries. The Principles of the 1992 Rio Declaration, which were reaffirmed at the 2002 Johannesburg World Summit on Sustainable Development, define the roles of the stakeholders in sustainable development, and rights and responsibilities in development processes. The Johannesburg Declaration went on to refer to "...the three components of sustainable development, economic development, social development and environmental protection as interdependent and mutually reinforcing pillars."

1.2 TOWARD A SUSTAINABLE COMMUNITY

There have been many papers proposing approaches to achieving sustainability in nations, in regions, and by individuals. Gallopin and Raskin (2002) compared a number of scenarios that have been used or may be used in the future for predicting the characteristics of a sustainable human society. They include *market forces, policy reform, eco-communalism,* through to *muddling through.* The global community is currently dominated by market forces in which there is often an absence of controls over development and therefore there are tensions between development and sustainability goals. The policy reform approach of Gallopin and Raskin (2002) was based on the assumption of consensus and strong political will to achieve a sustainable future. Radermaker (2004) compared the impact of several approaches to economic and political developments and concluded that a balanced philosophy must be based on the concept of a global *ecosocial* market, consensus, and respect for civil rights and human equity, where human behavior is agreed globally by social contract. It means that there needs to be a consensus on protection of resources and respecting the need for all humans to have an adequate quality of life with access to essential resources. It is a long-term view of how the global population should cooperate to secure environmentally sound developments, but one which he believed is essential to achieve a sustainable and equitable use of resources. Azapagic and Perdan (2005a,b) have presented a procedure for decision-making that includes all stakeholders in a development process defining their preferences and choosing the most suitable alternative for the development, implementation of the chosen alternative, and assessment of the outcomes. The decision-making process was based on the concept of Multiple Criteria Decision Analysis, which was described in some detail.

In order to assess the sustainability of a society and its consequent well-being, organizations in many countries, regional organizations, nongovernmental organizations, and private organizations have attempted to develop sustainability indicators (SIs). Hundreds of indicators have been produced under the three pillars of economic, social, and environmental. A fourth pillar, *institutional*, has also recently been added. In judging the importance of the three original pillars, Jesinghaus (2007) showed that economy is the most important pillar, with a weight of 45%, while social has a weighting of 35% and environment has a weighting of 20%. If these weights of importance lead to a degradation of the environment, then it seems a greater weight may be required for the environment.

While many of the concepts referred to above may be far removed from the practical implementation of remote sensing technologies, it can be argued that remote sensing and geographic information system (GIS) technologies have important roles to play in assisting in the understanding of the physical impacts of development and, therefore, by inference, their impacts on human well-being. Also, by virtue of information extracted from the analysis of geospatial data, unsustainable practices being undertaken may be identified and their likely consequences may be predicted if they are continued. Such analyses must involve experts in remote sensing and GIS, as well as those in ecology, biology, sociology, human resources, and politics.

A description of sustainability indicators (SIs) will be given Section 1.3, followed by those for urban environments and a demonstration of how remote sensing technologies can be incorporated for the practical assessment of sustainability of urban areas. This chapter will concentrate on applications of remote sensing technologies for assessing urban sustainability as well as land-use practices outside urban areas, since urban dwellers depend on land in the vicinity of urban environments for much of their well-being.

1.3   SUSTAINABILITY INDICATORS

1.3.1   DEFINING INDICATORS

Becker (1997) defined the approaches that can be taken for assessing sustainability using "an exact measurement of single factors and their combination into meaningful parameters" and indicators "as an expression of complex situations by a variable that compresses information into a more readily understandable form" (Harrington et al. 1993). Tanguay et al. (2010) stated that an observed datum or variable becomes an indicator, only when its role in the evaluation of a phenomenon has been established. An index (or composite indicator) is a synthesis of indicators. Moldan and Dahl (2007) stated that "Indicators of sustainability should measure characteristics of the human-environmental system that ensure its continuity and functionality far into the future."

Hák et al. (2007) provided a detailed scientific assessment of SIs, in the treatise implemented by SCOPE and UNEP, together with the IHDP and EEA, and sponsored by the ICSU.* This is a comprehensive coverage of SI from the perspective of the three pillars and includes methodological aspects, system and sectorial approaches, and case studies. A great deal can be learned from this volume about SIs. Rao (1998) stated that SIs are designed to monitor progress and assess the effectiveness and impact of policies on natural resource development. Becker (1998) reported on the proliferation of papers and recommendations on developing SIs since the UN Rio Earth Summit in 1992. The UN, Organisation for Economic Co-operation and Development (OECD), the World Bank, and many other organizations have developed sets of indicators. OECD countries use 23 so-called indices based on natural

---

* SCOPE, Scientific Committee on Problems of the Environment; UNEP, UN Environmental Programme; IHDP, International Human Dimensions Programme on Global Environmental Change; EEA, European Environmental Agency; ICSU, International Council for Science.

sciences, policy performance, accounting framework, and synoptic indices. There are also aggregate indices that cover countries or regions.

Petrosyan (2014) developed a new composite indicator, composite appraising supportive progress (CASP), derived from 12 other indicators, with the weighting based on the number of papers published on each indicator. While there are advantages to having a single measure of sustainability, there are considerable questions about the weights used to produce composite indices, which show little scientific justification.

Moldan and Dahl (2007) referred to the different time scales of indicators for the three pillars. Economic indicators will normally have a short-term effect, while the effects of environment indicators will be longer term. In addition, they suggested that SIs might be more easily understood if they are formed into frameworks, based on a hierarchy of subdomains, with the three pillars being the basis for one such framework. Brandon and Lombardi (2011) stated that frameworks have been developed in order to link indicators to policy processes and also for developing messages to decision-makers.

Olalla-Tárraga (2006) claimed that a reductionist approach in which the three areas, economic, social, and environmental, are separated has failed to provide a satisfactory set of SIs that can be practically implemented. He presented eight conceptual frameworks that have been published by various authors, namely, domain-based, issue-based, goal-based, sectoral, causal, comparative, ecosystemic, and combinations. His solution was the hierarchical concept shown in Figure 1.1, in which the three areas (economic, social, and environmental) are each subdivided into *area*, *objective*, *attribute*, and *indicators*. An attempt will be made to relate environmental

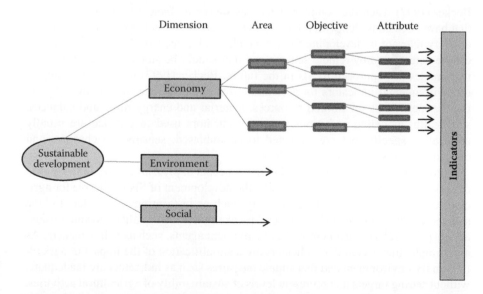

FIGURE 1.1 Hierarchical framework of sustainable indicator system. (From Olalla-Tárraga M. A. 2006. A conceptual framework for assessing urban ecological systems, *Int. J. Sustainable Dev. World Ecol.* 13: 1–15. With permission.)

indicators for urban areas to such a framework later in this chapter. Links may exist between the different criteria. This is a complex approach, but one which may be developed further in the future.

Olalla-Tárraga (2006) has also commented that a problem with SIs is the limited amount of data available to compile them. Becker (2007) provided a general discourse on frameworks for SIs that should lead to a more holistic approach to their implementation. She stated that limited tools are currently available for assessing the nonlinear multidimensional problem of sustainable development, but frameworks can assist in understanding the issues. Gutiérrez-Espeleta (2007) has also raised the issue that a new approach is necessary in the development of SIs. He has proposed a multilevel typology in which the indicators for *Environment* and *Society* can be classified into five generations. The first and second generations would measure a single characteristic of an environmental (or societal) issue, while the third, fourth, and fifth generations would be integrated measures. He has cited two indices—the Environmental Vulnerability Index (EVI) and Environmental Sustainability Index (ESI)—that have been developed as high-order indices and show promise. There are perhaps synergies with this approach and frameworks, but they have been developed along different lines. A major issue in demonstrating the usefulness of indices is the availability of data to test the use of indices (Brandon and Lombardi 2011). In summary, while frameworks have not yet resulted in a useful set of SI, the concept is believed by a number of researchers to lead to a better approach in the future.

1.3.2 Application of SIs

Becker (1997) listed the complex criteria as shown in Table 1.1 for developing SIs that have been compiled with reference to the work of a number of scientists. These are very relevant to developing SIs for the physical aspects of the environment that can be determined by remote sensing and that should be suitable for contributions to a framework approach to SIs in the future. Becker (1997) has shown examples of environmental indicators for agriculture that could include such items as "yield trends, coefficients for limited resources, material and energy flows and balances, soil health, modeling and bioindicators." Indicators used in practice are usually application specific but are expected to be unbiased, sensitive to changes, and convenient to communicate and collect.

Stevenson and Li (2001) referred to the concepts of *pressure*, *state*, and *response* developed by other authors in discussing the development of SIs applicable for agriculture. Pressure refers to human activity, such as in farming; state refers to "the state of the environment and resources," for example, water quality and soil erosion; and response refers to economic and environment agents, such as policy makers. As an example, they stated that indicators are a simplification of the impact of agriculture on the environment and that simple measures such as indicators are inadequate, without setting targets that represent levels of sustainability of agricultural activities. They argued that SIs must provide information on which to base decisions at the local or regional scale and must be a guiding framework for practitioners. They then derived a methodology for a meaningful set of indicators for agriculture that includes

TABLE 1.1

Criteria for the Selection of SIs Adapted for Their Relevance to Remote Sensing

1. Scientific Quality	2. Ecosystem Relevance	3. Data Management	4. Sustainability Paradigm
1.1 Indicator really measures what it is supposed to detect	2.1 Changes as the system moves away from equilibrium	3.1 Cost effective	4.1 What is to be sustainable?
1.2 Indicator measures significant aspect	2.2 Distinguishes agro-systems moving away from sustainability	3.2 Data available	4.2 Participatory definition
1.3 Problem specific	2.3 Identifies key factors leading to unsustainability	3.3 Quantifiable	4.3 Adequate rating of single aspects
1.4 Distinguishes between causes and effects	2.4 Warning of irreversible processes	3.4 Representative	4.4 Resource efficient
1.5 Can be reproduced and repeated over time	2.5 Proactive in forecasting future trends	3.5 Transparent	4.5 Carry capacity
1.6 Uncorrelated, independent	2.6 Covers full cycles through time	3.6 Geographically relevant	4.6 Health protection
1.7 Unambiguous	2.7 Corresponds to aggregation level	3.7 Relevant to users	4.7 Target values
	2.8 Highlights links to other system levels	3.8 User friendly	4.8 Time horizon
	2.9 Permits trade-off detection and assessment between system components and levels	3.9 Widely accepted	4.9 Social welfare
	2.10 Can be related to other indicators	3.10 Easy to measure	4.10 Equity
		3.11 Easy to document	
		3.12 Easy to interpret	
		3.13 Comparable across borders over time	

Source: Becker B. 1997. Sustainability assessment: A review of values, concepts, and methodological approaches, *Issues in Agriculture 10, CGIAR,* Washington, DC, The World Bank. With permission.

all stakeholders and goal setting. No actual values of indicators were given, which demonstrates that the concept of SI is only progressing slowly, and in many cases, no assessment of the sustainability of human activity is yet possible. In the future, values will be required for the indicators; otherwise, assessment of sustainability will not be achievable.

Hueting and Reijnders (2004) viewed the issue of SIs from an economic perspective, referring to Sustainable National Income (SNI) as the goal of countries. They also argued that physical aspects of the environment must be included in measures of sustainability and gave the example of the cod stocks in Canada in which physical aspects were inadequately considered before the industry collapsed. Hart (1999) demonstrated a new approach in SIs that included the interconnections between the three pillars of sustainability: social, environmental, and economic. She has stated that the indicators should be relevant, understandable, reliable, and timely, and has described a large number of indicators that cover many aspects of human activity, including production, energy, transport, education, health, recreation, ecosystem, land use, resource use, and many more. Most of these indicators are based on such measures as percentage of land covered by impervious materials, volumes of water used, harvest rates compared with growth rates, and so on. Indicators for sustainability of land practices include sustainable land use, sustainable resource use, amount of tree cover, sustained returns, impact on soil or water, diversity of products, and conservation of native habitats.

Böhringer and Jochem (2007) reviewed 11 indices and commented that they did not cover all factors related to sustainability. They remarked on the arbitrariness of the weighting systems in compiling the indicators. A comprehensive set of Biodiversity Indicators has been presented by Biggs et al. (2007) covering such topics as gene-level, species-based, population abundance, ecosystem-level, and composite indicators. They also provided a list of single-variable indicators in use that are related to a reference value. Many of these indicators should be measurable by earth observation technologies.

Van Woerden et al. (2007) described a set of indicators related to the Global Environmental Outlook covering indicators related to the atmosphere, land, and water. Barrios and Kimoto (2006) used an analytical approach, based on principal components and sparse principal component analyses of a number of core indicators of sustainable development for the Philippines for deriving relevant indicators.

In Phillis and Andriantiatsaholiniaina (2001) and Andriantiatsaholiniaina et al. (2004), fuzzy logic was used to determine a combined measure of sustainability and sensitivity to various indicators. They argued that the fragmentation of information about sustainability measures makes it difficult to assess the effects of developments on the environment. Hence, they attempted to use a systematic rule-based system that combined many measures of sustainability. Since these measures are never well defined, they have used fuzzy definitions of the measures, expressed verbally in such terms as *good*, *bad*, *medium*, and so on. A sensitivity value was determined for the factors that contribute to a lack of sustainability. The research concluded that there is no unique sustainable path, and hence policies need to be chosen to determine the most sustainable strategies. Cornelissen et al. (2001) also used fuzzy set theory to link human expectations expressed linguistically to SIs presented numerically. The conclusion from this review is that the determination of sustainability is complex and not clearly defined and that there is often a time lag between development and its negative impact.

It is well known that remote sensing technologies, because of the regional and multitemporal coverage of the images, can enable measurement of many variables more economically than other field-based or manual methods. It is necessary to identify those indicators that can be measured reliably by remote sensing on a regular basis and that are reproducible, without bias, and truly reflect the characteristics of the environment when it is changing. Appropriate indicators must be determined in association with experts in the particular fields in which the indicators are being developed. While there are redundancies in Table 1.1, many of the criteria for determining SIs are relevant to the capabilities of remote sensing, particularly those in columns 1, 2, and 3.

1.4 INDICATORS FOR URBAN SUSTAINABILITY

Sustainability of urban areas should define sustainable urban form, which can also be seen from both planning and landscape perspectives. From the planning viewpoint, a sustainable urban form is defined by its compactness, mixed use, density, sustainable transport, diversity, and greening (Jabareen 2006). Therefore, compactness is one of the design concepts for a sustainable city.

The growth in cities is causing increasing stress on many aspects of the urban environment. According to the World Health Organization (WHO), 54% of the global population in 2014 lived in cities and that percentage is growing annually at a rate of more than 1.5%. Rees and Wackernagel (1996) believed that the end of the 20th century was a turning point in the history of human civilization and provided many examples of how humans are increasingly consuming more of the Earth's resources, to the extent that a *full-world* assumption of economics should be adopted rather than that of an *empty world*. They then went on to discuss the ecological footprint and human load of cities, which is increasing with time for many developed countries. They concluded that "no city or urban region can achieve sustainability on its own" because cities depend on resources from the hinterland, stretching from nearby areas to globally through exports and imports. Therefore, "a prerequisite for sustainable cities is sustainable use of the global hinterland." They listed the advantages of city living, based on economies of scale, possibilities of reduced energy consumption, and recycling, among others. However, they were pessimistic about the ability of humans to make positive steps to achieve more sustainable living. Alberti (1996) stated that cities alter the local and global environment, because they alter the land, import vast amounts of food water and energy, and export waste. She stated that assessing sustainability of an urban environment requires assessment of the following aspects:

- Direct transformation of the physical structure and habitat
- Use of natural resources (renewable and nonrenewable)
- Release of emissions and waste
- Human health and well-being

Alberti (1996) also provided many examples of measures of urban sustainability, some of which are measurable by remote sensing technologies, for example, urban land cover, open areas, transport networks, and the effects of land use on accessibility to transport, urban renewal areas, derelict areas, air quality, and accessible green space.

Freedberg (2010) stated that there are six specific principles for urban sustainability:

- Greater viability of public transport
- Affordable housing that meets a range of demographic groups
- Greater access to quality education and jobs
- Supporting neighborhoods that are engaging in sustainable practices
- Engagement of governments in providing financial support for sustainable activities
- Investment in healthy, safe, and walkable neighborhoods

Several of these principles can be monitored by remote sensing technologies. Olalla-Tárraga (2006) stressed the need to take an integrated approach to developing indicators for urban environments that are a complex mixture of buildings, backyard gardens, open space, transport links, impervious surfaces, drainage systems, deteriorating atmosphere including heat sinks, and many more.

Shen et al. (2011) explored SIs used in a number of cities around the world, listing 37 indicators under the following headings: Environmental, Economic, Social, and Governance. The Environmental Indicators were as follows:

- En1—Geographically balanced settlement
- En2—Freshwater
- En3—Wastewater
- En4—Quality of ambient air and atmosphere (which would include the effect of heat sinks)
- En5—Noise pollution
- En6—Sustainable land use
- En7—Waste generation and management
- En8—Effective and environmentally sound transportation systems
- En9—Mechanisms to prepare and implement environmental plans

Tanguay et al. (2010) compiled 188 urban SIs derived from 17 studies in published literature. They noted that of the 188 indicators, only 72% were used in one or two studies, while very few were found in more than 5 of the 17 studies. The distribution of indicators in the social, environmental, and economic categories revealed the overlap between sustainability dimensions and their descriptions. By further aggregating the indicators and reviewing their significance, they concluded on a list of 29 indicators, only 3 of which were related to the environment: ecological footprint, space allotted to nature conservation relative to area of territory, and percentage of waterways with excellent water quality. While the authors rated the important economic and social aspects of sustainable city living, they did not consider the sustainability of city dwellers in terms of the input of environmental capital.

Brandon and Lombardi (2011) have adopted the framework for assessment of the sustainability of developments of the built environment, based on the concept of a *Cosmonomic Idea of Reality* incorporating 15 modalities in a specific order in which the latter modalities depend on the earlier ones. They comprise the following: numerical, spatial, kinematics, physical, biological, sensitive, analytic, historical, communicative, social, economic, aesthetic, juridical, ethical, and creedal. These modalities have been aggregated into two levels, where the highest level comprises physical environment, human cultural capital, and financial institutional capital. The second level of five main groups of urban policy activities comprises three each of the modalities listed above.

Goddard et al. (2010) reviewed the importance of neighborhood gardens and their effect on biodiversity in cities. They suggested that the scale of the gardens is important for different taxa, some being able to survive in small regions, whereas others require much larger areas than neighborhood gardens to thrive. They demonstrated that gardens are socioecological constructs at different scales depending on the location, neighborhood, and individuals responsible.

There is little guidance from the above discussion on which SIs should be adopted to assess the sustainability of urban environments based on remote sensing technologies, but some examples will be considered in the next section.

1.5 APPLICATION OF REMOTE SENSING FOR URBAN SUSTAINABILITY

1.5.1 AVAILABLE REMOTELY SENSED DATA

Foody (2003) stated that remote sensing can provide a wealth of environmental data over a range of spatial and temporal scales and therefore can play a major role in the provision of indicators of environmental conditions for sustainable development and associated decision-making. The reason why remote sensing technologies have not been presented as important for assessing the sustainability of urban environments may be partially attributed to the lack of data with suitable resolutions. However, with recent developments in high-resolution sensors, remote sensing technologies should now be able to make significant contributions to the assessment of sustainable or unsustainable practices. Also, the repeat coverage that is available with satellite remote sensing at the current high spatial resolutions should mean that they will be even more valuable for assessing trends in sustainability of urban areas. The characteristics of remote sensing systems that acquire data suitable for determining the sustainability of urban areas are presented in Table 1.2.

Optical systems that acquire high-resolution images from the air, either by RPAS (Remotely Piloted Aerial Systems, UAS, or drones) or by piloted aircraft, and imaging systems on Earth-orbiting satellites record reflected solar radiation from the Earth's surface. Because of the relatively low flying heights of airborne sensors, there is usually greater flexibility in decisions of selection of airborne sensors. While they are listed in Table 1.2, the following discussion on resolutions will refer to satellite sensors. Most modern satellites are referred to as *agile* and hence can point forward/backward to obtain stereo images and sideways to acquire images of

TABLE 1.2

Examples of Data Acquisition Systems for Urban Remote Sensing

Data Type	Characteristics	Examples
RPAS image data	Small cameras with spatial resolutions ranging from 1–2 cm to >10 cm Spectral resolution—RGB or CIR Temporal resolution—as required in daylight	Multipropeller copter or fixed wing RPAS under control of a remote operator
Manned aircraft image data	Digital aerial cameras Spatial resolution—3–5 cm to >50 cm Spectral resolution—RGB, CIR Hyperspectral sensor acquiring hundreds of bands Temporal resolution—as required in daylight	A range of commercial digital aerial cameras are available with resolutions ranging from 40 Mpixels to more than 250 Mpixels HyMap hyperspectral sensor with 128 bands with wavelengths from 0.45 to 2.48 μm and 3–10 m spatial resolution AVIRIS is an optical sensor with 4- to 20-m spatial resolution, 224 spectral bands, with wavelengths from 0.4 to 2.5 μm
Satellite images	High-resolution images with spatial resolutions from approximately 30 cm to several meters Spectral resolution—panchromatic (pan) with a single band to multiple bands (MSS) Temporal resolution—from many days to 1 or 2 days depending on location and capabilities of the satellite	IKONOS II with 0.80-m panchromatic images and 3.2 MSS WorldView-1 with 0.50-m panchromatic only CartoSat2B with 0.8-m panchromatic images GeoEye 1 with 0.41-m panchromatic images and 1.65-m MSS WorldView-2 with 0.4-m pan images and 1.85-m MSS with 8 bands Pléiades with 50-cm panchromatic and 2-m MSS WorldView-3 with 31-cm panchromatic and 8 bands MSS Multiple microsatellites
Synthetic aperture radar (SAR)	Microwave sensors installed in satellites or aircraft and operate in cloudy conditions and day or night Spatial resolution <1 m for airborne and ranging up to >10 m for space borne with various levels of polarization Spectral resolution—variable Temporal resolution—variable according to number of satellites available and orbits	SAR wavelengths used for satellite remote sensing X-band—3 cm C-band—8 cm L-band—25 cm Most satellite-borne SAR sensors acquire either X-band or C-band RADARSAT-2—C-band TerraSAR-X—X-band TanDEM-X—X-band COSMO-SkyMed—X-band ALOS-2—L-band Commercial airborne systems
Lidar	Currently airborne only. Elevation posts acquired at 2 to >10 posts/m^2 Multispectral sensors available	Many commercial systems are available with similar characteristics

the terrain surface from neighboring orbits, thus reducing the time between multiple imaging of the same area on the terrain.

The resolution of remote sensing images can be expressed in terms of spatial, spectral, and temporal resolution. A brief description of the three levels of resolution will be presented for completeness.

- Spatial resolution of data is generally described as the footprint of the sensor or the area that the sensor views instantaneously on the terrain surface, often referred to as ground sampling distance. This is a function of sensor altitude, detector size, sensor optics, and the system configuration. The dimensions of the smallest feature visible on the terrain surface are a function of the spatial resolution, the contrast and color of the object in relation to its background, and its shape (e.g., very narrow linear features are often detectable on satellite images when they are not expected to be). While sensors on satellite platforms are available with a very broad range of spatial resolutions, because of the requirements of urban remote sensing to detect buildings, pavements, roads, parking lots, the fragmentation of open space, and similar issues, only the so-called high-resolution satellite images with spatial resolutions on the order of 2 m and smaller will be discussed in this chapter. This is supported by Sliuzas et al. (2010) who stated that the identification of small urban objects or objects in complex environments requires a minimum spatial resolution of 5 m. They also argued that the required spatial resolution of images should be determined by the smallest object to be identified in the images. The availability of images with spatial resolutions smaller than 2 m over the past decade provides new opportunities for satellite remote sensing applications in sustainability studies in urban areas.
- Spectral resolution refers to the range and width of wavelengths that can be resolved by a sensor.
- Temporal resolution refers to the frequency of coverage of a certain area on the terrain surface by a sensor. The repeat cycle of orbiting satellites with optical systems viewing vertically only is usually many days, but this can be reduced by a constellation of satellites or by including tilt capability on the satellites so that they can view the same areas on the ground, weather permitting, from successive orbits.

In the design of sensors on satellites, there is necessarily a compromise between the demands for spatial, spectral, and temporal resolutions. For example, currently, there is the WorldView-3 satellite with a single panchromatic very high spatial resolution of 0.31 m and eight multispectral bands with spatial resolutions of 1.24 m. The spectral resolution of this satellite may be relatively coarse for analysis of some materials in urban areas. On the other hand, currently available airborne hyperspectral sensors have very high spectral resolutions in hundreds of bands but a relatively coarse spatial resolution (e.g., HyMap system referred to in Table 1.2).

Optical image satellite systems that are commonly used for urban applications have been described by Rashed and Jürgens (2010) and in the comprehensive review paper by Weng (2012). Gamba et al. (2011) referred to several problems with the geometry of high spatial resolution optical images, since a digital elevation model (DEM) with an accuracy on the order of 1 m is required to produce orthophotos from such data over urban areas. If high buildings exist in the image, they will be subject to relief distortions that can only be corrected if the DEM includes the heights of buildings. In addition, large distortions will occur in images that are acquired with the satellite tilted sideways to image areas at a specific time. Image interpretation may also be compromised for the tilted images.

Synthetic Aperture Radar (SAR) systems are active remote sensing systems that emit microwave radiation from an antenna and record the time of travel and the intensity of the radiation returned to the antenna. SAR systems have the advantage that they can be used day or night and in cloudy conditions, and the polarization of emitted and received radiation can be varied as well, providing additional potential for extracting information from the data. Soergel (2010) has described some applications of SAR systems for the analysis of urban areas.

Airborne lidar (Light Detection and Ranging—also written as LiDAR) data are based on a scanning laser that emits a pulse (with a wavelength in the infrared region of the electromagnetic spectrum) toward the terrain surface, and the distance to the terrain surface can be determined from the time of travel of the pulse to the terrain and back to the sensor.

Together with knowledge of the position and altitude of the aircraft, a dense point cloud on the terrain surface can be determined to represent the position and elevation of discrete posts at a density of two or more posts per square meter. In addition, the intensities of the returned laser pulses from the terrain or objects on the terrain are recorded, which can represent an infrared image. Airborne lidar can assist in the extraction of three-dimensional (3D) information about man-made features and therefore details of the built environment.

1.5.2 Urban Sustainability Measures Derived by Remote Sensing

Taking into consideration the above review of the literature on SIs for urban environments, the aspects of an urban environment that can be assessed for SIs by remote sensing technologies are presented in Table 1.3, where an attempt has been made to develop three levels of a framework for these environmental indicators. With reference to Figure 1.1, in Table 1.3, *Area* is the *Urban Sustainability Measure*, *Objective* is suggested as corresponding to the measurements made by the remote sensing technologies, while *Attributes* are the parameters derived by remote sensing technologies. While some *Indicators*, as shown in Figure 1.1, may be determined for the attributes, there is no adequate information available yet to present these details. The measures suggested in Table 1.3 would need to be considered in conjunction with other environmental, social, and economic SIs, which cannot be measured by remote sensing technologies, for the overall framework for assessing sustainability of urban areas.

In terms of the framework developed by Brandon and Lombardi (2011), while they are less specific to the measurements derived by remote sensing technologies, the first five modalities—*numerical*, *spatial*, *kinematics*, *physical*, and *biological*—would

TABLE 1.3
Assessable Sustainability Aspects in the Urban Environment by Remote Sensing Technologies

No.	Area—Urban Sustainability Measure	Objective—Measurement by Remote Sensing Technologies	Attribute—Parameters Derived by Remote Sensing Technologies
1	Balanced development—fraction of built versus open space	Measurement of impervious surfaces in relation to open space	Ratio of area of impervious surfaces to open spaces
2	Transformation of the physical structure and habitat from green space to impervious surfaces	Growth in fragmentation of open space versus impervious surfaces	Diversity, dominance, fragmentation
3	Effective and environmentally sound transportation systems	Mapping and analysis of transport systems to demonstrate effectiveness	Transport Mode Index
4	Consideration of healthy, safe, and walkable/cycle neighborhoods	Determine compactness of cities, mapping of walking and cycle paths, and township layout	Size Density Degree of distribution Clustering
5	Consumption of natural resources (renewable and nonrenewable), from hinterland and its impact	Measurement of deforestation and changes in land cover over time	Land use/land cover changes
6	Effects on biodiversity	Changes in local vegetation and native flora and potential habitats for fauna	Land-use conversion and loss of habitats
7	Release of emissions and waste, especially into waterways and the atmosphere	Determine surface water quality and chemical content of atmosphere	Not covered

apply to remote sensing technologies. *Numerical* refers to the accounting process that can be developed from remote sensing, *spatial* refers to the spatial details extracted, *kinematics* refers to transport and mobility, *physical* refers to the physical environment, and *biological* refers to health and ecological protection as well as biodiversity. While both frameworks should be relevant to urban sustainability, that shown in Figure 1.1 will be the basis of a demonstration of the contributions of remote sensing to urban sustainability in the following.

There is a need for greater attention to the future development of SIs for urban areas because more than half of the world's population lives in cities. Such indicators should also be used for monitoring whether an environment is becoming less sustainable as developments occur or populations grow, leading to a need for modification of decisions affecting an urban area. Each of the Indicators listed in Table 1.3 will be discussed below, except for Item 7, "Release of emissions and waste, especially into waterways and the atmosphere," which is discussed elsewhere in this book.

1.5.2.1 Measurement of Impervious Surfaces in Relation to Open Space

A great deal has been written on the measurement of impervious surfaces, which cover the full range of methods available for image classification of such surfaces, for example, in the review paper by Weng (2012). There are a number of issues raised in Weng's paper that are relevant to this chapter. Buildings and roads can be detected with optical sensors with spatial resolutions of 0.25 to 0.5 m (which is a much smaller resolution than referred to in Section 1.5.1), although shadows may cause problems with tall buildings. There is also the possibility of confusion of roads and buildings with surrounding features, and therefore, fusion of several forms of data of the area may prove beneficial. Weng (2012) has suggested that the investigation of the spectral diversity of impervious surfaces together with 3D characteristics and temporal changes should prove beneficial in urban studies. Salah et al. (2010) investigated various machine learning and ensemble learning approaches for the extraction of impervious surfaces created by buildings and roads and also extracted the ground and vegetation from high-resolution aerial photography and airborne lidar data over urban areas and found in excess of 90% accuracy for most methods.

The investigation by Wu and Yuan (2011) using high-resolution satellite images for extraction of impervious surfaces was based on pixel and object-oriented methods. They have estimated that approximately 40%–50% of pixels are mixed pixels, for high-resolution satellite data such as IKONOS and Quickbird multispectral data. They used a normalized spectral mixture analysis, regression trees, artificial neural networks, and object-oriented approaches for the extraction of impervious surfaces with overall accuracies of approximately 90%. Shadows from buildings and trees need to be differentiated since they hide different types of surfaces. Canters et al. (2011) also discussed the extraction of impervious surfaces from both high-resolution and medium-resolution images and suggested that the use of lower-cost medium-resolution images together with subpixel classification methods could be a better approach. However, this claim is not in agreement with other statements made above. Satellite SAR images have been used for extraction of buildings and roads in urban areas, sometimes with mixed success (Soergel 2010), but high-resolution airborne SAR images should provide better extraction capabilities than satellite SAR images.

While it was expressed earlier that spatial resolutions of 2 m or better are desirable for urban studies, the application of lower-resolution hyperspectral images has been extensively explored by Roessner et al. (2011) in studying impervious materials in urban areas. Images were derived from the HyMap sensor with pixel sizes ranging from 3 to 6 m, details of which are given in Table 1.2. The study extracted end members describing distinct spectral signatures of different surface materials in several German cities. The study revealed the increased information derived from the hyperspectral images compared with lower spectral resolution images used in previous studies, which assumed that the surfaces are completely impervious. The higher spectral resolution hyperspectral images, though lower in spatial resolution, allowed more detailed mapping of the surface materials in urban areas and the separation of semipermeable surfaces, which might include cobblestones or gravel, from completely impervious surfaces. This would affect the level of runoff and, thus, the hydrological cycle of urban areas. Also, there may be significant influences on the assessments of microclimates in urban areas. The availability of hyperspectral images should allow better definition of ground cover in urban areas and associated ecological factors and also allow for studies on the effects of developments on biodiversity. Time series would also allow for studies on changes in these factors. The definition of an attribute for impervious surfaces is complex, but the *ratio of areas of fully impervious surfaces to open spaces* is suggested as the *attribute* for this SI in Table 1.3. Further work would be required to determine the influence of partially impervious surfaces on urban sustainability.

1.5.2.2 Growth in Fragmentation of Open Space versus Impervious Surfaces

The growth in fragmentation in urban and suburban areas is suggested as a further SI for urban areas, since the transformation of fragments of open space into impervious surfaces is an important indicator of urban development. Greenhill et al. (2003) derived two parameters to act as ecological indicators for suburban areas:

- The *weighted mean patch size* (WMPS), which provides information about the size distribution of vegetation patches. The mean patch size is the average area of vegetative patches within a window. The weighted mean patch size includes information on both patch size and number.
- Lacunarity, which provides an indicator of the spatial clustering of such patches and is dependent on the number of interpatch nonvegetated pixels within a square box that are summarized in a histogram. The mean and variance of the counts are used to calculate the lacunarity.

These parameters could be used to assess the impact of fragmentation of open space caused by urban developments that change the distribution of vegetated versus nonvegetated areas. Certain values of WMPS and lacunarity could correspond to maintaining a relatively low density of housing and a good clustering of local green areas, which would be suitable for diversification of flora and fauna. On the other hand, high values of these parameters would demonstrate the primarily impervious environment, which provides little opportunity for diversification of flora and fauna, and the need for the provision of such environments elsewhere in the urban area for an adequate lifestyle for the inhabitants.

Lein (2014) argued that limiting sustainability studies to regional scales enables the linking of the economic, social, and environmental factors of sustainability to the landscape unit. In addition, the revelation of unsustainable trends in aspects of the environment should lead to a change in behavior or development programs. He then presented advantages of remote sensing methods for assessing sustainability: the capacity of obtaining unique measurements of reflectances from the Earth's surface, the repeatability of the observations, and the archival capacity of the data. He espoused the possibility of developing indices for sustainability, which should have the following attributes: relevance, concept integrity, reliability, scale appropriateness, scale sensitivity, and robustness, many of which are similar to those recommended by Becker (1997), as shown in Table 1.1. Lein thus provided indices that can be based on metrics derived from remote sensing, namely, impervious surfaces, fragmentation, diversity, and dominance. The estimates of *impervious surfaces* are based on NDVI. *Fragmentation* of the landscape is dependent on the following formula: $F = (n - 1)/(c - 1)$, where n is the number of patches in a kernel and c is the number of cells considered. Diversity $= (\Sigma p \times \ln(p))$, where Σ is the summation of all land types in the study area, p is the proportion of each land type in the spatial unit of measure (pixel), and ln is the natural logarithm. *Dominance* is defined by the most abundant land type by $\ln S + \Sigma p_k \times \ln p_k$, where S is the number of habitat types and p_k is the proportion of area in habitat k. As these metrics were assessed using Earth observation data, monitoring could lead to a dynamic management process. Lein demonstrated an ecological integrity component that is dependent of the *diversity, dominance, fragmentation*, and *impervious surfaces*. These components could be monitored to assess trends in sustainability. The use of high-resolution satellite data could provide greater granularity to the information extracted in this research, thus enabling more detailed analysis of *fragmentation, diversity*, and *dominance*.

Sapena and Ruiz (2015) have defined several metrics to determine the rate of growth patterns on the European Urban Atlas databases in 2006 and 2012. The measures used to describe fragmentation are as follows: urban density, which estimates the proportion of developed areas, comprising housing, commercial, industrial and landmark buildings, roads, barren land, and leisure areas, as a ratio of the classified urban areas in the region; weighted standard distance, which measures concentration or scattering around the centroid of the objects of a given class, weighted by the size of each object; Euclidean nearest neighbor (ENN) mean distance, which is also a measure of scattering, being the mean of the distances between the edges of the objects of the same class; Shannon diversity, which represents the abundance and evenness of the classes; and edge contrast ratio, which quantifies the degree of contrast between objects from different land uses, excluding the road network.

On the basis of the above discussion, the primary *attribute* for this indicator is *fragmentation* together with *diversity and dominance*, which have been added to Table 1.3. The measures proposed by Lein (2014) and Sapena and Ruiz (2015) are demonstrations of how such attributes could be measured using remote sensing data.

1.5.2.3 Mapping and Analysis of Sustainable Transport Systems

The task of determining the overall effectiveness of transport routes is beyond the capacity of remote sensing. However, an essential task is to provide data for the spatial analysis, together with GIS tools that enable transport experts to determine the effectiveness of the transport infrastructure. Zhang and Guindon (2006) and Guindon and Zhang (2007) have described the process for developing a set of *Sustainable Transportation Performance Indicators*, in which the basis for land-use mapping were combined with socioeconomic data derived from the national census, for the determination of such parameters as urban population density and compactness of cities. The *Transport Mode Index*, which was adopted as the *attribute* in Table 1.3, measures the impact of the land-use mix and urban form on the feasibility of various modes of transportation. The authors stated that while indicators are important for demonstrating the sustainability of transport systems in

Canadian cities, simple measures such as density and compactness are poor indicators of energy consumption, whereas land-use mix is a better indicator. This would be the cue for the application of remote sensing technologies since land-use analysis can be undertaken economically over urban areas using medium-resolution images. However, they argued that more research is required to develop better indicators that incorporate median travel distance.

1.5.2.4 Compactness of Cities and Mapping of Township Layout

Compactness is one of the "design concepts of sustainable urban form" (Jabareen 2006). The logic behind it is that a more compact city results in less travel and therefore lower energy consumption, leading to environmentally sustainable transport systems, less fragmentation of neighboring lands, walkable environments, and elimination or significant reduction in urban sprawl. The scale of analysis of compactness of urban areas may determine the methods and resolutions of images used for measurement by remote sensing technologies. Large metropolitan areas may require regional-scale measurements while walkable cities may require larger-scale images and manual mapping operations.

A number of parameters have been used by Tsai (2005) to define compactness versus urban sprawl. These include size of the metropolitan area, which varies according to the extent of sprawl; density, expressed in terms of land occupation per capita; degree of distribution of development in a metropolitan area; and clustering or centralization of the metropolis.

Tsai (2005) then discussed the Moran, Geary, and Gini coefficients, including simulation of various forms of urban areas based on population or employment that can be assessed as to their suitability for describing compactness. He found that the Moran coefficient was useful as a metric for distinguishing between compact urban areas and urban sprawl, but it was not able to differentiate between a circular and a linear shape of cities. There was no index that completely described compactness versus sprawl, but the Moran and Gini coefficients were suitable for some examples of urban form. Population or employment cannot be determined by remote sensing technologies. However, the 3D form determined by airborne lidar may be a surrogate for population required by the Moran coefficient, although it is unlikely to be able to distinguish between actual population and employment locations. The 2D geometry of urban areas required for the Gini coefficient may be determined from medium- to high-resolution images, although it would require manual interpretation of the images.

Herold et al. (2003) analyzed four urban spatial metrics using high-resolution satellite data to determine their characteristics for several types of land covers ranging from forests to high-density single-unit residential areas. They supported the use of spatial metrics as descriptors of built-up structures and open areas. Huang et al. (2007) suggested additional measures of compactness that included an area weighted average shape index, a patch fractal dimension, centrality, several compactness indices, ratio of open space, density, and several measures related to social indices for comparisons of the form of many cities around the world using medium-resolution remotely sensed data. They found significant differences between developed world cities in America, Australia, and parts of Europe, which tend to be subject to sprawl, and those generally more compact cities in Asia.

Sim and Mesev (2011) extracted similar parameters as Tsai (2005) (size, density, continuity of development, and scattering of developments) but also added shape and loss of green space as indicators of urban sprawl. They used different statistical methods for indicators, including entropy and ENN. Therefore, from the above discussion as shown in Table 1.3, the *attributes* for urban compactness are assumed to be those defined by Tsai above, namely, *size of the metropolitan area*, *density*, *degree of distribution*, and *clustering*.

Locating and mapping the lengths of walking paths and mapping township layouts are primarily a manual mapping task that can be undertaken with medium- to high-resolution optical images for the production of digital orthophotos.

1.5.2.5 Deforestation and Changes in Land Cover

Since urban areas depend on imports of products derived from outside the urban areas, and further in some cases, a considerable amount of land clearing and deforestation occurs to service urban areas. In order to define the effects of these land-use changes, a detailed analysis of particular cities and the resulting changes in surrounding areas would have to be undertaken. Medium-resolution remote sensing technologies should be applicable for these purposes. For example, NASA (2010) revealed, based on the application of the QuickScat satellite radar system, that Beijing has quadrupled its size in 9 years. Foody (2003) discussed aspects of maps of forest and forest change, estimation of forest biomass, biodiversity, and drought, showing how remote sensing can satisfy some of the broadly based indicators of changes in open space areas, and that satellite remote sensing would be the most economical approach. As an example, Brazil's National Institute for Space Research (INPE) has used satellite data for monitoring deforestation for more than 15 years based on a range of medium-resolution satellite images, demonstrating the conversion of forests to farming in the Amazon region. INPE has developed an almost real-time monitoring system to detect illegal land clearing. High-resolution images are required to determine the exact area of clearing and for extracting small areas of clearing. Therefore, the study of land-use changes over the period due to urbanization could be undertaken economically by processing time series remote sensing images.

The information derived from remote sensing can also be directly related to measuring important socioeconomic impacts. Rates of land cover change and drought will strongly influence vegetation yield, which could substantially affect human health and well-being of neighboring cities. These factors will, for example, influence the demand for and rate of fertilizer application, which may be associated with downstream pollution. Issues such as soil erosion are a major concern for land users and are also strongly associated with consequential impacts, including the silting of lakes and damage to hydroelectric power stations.

Bacchus et al. (2000, 2003) investigated the detrimental effects on the health of the vegetation, in the case of pond cypress, caused by the withdrawal of groundwater from aquifers in Florida. Their investigations were based on laboratory spectrometry studies (in the visible, near-infared [NIR], and mid-infrared regions of the spectrum) of dried milled branch tips collected from natural stands of pond cypress stands, both in summer and in winter before bud-break. They found that the NIR spectral response was more affected by stress than by site-related factors. Visual effects of stress that were found to be well correlated with the spectrometry studies were also evident. They believed that the chemical changes in the vegetation that were revealed in the spectrometry studies could be used as an indicator of unsustainable withdrawal of water from the aquifers. This research is an important development in the use of indicators for detecting unsustainable practices leading to stress on vegetation caused by inadequate water. In a similar manner, Chisholm et al. (2003) studied moisture stress on *Eucalyptus camaldulensis* (River Red Gum) in Australia, using high-resolution spectral data at the leaf level. Their results indicate that even low levels of stress can be detected from such data before they become visible, and therefore spectral reflectance regions that would be indicators of moisture stress in vegetation and hence act as appropriate SIs may be developed. Withdrawal of water from aquifers is often used as a source of water for urban areas and could be a further manifestation of the impacts of urbanization on surrounding areas. These impacts need to be assessed by appropriate remote sensing technologies. A general term for the *attributes* that could be adopted is *land use/land cover changes* but there are many impacts related to the effects of urbanization on the hinterland that should be included in a detailed set of indicators.

1.5.2.6 Biodiversity

Land cover change threatening biodiversity, and a major variable in the loss of nutrients from productive lands, can be mapped and monitored by a range of remote sensing data sources. This may require high-resolution images or spectrally unmixing approaches to determine the class composition of mixed pixels to capture land cover modifications systematically and on a repetitive basis. Vegetation indices and change detection techniques derived from images from high-resolution optical sensors permit the mapping, monitoring, and measurement of the areal extent of the change.

Hepinstall-Cymerman (2011) demonstrated that analyzing biodiversity requires knowledge of the existing fauna and vegetation, which also involves field surveys and documentation. He described a land cover change avian biodiversity model based on remote sensing, which, together with adequate knowledge of the avian fauna, provided details of the impacts on the species. *Land-use conversion and loss of habitats* are shown as *attributes* for the effects on biodiversity.

1.6 CONCLUSIONS

This chapter aims to present some principles for assessing sustainability of development in urban areas and to describe the ways in which remote sensing can be used in this process. Definitions of sustainable development have been given, and the approach to its assessment based on SIs is described. While a number of different indicators are currently available for the three pillars (social, economic, and environmental), there appears to be no consensus on the most appropriate indicators for sustainability of the environment, and especially for urban areas. The list suggested in Table 1.3 could represent some components of a framework for SIs in urban areas, but considerable work would still be required to further develop and test the set of SIs.

Since more than 50% of the global population now lives in urban areas and this percentage is increasing, there is urgency in determining and assessing SIs for urban areas. Unless the sustainability of urban areas is addressed, there may be little chance of real sustainability of the environment being achieved. Decision-making, which affects urban environments, in many cases, is dominated by market forces often with inadequate consideration of effects on the environment. Interdisciplinary collaborations between the remote sensing community and experts in a range of scientific fields such as ecology, biology, sociology, human resources, and politics, who can take responsibility for assessing the sustainability of urban communities, should be developed so that their combined expertise can determine the impacts of unsustainable practices in urban areas and the hinterland, and decision-makers can be notified of the need to change current practices. While remote sensing technologies will not be the only tools for assessing sustainability, they should make an important contribution to this multidisciplinary process, provided they satisfy scientific criteria, such as being subject to strict calibration and validation. A great deal has yet to be learned about these processes and how the full potential of remote sensing can be achieved in this very important issue of environmental sustainability of urban areas.

REFERENCES

Alberti M. 1996. Measuring urban sustainability, *Environ. Impact Assess. Rev.* 16: 381–424.
Andriantiatsaholiniaina L. A., Kouikoglou V. S., and Phillis Y. A. 2004. Evaluating strategies for sustainable development: Fuzzy logic reasoning and sensitivity analysis. *Ecol. Econ.* 48: 149–172.
Azapagic A. and Perdan S. 2005a. An integrated sustainability decision-support framework. Part I: Problem structuring. *Int. J. Sustainable Dev. World Ecol.* 12: 98–111.
Azapagic A. and Perdan S. 2005b. An integrated sustainability decision-support framework. Part II: Problem analysis. *Int. J. Sustainable Dev. World Ecol.* 12: 112–131.
Bacchus S. T., Archibald D. D., Brook G. A. et al. 2003. Near-infrared spectroscopy of a hydroecological indicator: New tool for determining sustainable yield for Floridan aquifer system. *Hydrol. Processes* 17: 1785–1809.
Bacchus S. T., Hamazaki T., Britton K. O., and Haines B. L. 2000. Soluble sugar composition of pond-cypress: A potential indicator of ground-water perturbations. *J. Am. Water Resour. Assoc.* 36(1): 55–65.
Barrios E. and Kimoto K. 2006. Some approaches to the construction of sustainable development index for the Philippines, *Int. J. Sustainable Dev. World Ecol.* 13: 277–288.
Becker B. 1997. Sustainability assessment: A review of values, concepts, and methodological approaches, *Issues in Agriculture 10, CGIAR*, Washington, DC, The World Bank.
Becker J. 1998. Sustainable development assessment for local land uses. *Int. J. Sustainable Dev. World Ecol.* 5: 59–69.
Becker J. 2007. How frameworks can help operationalize sustainable development indicators, *World Futures.* 63: 137–150.
Biggs R., Scholes R. J., ten Brink J. E., and Vačkář D. 2007. Biodiversity indicators. In Hák T., Moldan B., and Dahl A. J. (Eds.), *Sustainability Indicators: A Scientific Assessment*, pp. 249–270, Island Press, Washington, DC.
Blanco H., Wautiez F., Llavero A., and Riveros C. 2001. Sustainable development indicators in Chile: To what extent are they useful and necessary? *Eure-Revista Latinoamericana de Estudios Urbano Regionales.* 27: 85–95.
Böhringer C. and Jochem P. E. P. 2007. Measuring the immeasurable—A survey of sustainability indices. *Ecol. Econ.* 63: 1–8.

Brandon P. and Lombardi P. 2011. *Evaluating Sustainable Development in the Built Environment, Second Edition*, Wiley-Blackwell, Hoboken, New Jersey.

Canters F., Batelaan O., van de Voorde T. et al. 2011. Use of impervious surface data obtained from remote sensing in distributed hydrological modeling of urban areas, In Yang X. (Ed.), *Urban Remote Sensing*, pp. 255–273, Wiley-Blackwell, Hoboken, New Jersey.

Chisholm L. A., Cooke J., Erdmann B. et al. 2003. Preliminary Investigations into Observed River Red Gum Decline along the River Murray below Euston: *Technical Report 03/03. Australia: Vic: Murray-Darling Basin Commission.*

Cornelissen A. M. G., van den Berg J., Koops W. J. et al. 2001. Assessment of the contribution of sustainability indicators to sustainable development: A novel approach using fuzzy set theory. *Agric. Ecosyst. Environ.* 86: 173–185.

Foody G. M. 2003. Remote sensing of tropical forest environments: Towards the monitoring of environmental resources for sustainable development, *Int. J. Remote Sens.* 24(20): 4035–4046.

Freedberg M. 2010. In Schaffer D., Vollmer D. (Eds.), *Pathways to Urban Sustainability: Research and Development on Urban Systems: Summary of a Workshop by Committee on the Challenge of Developing Sustainable Urban Systems*, pp. 9–10, National Research Council, Washington, DC.

Gallopin G. C. and Raskin P. D. 2002. *Global Sustainability Bending the Curve.* Routledge, London.

Gamba P., Erll'Acqua F., Stasolla M. et al. 2011. Limits and challenges of optical very-high-resolution satellite remote sensing for urban applications, In Yang X. (Ed.), *Urban Remote Sensing*, pp. 36–47, Wiley-Blackwell, Hoboken, New Jersey.

Greenhill D. R., Ripke L. T., Hitchman A. P. et al. 2003. Characterization of suburban areas for land use planning using landscape ecological indicators derived from Ikonos-2 multispectral imagery, *IEEE Trans. GRSS* 41(9): 2015–2023.

Goddard M. A, Dougill A. J., and Benton T. G. 2010. Scaling up from gardens: Biodiversity conservation in urban environments, *Trends Ecol. Evol.* 25(2): 90–98.

Guindon B. and Zhang Y. 2007. Using satellite remote sensing to survey transport-related urban sustainability. Part II. Results of a Canadian urban assessment, *Int. J. Appl. Earth Observ. Geoinf.* 9: 276–293.

Gutiérrez-Espeleta E. E. 2007. Further work needed to develop sustainable development indicators. In Hák T., Moldan B., and Dahl A. J. (Eds.), *Sustainability Indicators: A Scientific Assessment*, pp. 351–360, Island Press, Washington, DC.

Hák T., Moldan B., and Dahl A. J. (Eds.) 2007. *Sustainability Indicators: A Scientific Assessment*, Island Press, Washington, DC.

Harrington L., Jones P. G., and Winograd M. 1993. Measurements and indicators of sustainability. *Report of a Consultancy Team*, Centro Internacional de Agricultura Tropical (CIAT), Cali, Colombia.

Hart M. 1999. *Guide to Sustainable Indicators*. 2nd Ed., Hart Environmental Data, North Andover, MA.

Hepinstall-Cymerman J. 2011. Ecological modeling in urban environments: Predicting changes in biodiversity in response to future urban development. In Yang X. (Ed.), *Urban Remote Sensing*, pp. 359–370, Wiley-Blackwell, Hoboken, New Jersey.

Herold M., Couclelis H., and Clarke C. C. 2003. The role of spatial metrics in the analysis and modeling of urban land use change, *Comput. Environ. Urban Syst.* 29: 369–399.

Huang J., Lu X. X., and Sellers J. M. 2007. A global comparative analysis of urban form: Applying spatial metrics and remote sensing, *Landscape Urban Plann.* 82: 184–197.

Hueting R. and Reijnders L. 2004. Broad sustainability contra sustainability: The proper construction of sustainability indicators, *Ecol. Econ.* 50: 249–260.

Jabareen Y. R. 2006. Sustainable urban forms, *J. Plann. Educ. Res.* 26: 38–52.

Jesinghaus J. 2007. Indicators: Boring statistics or the key to sustainable development? In Hák T., Moldan B., and Dahl A. J. (Eds.), *Sustainability Indicators: A Scientific Assessment*, pp. 351–360, 83–96, Island Press, Washington, DC.

Karlsson S., Dahl A. L., Biggs R. et al. 2007. Meeting conceptual challenges, In Hák T., Moldan B., and Dahl A. J. (Eds.), *Sustainability Indicators: A Scientific Assessment*, pp. 27–48. Island Press, Washington, DC.

Kates R. W. 2000. Population and consumption: From more to enough. In Schmandt J. and Ward C.H (Eds.), *Sustainable Development: The Challenge of Transition*, pp. 79–99, Cambridge University Press: Cambridge.

Lein J. K. 2014. Toward a remote sensing solution for regional sustainability assessment and monitoring, *Sustainability* 6: 2067–2086.

Mahi P. 2001. Developing environmentally acceptable desalination projects. *Desalination* 138: 167–172.

McGlade J. 2007. Foreword: Finding the right indicators for policymaking, In Hák T., Moldan B., and Dahl A. J. (Eds.), *Sustainability Indicators: A Scientific Assessment*, Island Press, Washington, DC.

Moldan B. and Dahl A. J. 2007. Challenges to sustainability indicators. In Hák T., Moldan B., and Dahl A. J. (Eds.), *Sustainability Indicators: A Scientific Assessment*, pp. 1–24, Island Press, Washington, DC.

NASA. 2010. Beijing Quadrupled in Size in a Decade, NASA Finds. http://www.jpl.nasa.gov /news/news.php?feature=4641, accessed July 31, 2015.

Olalla-Tárraga M. A. 2006. A conceptual framework for assessing urban ecological systems, *Int. J. Sustainable Dev. World Ecol.* 13: 1–15.

Petrosyan A. 2014. Proposal of composite appraising supportive progress beyond twelve (12) economic sustainability indices, *J. Econ. Dev. Stud.* 2(2): 547–569.

Phillis Y. A. and Andriantiatsaholiniaina L. A. 2001. Sustainability: An ill-defined concept and its assessment using fuzzy logic. *Ecol. Econ.* 37: 435–456.

Radermaker F. J. 2004. *Balance or Destruction*, Oekosoziales Forum Europa, Vienna, Austria.

Rao D. P. 1998. Remote sensing and GIS for sustainable development: An overview, *Int. Arch. Photogramm. Remote Sens.* XXXII(7): 156–163.

Rashed T. and Jürgens C. 2010. (Eds) *Remote Sensing of Urban and Suburban Areas*, Heidelberg, Springer.

Rees W. and Wackernagel M. 1996. Urban ecology footprints: Why cities cannot be sustainable—And why they are the key to sustainability, *Environ. Impact Assess. Rev.* 16: 223–248.

Roessner S., Segl K., Bochow M. et al. 2011. Potential of hyperspectral remote sensing for analysing the urban environment. In Yang X. (Ed.), *Urban Remote Sensing,* pp. 50–61, John Wiley & Sons, Hoboken, New Jersey.

Salah M., Trinder J., Shaker A. et al. 2010. Integrating multiple classifiers with fuzzy majority voting for improved land cover classification, *Int. Arch. Photogramm. Remote Sens. Spat. Inf. Sci.* 39(3A): 7–12.

Sapena M. and Ruiz L. A. 2015. Analysis of urban development by means of multi-temporal fragmentation metrics from LULC data, *Int. Arch. Photogramm. Remote Sens. Spat. Inf. Sci.* XL-7/W3: 1411–1418.

Shen L.-Y., Ochoa J. J., Shah M. N., and Zhang X. 2011. The application of urban sustainability indicators—A comparison between various practices, *Habitat Int.* 35:17–29.

Sim S. and Mesev V. 2011. Measuring urban sprawl and compactness: Case study Orlando, *International Cartographic Conference*, CO-437, pp. 1–10, Paris, France.

Sliuzas R., Kuffer M., and Masser I. 2010. In Rashed T. and Jürgens C. (Eds.), *Remote Sensing of Urban and Suburban Areas*, pp. 67–84, Heidelberg, Springer.

Soergel U. (Ed.) 2010. *Radar Remote Sensing of Urban Areas*, Heidelberg, Springer.

Stevenson M. and Li H. 2001. Indicators of sustainability as a tool min agricultural development: Partitioning scientific and participatory process, *Int. J. Sustainable Dev. World
Ecol.* 8: 57–65.

Tanguay G. A., Rajaonson J., Lefebvre J.-F. et al. 2010. Measuring the sustainability of cities:
An analysis of the use of local indicators, *Ecol. Indic.* 10: 407–418.

Tsai Y.-H. 2005. Quantifying urban form: Compactness versus 'sprawl', *Urban Stud.* 42(1):
141–161.

Turner R. K. 1993. Sustainability: Principles and practices. In Turner R. K. (Ed.), *Sustainable
Environmental Economics and Management: Principles and Practices*, pp. 3–36,
London, Belhaven Press.

Van Woerden J., Singh A., and Demkine V. 2007. Core set of UNEP GEO indicators among
global environmental indices, indicators and data, In Hák T., Moldan B., and Dahl
A. J. (Eds.), *Sustainability Indicators: A Scientific Assessment*, pp. 343–350, Island
Press, Washington, DC.

Weng Q. 2012. Remote sensing of impervious surfaces in the urban areas: Requirements,
methods, and trends, *Remote Sens. Environ.* 117: 34–49.

Wu C. and Yuan F. 2011. Remote sensing of high resolution urban impervious surfaces,
In Yang X. (Ed.), *Urban Remote Sensing*, pp. 241–254, Wiley-Blackwell Hoboken,
New Jersey.

Zhang Y. and Guindon B. 2006. Using satellite remote sensing to survey transport-related
urban sustainability. Part 1: Methodologies for indicator quantification, *Int. J. Appl.
Earth Observ. Geoinf.* 8: 149–164.

2 Earth Observation for Urban and Spatial Planning

Mattia Marconcini, Annekatrin Metz, Thomas Esch, and Julian Zeidler

CONTENTS

2.1 EARTH OBSERVATION APPLICATIONS IN URBAN AND SPATIAL PLANNING

At present, the two most critical phenomena affecting cities worldwide are urbanization and climate change. Indeed, on the one hand, the United Nations (UN 2014) estimate that nowadays 54% of the human population is living in urban areas (up from 34% in the 1960s); on the other hand, global climate changes are directly affecting the economy of cities as well as the quality of urban environments.

In such context, cities play a dual role: they are part of the problem and a key part of the solution (Kamal-Chaoui and Alexis 2009). In particular, despite covering ~2%–3% of the emerged land, cities are responsible for 30%–40% of greenhouse gas emissions (or 70% if all the human activities are taken into account) (UN 2011), which directly affect population health and result in flooding, storms, heat, drought, sea-level rise, and damages to infrastructures and buildings. Nevertheless, by means of proper managing strategies, they can also lessen the adverse impacts of climate change through a wide range of system-specific actions and become sustainable and livable environments. However, this is a challenging task for urban planners as they must take into consideration all the complex and interconnected features of urban systems for properly balancing their socioeconomic and environmental dynamics.

In this framework, it is of paramount importance to provide the decision makers with proper and reliable information and to directly engage the citizenry in

29

implementing suitable adaptation and mitigation actions. To this aim, in addition to in situ measurements as well as socioeconomic and environmental variables, Earth Observation (EO)–based products are also increasingly used at present as part of the planning intelligence; indeed, they proved to be of great support to environmental and climate managers in a variety of applications. Nevertheless, there still exist several methodological and technological gaps between the pressing requirement from the users and the current state-of-the-art methodologies. As an example, there is a growing need for assimilating EO-based products into urban models or for facilitating the use of geo-spatial products by nonexperts as well.

This contribution introduces selected geo-information products and their related application in support of local and regional urban planning, such as the assessment of impervious urban surface, the mapping of urban growth, and the characterization of settlement patterns. The corresponding studies were conducted for five regions of interest, including Antwerp (Belgium), Helsinki (Finland), London (United Kingdom), Madrid (Spain), and Milan (Italy).

2.2   SELECTED APPLICATIONS

2.2.1   Aseessment of Impervious Urban Surface

Urban growth is associated not only with the construction of new buildings but also (and in more general terms) with a consistent increase of all the impervious surfaces (hence including roads, parking lots, squares, pavements, and railroads as well), which do not allow water to penetrate, forcing it to run off. Effectively mapping the impervious surface area is then of high importance because it is related to the risk of urban floods, the urban heat island phenomenon, and the reduction of ecological productivity. However, to date, this task has been mostly carried out by photointerpretation of very high resolution (VHR) airborne optical imagery or in situ surveys, which are generally costly and time-consuming and hence forbid a systematic regional-scale mapping as well as regular updates. To this purpose, EO data have started being used since they proved capable of improved detection capabilities, larger-scale analysis, and lower costs. In this context, one of the current state-of-the-art methodologies has been presented by Esch et al. (2009), which allows one to automatically estimate the percentage of impervious surface (PIS) by analyzing the Normalized Difference Vegetation Index (NDVI) calculated from single-date Landsat Thematic Mapper scenes. Nonetheless, the performances might strongly vary depending on (i) the quality of the Landsat image used for the analysis and (ii) the availability of local railway and road network vector layers, as well as VHR optical imagery needed to train the employed empirical regression model based on support vector regression (SVR). To overcome such drawbacks, we improved the above-described technique by (i) considering the mean temporal NDVI calculated from a series of Landsat-8 (LS8) scenes acquired over the area of interest in a 1-year time frame (which allows a drastic reduction in the effect of vegetation phenology and prevents problems related to cloud coverage and shadow in specific scenes) and (ii) deriving the training points for the SVR model starting from OpenStreetMap data (OSM 2015). In particular, samples are extracted in selected areas from OpenStreetMap layers corresponding to impervious surfaces. Next, they are first rasterized and then aggregated at the 30-m spatial resolution of Landsat imagery. An SVR model is finally used to correlate the resulting training information with the mean temporal NDVI and hence estimate the PIS for the whole study area. Specifically, only pixels denoted as urban in the Global Urban Footprint (GUF) mask described by Esch et al. (2012) are preserved. An example of the estimated PIS for Helsinki (Finland) based on a series of 25 LS8 scenes collected for two path-row combinations covering the city in the entire 2014 is illustrated in Figure 2.1.

A quantitative assessment of the effectiveness of the obtained results has been carried out by means of WorldView-2 (WV2) multispectral scenes acquired at 2-m spatial resolution and available for the five considered study cases (see Table 2.1). In particular, given the spatial detail offered by the WV2 images, we could mark with a very high degree of confidence all the impervious structures included in the different study regions. To this aim, we first computed for each scene the NDVI and manually identified the most suitable threshold that allows the exclusion of all the green areas (i.e., nonimpervious); then, we refined the resulting mask by extensive photointerpretation and aggregated it to 30-m spatial resolution. Finally, we compared the resulting WV2-based reference PIS to the corresponding portion of the 2014 PIS products obtained with the proposed method.

To this aim, three different measures have been considered:

- The Pearson's correlation coefficien, which measures the strength of the linear relationship between two variables, and it is defined as the covariance of the two variables divided by the product of their standard deviations; in particular, it is largely used in the literature for validating the output of regression models (as in our case).
- The Mean Error (ME), which is calculated as the difference between the estimated value (i.e., the 2014 LS8-based PIS) and the reference value (i.e., the WV2-based reference PIS) averaged over all the pixels of the image.
- The Mean Absolute Error (MAE), which is calculated as the absolute difference between the estimated value (i.e., the 2014 LS8-based PIS) and the reference value (i.e., the WV2-based reference reference) averaged over all the pixels of the image.

The results of this comparison are reported in Table 2.2 and are extremely promising. Indeed, we obtained a mean correlation of 0.8271 and average ME and MAE equal to −0.09 and 13.33, respectively, which confirms the great effectiveness of the LS8-based PIS products. However, it is also worth pointing out that because of the different acquisition geometries, the WV2 and LS8 images generally exhibit a very small shift. Nevertheless, despite being limited, such displacement often results in a one-pixel shift between the LS8-based PIS and the WV2-based reference PIS aggregated at a 30-m resolution. This somehow affects the computation of the MAE and of the correlation coefficient (which, however, resulted in highly satisfactory values). Instead, the bias does not alter the ME, which always exhibited values close to 0, thus confirming the capabilities of the implemented technique.

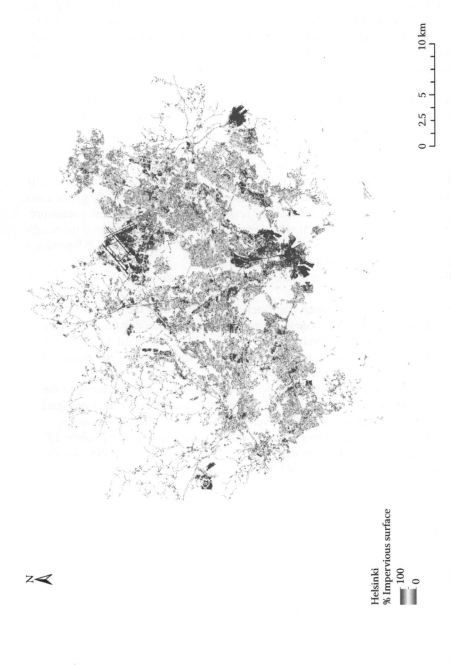

FIGURE 2.1 PIS for Helsinki (FI), modeled on the basis of multitemporal indices derived from Landsat-8 TM data.

TABLE 2.1

Acquisition Dates and Size of the WV2 Images Available for the Five Study Sites

	Acquisition Date (DD.MM.YYYY)	Size (Pixel)
Antwerp	31.07.2014	5404 × 7844
Helsinki	21.04.2014	12,468 × 9323
London	28.08.2013	7992 × 8832
Madrid	20.12.2013	10,094 × 13,105
Milan	14.05.2014	8418 × 7957

TABLE 2.2

Pearson's Correlation Coefficient, Mean Error (ME), and Mean Absolute Error (MAE) Obtained from the Comparison of the 2014 PIS Obtained with the Proposed LS8-Based Method and the WV2-Based Reference PIS

	Pearson's Correlation Coefficient	ME	MAE
Antwerp	0.8713	−2.69	12.07
Helsinki	0.7847	0.44	15.60
London	0.8094	−1.32	13.79
Madrid	0.8079	2.45	12.43
Milan	0.8623	0.67	12.76
Mean	0.8271	−0.09	13.33

2.2.2 MAPPING OF URBAN GROWTH

Reliably delineating the urban growth that occurred in the last decades is of great importance to properly model the temporal evolution of urbanization and, hence, to better estimate future trends and implement suitable planning strategies. In Europe, all major cities have access nowadays to digital databases with highly detailed information about the urban extent often down to the single-building level (mostly derived from VHR optical airborne imagery taken every 2 to 5 years). Nonetheless, the information available for the past 20 years is often of poor quality because urban extent maps were mostly generated manually by in situ surveys at that time.

For this purpose, we developed a novel approach for delineating the urban area extent from ESA radar imagery, namely, ERS-1/2 SAR Precision Image and Envisat ASAR Image Mode Precision products acquired between 1992 and 2012 at 30 × 30 m spatial resolution with 12.5 m pixel distance (Marconcini et al. 2014). The corresponding block scheme is reported in Figure 2.2.

There are two main assumptions of the implemented method. On the one hand, we suppose that for the investigated region of interest, a binary mask M is available, which delineates the current extent of urban areas, for example, from local cadastral data or, alternatively, from GUF (Esch et al. 2012) or OpenStreetMap

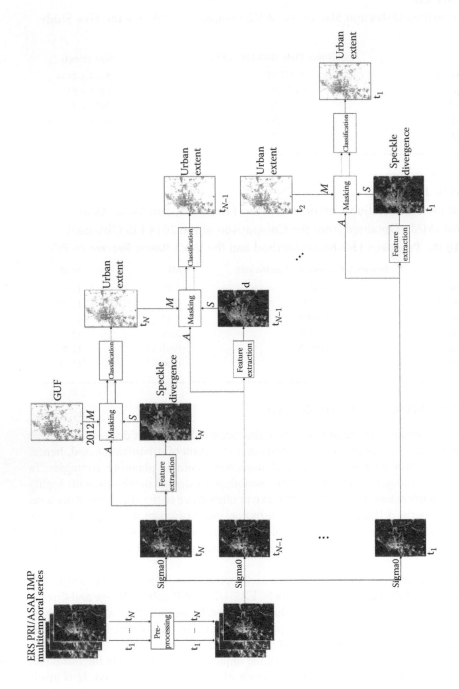

FIGURE 2.2   Schematic view of the work flow for the spatiotemporal mapping of urbanization based on ERS/ASAR time series data.

data (OSM 2015). Specifically, we solely consider built-up structures without linear elements as streets, roads, or railways because it is not feasible to properly identify them by means of radar data. On the other hand, we assume that human settlements in the study area experienced urban expansion rather than shrinkage in the last decades, which is generally reasonable given the current global trend of urbanization.

Given a multitemporal sequence of backscattering ERS/ASAR images available for the investigated region in the periods of interests, we sort them by acquisition date from the newest to the oldest; then, we apply calibration and terrain correction to each scene and finally co-register the entire series. Next, we consider the first item of the series A and, in line with the strategy adopted for deriving the GUF, we extract its speckle divergence texture feature S, which represents an estimate of the local true image texture and exhibits high values solely in correspondence of heterogeneous and highly structured built-up areas.

Under the considered working hypothesis, one pixel cannot be categorized as belonging to the built-up class earlier in the time series, if this does not occur at a later time. Accordingly, we mask A and S by using M and only keep those pixels labeled as urban in M, which are then provided as input to the unsupervised classification scheme presented in Esch et al. (2013), which allows one to automatically derive the corresponding built-up extent. This process is then iteratively applied to the next images of the multitemporal sequence and, for each scene, the corresponding A and S are masked by means of the built-up mask obtained for the previous item of the series.

It might occur that, for specific periods of interest (e.g., 1 or 2 years), several ERS/ASAR scenes have been acquired over the study area. In this case, to properly account for the generally stable behavior of the urban areas (mostly associated with high backscattering values) compared to the other information classes (which might show high values only under specific conditions), it might be beneficial to first calculate the mean temporal backscattering and then derive its speckle divergence. This allows an improvement in the performances and a reduction of the effect of the speckle noise.

The outcome of the spatiotemporal urbanization mapping for Madrid (Spain) is presented in Figure 2.3, showing the urban area extent derived from ERS-1, ERS-2, and ASAR imagery acquired in 1992 (light brown), 2000 (red), 2006 (blue), and 2010 (dark blue). Moreover, the change polygons provided by the European CORINE Land Cover data set (Figure 2.3a) are compared to the changes identified based on the SAR time series (Figure 2.3b). Considering the building block–related change indication, a block was only assigned as change area if a sufficient proportion of change pixels were identified within the corresponding area.

To assess the effectiveness of the proposed methodology, we considered the city of Madrid as the representative validation case. Indeed, the municipality has a surface of ~600 km^2 and includes both high- and low-density residential areas, as well as large portions of rural and forest areas. Accordingly, we evaluated the accuracies of the corresponding built-up extent products generated for the years 2000, 2006,

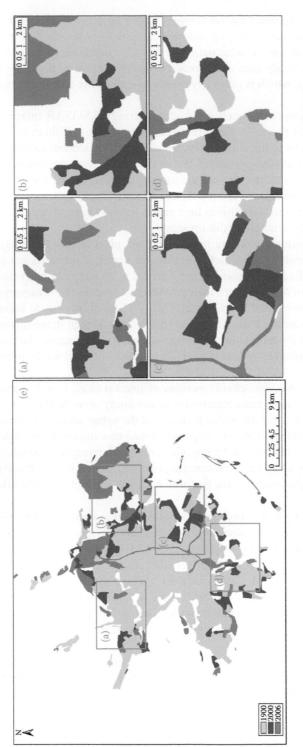

FIGURE 2.3 Urban growth estimated for specific hot spots (a–d) and the entire city (e) of Madrid (Spain) as defined from CORINE data.

(Continued)

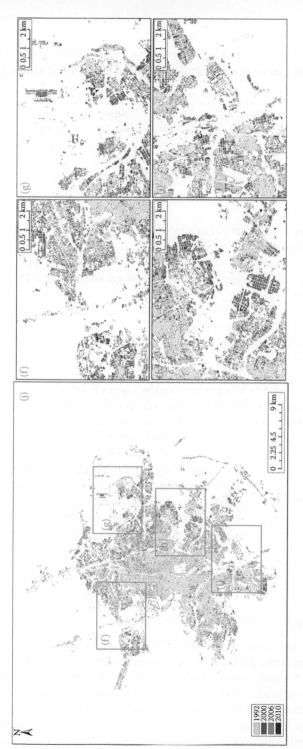

FIGURE 2.3 (CONTINUED) Urban growth estimated for specific hot spots (f–i) and the entire city (j) of Madrid (Spain) as defined from the presented methodology based on ERS/ASAR SAR time series data (12.5 m spatial resolution).

TABLE 2.3
OA%, PA%, UA%, and Kappa Coefficient Computed for the 2000, 2006, and 2010 Built-Up Extent Products Generated for the City of Madrid (Spain)

		PA%		UA%		
Year	OA%	Non–Built-Up	Built-Up	Non–Built-Up	Built-Up	Kappa
2010	89.40	89.30	89.51	89.57	89.24	0.7880
2006	85.98	85.70	86.28	86.30	85.67	0.7197
2000	87.29	86.60	87.99	87.92	86.68	0.7459

and 2010, by means of VHR optical imagery from Google Earth. In particular, VHR data are available for the following dates:

- September 15, 2000—suitable for the assessment of the year 2000 built-up extent product
- July 5, August 5, and September 10, 2006—suitable for the assessment of the year 2006 built-up extent product
- October 2, 2009—suitable for the assessment of the year 2010 built-up extent product

In our analysis, we used a blind interpretation approach, where randomly selected points have been labeled as urban/nonurban via photointerpretation before generating the built-up extent maps with the presented technique. For each of the three considered years, we labeled 1000 built-up and 1000 non–built-up points randomly distributed within the boundaries of the Madrid municipality. Specifically, these validation points were chosen not to be falling within a buffer of 10 m between built-up and non–built-up areas. Then, we compared the labels manually associated to the resulting set of 2000 points against the corresponding ones of the built-up extent products. Finally, different statistics were computed, namely, percentage overall accuracy (OA%), percentage producer's accuracy (PA%), percentage user's accuracy (UA%), and kappa coefficient (which also takes into consideration errors and their type). Results reported in Table 2.3 assess the capabilities of the proposed technique, which always exhibited very high accuracies.

2.2.3 CHARACTERIZATION OF SETTLEMENT PATTERN

In order to define and implement effective adaptation and mitigation strategies for a certain study area (from the local to the regional scale), it is important to properly characterize its settlement pattern. This involves not only gathering a comprehensive knowledge about form, size, and distribution of the corresponding types of settlements but also quantitatively assessing their significance with respect to each other. As an example, the local impact of a big but isolated city might be lower than that of a smaller city connected to many others in its surrounding. In this framework, we implemented a novel technique that, by means of graph theory and spatial network analysis, allows one to properly model the relationships between different settlements and to quantitatively describe their mutual relevance (Esch et al. 2014). In particular, given a binary mask delineating urban and nonurban areas within the region of interest, we first identify all settlement objects (defined as clusters composed of pixels labeled as urban and connected via at least one edge or corner). Then, a spatial network is created where the nodes are associated with the centroids of the different objects, while the edges connect neighboring settlements within a predefined Euclidean distance from each other. Next, several attributes are computed for each node describing the geometrical properties of the corresponding settlement, for example, area, perimeter, solidity, equivalent diameter, shape index, and eccentricity. Likewise, weights are calculated for each edge characterizing the link between the two connected objects, for example, minimum Euclidean distance, number of crossed edges, and cohesion index (see Esch et al. 2014).

To characterize the impact of each node (and hence of the corresponding settlement object), we finally compute different relevance indices that jointly take into account its attributes as well as the weights associated with all its edges. For instance, we considered the degree centrality (defined as the total number of edges connected to a given node), the betweenness centrality (defined as the number of times that one node is included in the shortest path between any two other nodes in the network), or the local dominance (defined as the ratio between the degree centrality and the number of edges for which the given node has a size greater than that of the neighbor to which it is connected).

Many experimental trials have been carried out at different scales for assessing the performances of the presented method, which proved to be an effective and promising tool for supporting both quantitative and qualitative settlement pattern analyses.

In Figure 2.4, we report the spatial network obtained for the greater London area starting from the corresponding portion of the GUF data set (Esch et al. 2012) further split up and subdivided by the administrative district boundaries (for this reason, the city of London consists of different polygons associated with its 33 boroughs). A minimum Euclidean distance of 1 km has been considered when computing the edges. The size of different nodes is proportional to the area of the corresponding settlement, whereas their color varies based on the resulting degree centrality. Here, it is worth noting that nodes appearing in darker tones are those involved in the highest number of interactions and that most of them are located at the western side of London (i.e., where then the city receives more pressure from the outside).

2.3   CONCLUSIONS AND OUTLOOK

The constantly increasing availability and accessibility of modern remote sensing technologies provides new opportunities for a wide range of urban applications such as mapping and monitoring of the urban environment (land cover, land use, morphology, urban structural types), socioeconomic estimations (population density), characterization of urban climate (microclimate, human health conditions), analysis of regional and global impacts (groundwater and climate modeling, urban heat

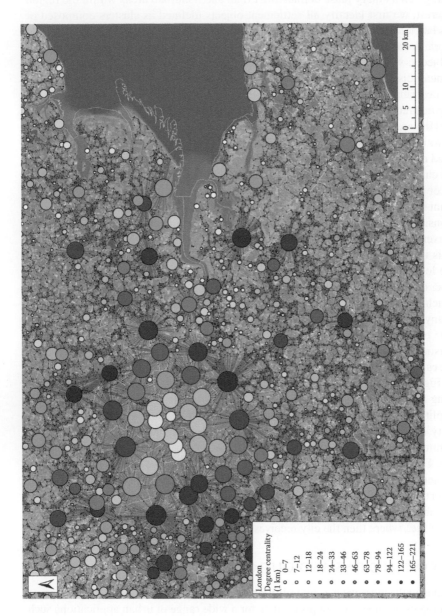

FIGURE 2.4 Spatial network obtained for the Greater London area. The size of the nodes is proportional to the area of the corresponding settlement, whereas their color varies depending on the computed degree centrality. Edges connect objects within a minimum Euclidean distance of 1 km from each other.

islands), or urban security and emergency preparedness (sustainability, vulnerability). In this contribution, we have introduced a selection of applications and example products that have been developed for providing additional and innovative data that can support day-to-day decision-making in urban and regional planning.

For most of the applications dealing with urban environments, the basic challenge is related to the spectral heterogeneity and morphological complexity of built-up areas. The spectral heterogeneity originates from the enormous diversity of different materials forming the urban landscape. Thereby, some land cover types such as vegetation, bare soil, or water are also found in nonurban environments. Moreover, certain surfaces (e.g., bare soil and specific construction materials of buildings or pavements) can hardly be differentiated from each other through their spectral signature. Regarding the morphological complexity, urban areas are characterized by structural elements featuring diverse scales and shapes. In order to accurately capture the morphological properties of urban objects, a very high spatial resolution of the sensor system is required. However, although an increased spatial resolution certainly expands the spectrum of urban applications, this development comes along with new challenges in terms of an automated image analysis. On the one hand, the observable heterogeneity within the specific object types increases significantly since many local but often nonrelevant characteristics appear (e.g., roof lights and chimneys on top of buildings or cars, street furniture, and sign postings on streets). On the other hand, urban features are hence formed by groups of pixels with similar spectral signatures.

To address these challenges arising from an improved spatial resolution, recent studies have increasingly used object-oriented analysis approaches. Compared to the established pixel-based approaches, these techniques facilitate an improved consideration of spectral, geometric, textural, contextual, and hierarchical characteristics.

The previous remarks regarding urban remote sensing stress that the suitable approach, technology, and data are highly dependent on the thematic focus and the spatial scale of the analysis. Medium-resolution multispectral data (e.g., Landsat, Spot, and IRS-P6) are best suited for regional analyses since they cover areas of up to 32,000 km² with one image ensuring cost-effective analyses. At the same time, the spatial resolution is still sufficient to discriminate built-up areas from nonurban regions based on spectral and textural characteristics. Because of their direct link to morphological properties, high- and medium-resolution SAR images provide particularly robust features for the detection of settlements. However, the applicability of SAR data for local analysis of the urban structures is still limited since the complex geometrical and physical characteristics of metropolitan areas and the varying appearance and visibility of objects subject to the line of sight lead to significant distortions of and ambiguities in the resulting radar images.

To cope with the heterogeneity and complexity of urban areas, VHR multispectral systems such as IKONOS or QuickBird are required. Indeed, their sensors provide images in four spectral bands featuring a ground resolution of 4 m (IKONOS) and 2.44 m (QuickBird) supplemented by a panchromatic channel with a geometric resolution of 1 m (IKONOS) and 61 cm (QuickBird). Some drawbacks of these data are the limitation of the spectral resolution to four bands—facilitating only a very rough reconstruction of the spectral signature—and the limited spatial coverage of a few hundred square kilometers by one image. Hence, analyses of complete metropolitan areas and major or mega cities demand a data volume that significantly increases the complexity and expense for image processing and classification.

The immense spectral resolution of hyperspectral sensor systems enables thematically comprehensive and spatially detailed characterizations of the urban environment. However, current hyperspectral sensor systems showing a spatial resolution that is useful for urban applications are limited to airborne platforms. The first high-resolution hyperspectral satellite sensor EnMAP is supposed to be launched by 2017. This system will feature a spatial resolution of 30 m and cover the spectral range of 420–2450 nm with approximately 200 bands. Heldens et al. (2011) reviewed 146 publications to give an outlook on the capabilities of the EnMAP mission for urban applications.

The synchronism and coexistence of economic activities, environmental threats, infrastructural deficits, poverty, and population growth mark a significant challenge to urban planning. Therefore, future research has to focus on integrated interdisciplinary studies to understand the multidimensional and complex interactions of urban systems and to analyze and assess the effects of plans, actions, and concepts. An important step toward the improvement of the generated information products and their acceptance by decision makers consists in the adaptation to holistic approaches on complex urban systems. Hence, the appropriate concepts have to integrate and correlate multiple analysis tools (image analysis software, geographic information system), data types (satellite images, vector data and statistics), and data sources (EO, in situ survey, census). The synergetic use of various data sources and their combined analysis increases the quality and information content of the resulting products, opens new levels of information, and enhances the possibilities of integrating the resulting data and information into existing systems and concepts. However, in view of regional, national, or even global monitoring tasks, there is still some effort needed with respect to the availability and accessibility of remote sensing data and the operationalization of image processing in order to allow for cost- and time-efficient analyses and a rapid provision of the required information. Thereby, new sensor systems such as RapidEye and GeoEye will improve the capabilities of urban remote sensing application, particularly in terms of providing detailed time series of multispectral imagery.

ACKNOWLEDGMENTS

The results of the studies introduced in this chapter were generated in the context of the FP7 DECUMANUS project. The SAR data used in this study was provided by ESA based on the Category-1 proposal no. 16969.

REFERENCES

Esch, T., V. Himmler, G. Schorcht, M. Thiel, T. Wehrmann, F. Bachofer, C. Conrad, M. Schmidt, and S. Dech, Large-area assessment of impervious surface based on integrated analysis of single-date Landsat-7 images and geospatial vector data. *Remote Sensing of Environment*, 113, 1678–1690, 2009.

Esch, T., H. Taubenböck, A. Roth, W. Heldens, A. Felbier, M. Thiel, M. Schmidt, A. Müller, and S. Dech. TanDEM-X mission-new perspectives for the inventory and monitoring of global settlement patterns. *Journal of Applied Remote Sensing*, 6(1), 061702, 2012; 21 pp., doi: 10.1117/1.JRS.6.061702.

Esch, T., M. Marconcini, A. Felbier, A. Roth, W. Heldens, M. Huber, M. Schwinger, H. Taubenböck, A. Müller, and S. Dech, Urban footprint processor—Fully automated processing chain generating settlement masks from global data of the TanDEM-X Mission. *IEEE Geoscience and Remote Sensing Letters*, 10(6), 1617–1621, 2013.

Esch, T., M. Marconcini, D. Marmanis, J. Zeidler, S. Elsayed, A. Metz, and S. Dech. Dimensioning urbanization—An advanced procedure for characterizing human settlement properties and patterns using spatial network analysis. *Applied Geography*, 55, 212–228, 2014.

Heldens, W., U. Heiden, T. Esch, E. Stein, and A. Müller. Can the future EnMAP mission contribute to urban applications? A literature survey. *Remote Sensing* 3(9), 1817–1846, 2011.

Kamal-Chaoui, L. and R. Alexis. Competitive Cities and Climate Change, OECD Regional Development Working Papers no. 2, OECD Publishing, 2009.

Marconcini, M., A. Metz, T. Esch, and J. Zeidler, Global urban growth monitoring by means of SAR data. *Proceedings of the IEEE International Geoscience and Remote Sensing Symposium* (IGARSS 2014), Quebec, Canada, July 13–18, 2014.

OpenStreetMap: http://www.openstreetmap.org/ December 21, 2015.

United Nations, Human Settlements Programme, Cities and Climate Change: Global Report on Human Settlements 2011, Earthscan, 2011.

United Nations, Department of Economic and Social Affairs, Population Division. World Urbanization Prospects: The 2014 Revision, Highlights (ST/ESA/SER.A/352), 2014.

3 Assessment of Urban Growth in the Pearl River Delta, China, Using Time Series Landsat Imagery

Lei Zhang and Qihao Weng

CONTENTS

3.1 INTRODUCTION

Urban growth has a significant influence on urban environments, including climate change (Liao et al. 2015; Pathirana et al. 2014), biochemical cycles (Hutyra et al. 2014), and environment quality (Panagopoulos et al. 2015; Zhao et al. 2015). Assessment of urban growth is needed for sustainable development and studies on ecological consequences. Since remote sensing technology provides spatially consistent data with high spatial resolution and high temporal frequency, remote sensing imagery makes it possible to analyze and model urban growth over long periods at various scales in a timely and cost-effective manner. Multitemporal coarse or medium spatial resolution imagery has been commonly applied for analysis of urban growth, such as DMSP/OLS (Defense Meteorological Satellite Program/Operational Linescan System) nighttime light data (Liu et al. 2012; Ma et al. 2012; Zhang and Seto 2011), Landsat archive (Bagan and Yamagata 2012; Michishita et al. 2012; Sun et al. 2013; Xian and Crane 2005), China–Brazil Earth Resources Satellites images, and HJ-1 images (Du et al. 2015). The main drawback to using coarse or moderate spatial resolution imagery is the mixed pixel problem, which leads to a salt-and-pepper effect caused by spectral heterogeneity. Multitemporal high spatial resolution imagery has also been used to extract urban areas, such as Spot 5 imagery (Durieux et al. 2008; Jacquin et al. 2008) and RapidEye and IRS data (Dupuy et al. 2012). Although these methods have been applied to analyze urban growth successfully, the successive imagery with high temporal resolution was difficult to obtain. Additionally, the intraclass spectral variability problem is inevitable in high spatial resolution imagery.

Multitemporal analysis for urban growth usually requires single classification or segmentation of all the stacked images and is limited to provide detailed change information because of low temporal resolution. Therefore, time series imagery applied in differentiating land cover has attracted increased attention from researchers in recent years, because temporal domain has showed its advantages in resolving class confusion between classes with similar spectral characteristics (Bhandari et al. 2012; Schneider 2012). Specifically, Landsat time series have been successfully applied to map dynamics of urban areas because of their long record of continuous measurement at effective spatial resolution and temporal frequency (Gao et al. 2012; Li et al. 2015; Sexton et al. 2013a,b). However, these methods focused on spectral differences or temporal consistency after classification. Little attention was paid to temporal data mining method to differentiate urban areas from other land cover using dense time series Landsat images.

Since time series clustering has been shown to be effective in time series data mining (Fu 2011; Liao 2005), in this study, we aimed at extracting urban areas using a semi-supervised fuzzy time series clustering method through the Biophysical Composition Index (BCI) (Deng and Wu 2012) and Land Surface Temperature (LST) time series and applied the method to the Pearl River Delta, China, from 1987 to 2014. Kernel fuzzy C-means (KFCM), proposed by Zhang and Chen (2003), was introduced in this study, because it could provide a more robust signal-to-noise ratio and is less sensitive to cluster shapes in comparison to other clustering algorithms (Du et al. 2005). BCI and LST time series images were derived because of their strong correlation with urban areas. BCI aimed to identify different urban biophysical compositions and has been demonstrated to be effective in identifying the characteristics of impervious surfaces and vegetation and in distinguishing bare soil from impervious surfaces. LST, as a significant parameter in urban environmental analysis, tends to be positively correlated with urban expansion (Weng and Hu 2008; Yuan and Bauer 2007).

3.2 CASE STUDY

3.2.1 Study Area

The Pearl River Delta, as the third most important economic district of China, is located in the developmental core region of Guangdong Province, between 21°N–23°N and 111°E–115°E. It has experienced rapid urbanization since the reform process started in the late 1970s in China. The Pearl River Delta has a subtropical climate with an average annual temperature of 21°C–23°C, including a dry season from October to April and a wet season from May to September.

Quantifying and analyzing the urban growth in the Pearl River Delta are important to characterize the effects of anthropogenic activities on urban environments during the past years. In our study, 239 Landsat images, covering the period from 1987 to 2014, were used to monitor urban area dynamics. Landsat imagery, including TM, ETM+ (including SLC-off data), and OLI data with cloud cover less than 50%, was ordered and downloaded from the USGS Earth Explorer (Reference system: WRS-2, Path: 122, Row: 44). A clipped region from Landsat imagery with an area of 16,824 km², covering five cities, Guangzhou, Foshan, Zhongshan, Dongguan, and Shenzhen, was used to monitor urban growth. These five cities were the most developed cities in the Pearl River Delta. The geographic location of the study area is shown in Figure 3.1.

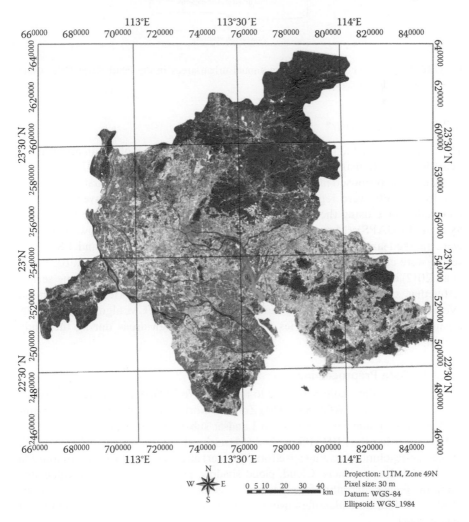

FIGURE 3.1   Geographical location of the case study area, Pearl River Delta.

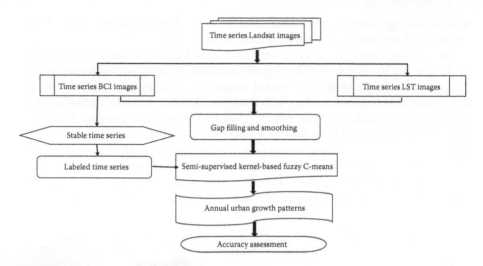

FIGURE 3.2 Procedures for mapping annual urban areas in the Pearl River Delta using time series Landsat images.

3.2.2 METHODOLOGY

This study intended to quantify the spatiotemporal patterns of urban areas in the Pearl River Delta using dense time series Landsat images from 1987 to 2014. In this study, the proposed method included five steps (Figure 3.2): First, time series Landsat subsets were registered in the same projection and converted to surface reflectance using the Landsat Ecosystem Disturbance Adaptive Processing System (LEDAPS) (Masek et al. 2006). Then, cloud, cloud shadow, and snow were masked and their values were set to null. Next, the BCI and LST time series were computed from preprocessed data. Gap filling (Garcia 2010; Wang et al. 2012) and smoothing were adopted to fill missing values in the time series. After that, stable time series were selected based on time series decomposition (Verbesselt et al. 2010). Finally, the semi-supervised KFCM algorithm was performed to clustering time series to map annual urban areas during the study period.

3.2.2.1 Data Preprocessing

All the Landsat data were registered to the 1984 World Geodetic System Universal Transverse Mercator (WGS-84 UTM) Zone 49 North projection system and resampled at 30-m spatial resolution. Each Landsat subset was standardized to surface reflectance using the LEDAPS method (Masek et al. 2006), which applied MODIS (moderate-resolution imaging spectroradiometer) atmospheric correction routines to Landsat L1 data products. Cloud, cloud shadow, and snow mask were calculated using the Fmask algorithm for all scenes (Zhu and Woodcock 2012). The locations of SLC-off data were identified using band-specific gap mask files in each SLC-off data product.

3.2.2.2   Calculation of BCI and LST

BCI (Deng and Wu 2012), involving Tasseled Cap (TC) transformation and the V–I–S triangle model, was given as

$$BCI = \frac{\dfrac{TC_1 + TC_3}{2} - TC_2}{\dfrac{TC_1 + TC_3}{2} + TC_2},$$

(3.1)

where TC_i ($i = 1$, 2, and 3) were three normalized TC components: TC_1 was *high albedo*, TC_3 was *low albedo*, and TC_2 was *vegetation*. Each derived TC component was then linearly normalized to the range from 0 to 1.

The LST was calculated using the radiative transfer equation method (Sobrino et al. 2004):

$$B_T = \frac{[L_\lambda - L\!\uparrow - \tau(1-\varepsilon)L\!\downarrow]}{\tau\varepsilon},$$

(3.2)

where B_T was the atmospherically corrected pixel value as brightness temperature. $L\!\uparrow$, $L\!\downarrow$, and τ, derived from an atmospheric correction tool (Barsi et al. 2005), were upwelling radiance, downwelling radiance, and transmittance, respectively. ε was emissivity, derived from NDVI. L_λ was the pixel values as radiance derived from digital numbers DN:

$$L_\lambda = \text{gain} \times DN + \text{bias},$$

(3.3)

where gain and bias were gain value and bias value for the specific band, respectively.

Then, the radiance was converted to surface temperature (Chander and Markham 2003):

$$T = \frac{K_2}{\ln\left(\dfrac{K_1}{B_T} + 1\right)},$$

(3.4)

where T was the temperature in Kelvin (K) and K_1 and K_2 were prelaunch calibration constants. For Landsat 5 TM, $K_1 = 607.76$ W/(m^2 sr μm) and $K_2 = 1260.56$ K; for Landsat 7 ETM+, $K_1 = 666.09$ W/(m^2 sr μm) and $K_2 = 1282.71$ K; for Landsat TIRS 10, $K_1 = 774.89$ W/(m^2 sr μm) and $K_2 = 1321.08$ K.

3.2.2.3   Gap Filling and Smoothing

Discontinuities existed in the time series Landsat data due to missing values caused by cloud cover and SLC-off data, which made uncertainties under incomplete time series in a subsequent analysis. Gap filling and smoothing were needed to improve the continuity and consistency in the time series. The gap filling method (Garcia 2010;

Wang et al. 2012), as a penalized least square regression based on three-dimensional discrete cosine transform (DCT-PLS), was adopted to predict missing values. It used information from both spatial and temporal variability to provide robust gap filling and it required no ancillary data sets such as alternative geospatial data sets or digital elevation models to model missing values. A smoothing gap filled time series was essential to improve the accuracy of the phenology derived from the reconstructed time series (Atkinson et al. 2012). Fourier fitting, as a most common and useful method, was adopted for smoothing gap filled time series BCI and LST data. Fourier analysis has showed promise in monitoring interannual vegetation changes (Geerken 2009). For the BCI and LST time series, Fourier fitting reduced the amount of noise, which mitigated the effects of outliers, anomalies, and spurious values in the time series.

3.2.2.4   Selection of Stable Time Series

Stable time series were derived from BCI time series. Stable time series means time series of labeled land covers, which have not experienced land cover transition during the study period. Since there was no exponential growth in the time series, an additive decomposition model was applied to separate BCI time series into three distinct components:

$$T = T_t + S_t + I_t, \tag{3.5}$$

where T was the observed data at time t, T_t was a nonseasonal secular trend component, S_t was a seasonal component, and I_t was an irregular component. T_t was estimated using the regression model as

$$T_t = \beta_0 + \beta_1 t + \beta_2 t. \tag{3.6}$$

β_0, β_1, and β_2 were regression coefficients. The seasonal component S_t was derived from the detrend time series using a parametric regression model. Detrend time series were computed by subtracting the trend component T_t from the original time series. Given the trend component T_t and the seasonal component S_t, the irregular component was estimated as

$$I_t = T - T_t - S_t. \tag{3.7}$$

The time series with the trend component as constant and without obvious phenology circle in seasonal variables were selected as stable time series.

3.2.2.5   Semi-Supervised KFCM Algorithm

Time series BCI and LST images were processed using a semi-supervised KFCM to obtain clustering results. Given time series data $X = \{x_1, x_2,..., x_n\}$, $x_k \in R^d (k = 1, 2,...,n)$, where d was temporal dimension and n was the number of samples.

KFCM partitions X into c fuzzy subsets by minimizing the following objective function:

$$J_m(U,V) = 2\sum_{i=1}^{c}\sum_{k=1}^{n} u_{ik}^m (1 - K(x_k, v_i)),\qquad (3.8)$$

where c was the number of clusters, v_i was the ith cluster centroid, u_{ik} was the membership of x_k in class i, and $\Sigma_i u_{ik} = 1$; $m \in [1, +\infty]$ was the weighting exponent determining the fuzziness of the clusters. $K(x_k, v_i)$ was the kernel function, aiming to map x_k from the input space X to a new space with higher dimensions. In this study, radial basis function kernel was adopted:

$$K(x_k, v_i) = \exp\left(-\|x_k - v_i\|^2 / \sigma^2\right),\qquad (3.9)$$

where the parameter σ was computed by

$$\sigma = \frac{1}{c}\left(\sqrt{\frac{\sum_{i=1}^{n}\|x_i - m\|^2}{n}}\right).\qquad (3.10)$$

In order to search for new clusters, the objective function was minimized:

$$\text{Min } J_m(U,V) = 2\sum_{i=1}^{c}\sum_{k=1}^{n} u_{ik}^m (1 - K(x_k, v_i))\qquad (3.11)$$

$$s.t. \sum_{i=1}^{c} u_{ik} = 1, k = 1,2,\cdots,n.\qquad (3.12)$$

The Lagrange function converted the constrained objective as an unconstrained optimization model. By optimizing the objective function, the membership u_{ik} and centroid v_i could be updated:

$$u_{ik} = \frac{\left(\dfrac{1}{(1 - K(x_k, v_i))}\right)^{\frac{1}{m-1}}}{\sum_{j=1}^{c}\left(\dfrac{1}{(1 - K(x_k, v_i))}\right)^{\frac{1}{m-1}}}\qquad (3.13)$$

$$v_i = \frac{\sum_{k=1}^{n} u_{ik}^m K(x_k, v_i) x_k}{\sum_{k=1}^{n} u_{ik}^m K(x_k, v_i)}.$$ (3.14)

Labeled time series samples were derived from stable time series, and unlabeled samples were derived from the remaining time series. Given time series data X consisted of X_l and X_u, X_l was labeled samples and X_u was unlabeled samples. "l" and "u" indicate labeled or unlabeled data, respectively.

The whole process of semi-supervised KFCM algorithm was shown as follows:

1. Initialize the values of σ and u_{ik} using X_l and X_u. For X_l, the value of component u_{ik} was set to 1 if the data x_k were labeled with class i, and 0 otherwise. For X_u, positive random values within [0,1] were set to unlabeled data. The initial set of centroid v_i was calculated as $v_i^0 = \frac{\sum_{k=1}^{n'} \left(u_{ik}^l \right)^m x_k^l}{\sum_{k=1}^{n'} \left(u_{ik}^l \right)^m}$, where n' was the number of labeled data.

2. Update the membership u_{ik} in X_u and centroid v_i until the objective function was minimized.

Finally, inconsistent labeled pixels were mapped comparing the LST L and BCI B clustering results. For those pixels, if the maximum membership $\max(u_{ik})^L$ of the pixel k in L was higher than $\max(u_{ik})^B$ in B, the pixel was labeled as the class with $\max(u_{ik})^L$ in L, and vice versa. However, if the values were equal, the pixel was labeled as the class with $\max(u_{ik})^L$.

3.3   RESULTS

3.3.1   QUANTITATIVE CHARACTERISTICS OF URBAN GROWTH

The urban area distributions for the Pearl River Delta from 1987 to 2014 are shown in Figure 3.3. The dark gray represents urban area, the medium gray shows water bodies, and the light gray shows nonurban area. Because image numbers in 1989, 1992, 1997, and 1998 were fewer than 3 and cloud cover was more than 50% for all images, urban areas in these 4 years were not analyzed.

The annual urban area maps in the Pearl River Delta in Figure 3.3 show a dramatic urban expansion from 1987 to 2014. Urban areas increased from 598 km² in 1987 to 5768 km² in 2014. To evaluate urban growth in the study area, urban areas by year were calculated from annual clustering maps in Figure 3.4. The spatial distributions of annual urban areas could be divided into four periods.

1. For the period 1987–1991, the Pearl River Delta experienced no significant change in urban areas. The Pearl River Delta was in the early phases of development during this period. Urban areas increased from 3.56% of the study area in 1987 to 5.97% in 1993, with an annual average rate of 13.57%. The average increased urban area per year was 101.53 km²/year.

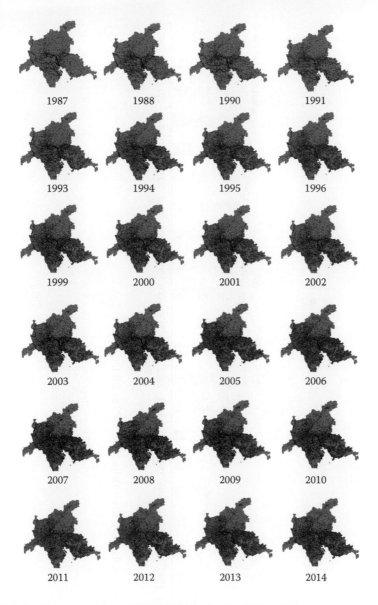

1987	1988	1990	1991
1993	1994	1995	1996
1999	2000	2001	2002
2003	2004	2005	2006
2007	2008	2009	2010
2011	2012	2013	2014

FIGURE 3.3 Urban areas from 1987 to 2014 in the Pearl River Delta.

2. For the period 1993–2000, large urban areas formed within and around existing urban areas. The Pearl River Delta experienced its first rapid growth period as a result of the deep development of China's reform and opening up. Urban areas increased from 6.81% in 1993 to 19.07% in 2000 with an annual average rate of 22.52%. The average increased urban area per year was 412.62 km²/year.

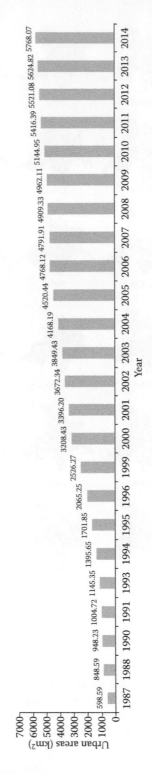

FIGURE 3.4   Annual urban areas in the Pearl River Delta, from 1987 to 2014.

3. For the period 2001–2006, the Pearl River Delta was still experiencing a rapid growth, but the urban expansion was slower than that in the previous period. In this period, the Southeast Asian economic crisis affected urban development to some degree. The driving forces of continuous urban expansion were mainly because China had a fast-growing economy since it joined the World Trade Organization in 2001. Urban areas increased from 20.19% in 2000 to 28.34% in 2006, yielding an annual average rate of 6.73%. The average increased urban area per year was 228.65 km²/year.

4. For the period 2007–2014, Pearl River Delta entered a period of stable development and urban development was slowed down. Urban areas increased from 28.48% in 2007 to 34.28% in 2014 with an annual average rate of 2.55%. The average increased urban area per year was 122.02 km²/year.

3.3.2 Assessment of Urban Growth

Since the 1980s, rapid urbanization in the Pearl River Delta was strongly linked to economic growth and population attributed to reform policies. From 1980 to 2000, the average annual gross domestic product (GDP) growth of the Pearl River Delta grew up to 16.9%. In this period, the driving force of urban growth in the study area was the establishment of Special Economic Zones of Shenzhen, followed by foreign investment from overseas investors. Since the new century, the Pearl River Delta has been continuing to experience strong economic growth with an average annual GDP growth of up to 15% between 2000 and 2007. However, the GDP growth has slowed down since 2008 because of the global financial crisis and industrial structure adjustment. The GDP growth particularly slowed down to below 10% since 2011. According to the Hong Kong Trade Development Council, the Pearl River Delta enjoyed a per-capita GDP of RMB 93,114 in 2013 (approximately $15,000), with real GDP growing by an average of 9.4%.

The urban growth types differed during four periods: 1987–1991, 1993–2000, 2001–2006, and 2007–2014. In the 1987–1991 period, urban growth was dominated by scattered development. In the 1993–2000 period, scattered development was decreased and strip development was increased. Urban areas mainly expanded along transportation networks. According to the fifth national population census, the population of the Pearl River Delta was 4 million in 2000, rising by 1.94 million when compared to the population in 1990. From 2001 to 2014, strip development and compact development became dominant. City clusters and metropolitan stretches came into being in the Pearl River Delta. Accompanying the urbanization process was population growing to 47.72 million in 2008. Since 2011, the population growth affected by economic growth and industrial structure adjustment began to stabilize (Pearl River Delta Region Planning and Guangdong Statistical Yearbook 2014).

3.3.3 Clustering Accuracy Assessment

In this study, historical imagery acquired from Google Earth was used as reference data for each year. Since available historical imagery cannot cover the whole study

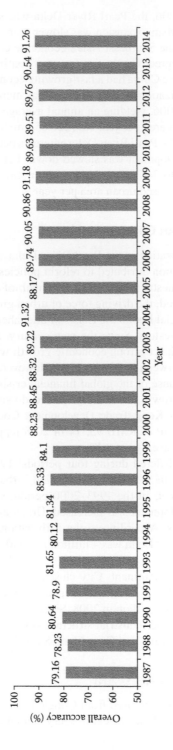

FIGURE 3.5   Annual clustering accuracy of the proposed method.

area and historical imagery from Google Earth for some period was few, reference samples were also selected from time series BCI images using visual judgments. Stable time series derived from stacked cloud free BCI images were used as reference data. Then, the stratified random sampling method was employed for selecting reference data for each year. Five hundred samples were randomly selected to each class and divided into two subsets. One subset was used for classifier training and the second was used for accuracy assessment. It would help eliminate the bias resulting from using the same samples for both training and testing. The annual clustering accuracy is shown in Figure 3.5.

The annual clustering accuracy yielded 78.23% to 91.32%, which shows the effectiveness and feasibility of the time series clustering method. However, the average clustering accuracy in the 1987–1999 period was 81.05%, and the average clustering accuracy in the 2000–2014 period was 89.75%. The reason was that different temporal dimensions in each year led to different clustering accuracy. The available annual average image number before 2000 was smaller than 3, but the available annual average image number after 2000 was larger than 14. The main reason for this phenomenon was that the low temporal resolution in each year could obscure land cover changes and reduce the separability of temporal characteristics for urban areas and nonurban areas. The vegetation phenology characteristic would particularly be weakened. Time series clustering also showed values of imagery with cloud contamination or SLC-off data in identifying urban areas. Although cloud contamination and SLC-off data caused significant temporal noise and resulted in incomplete time series, gap filling and smoothing were helpful for solving missing data problems through enhancing temporal resolution of time series Landsat imagery.

3.4 CONCLUSIONS

This chapter explores urban growth in the Pearl River Delta using time series Landsat imagery from 1987 to 2014 based on the time series clustering method. The annual spatiotemporal patterns of urban areas in the Delta were quantified. The results indicated that urban areas in study areas increased rapidly from 3.56% to 34.28% during the 1987–2014 period. The method proposed in this chapter verifies the feasibility and effectiveness of the time series clustering method in assessing urban growth patterns using time series Landsat imagery. Results from this study can be used by policymakers for urban planning and management and for hydrological modeling to determine the effect of increasing urban areas on urban environments of the study area. This study can also be valuable for exploring the mechanisms of urban areas and environmental relationship for sustainable urban planning and management. However, how to evaluate the differences between two time series and how time series components affect time series clustering remain unanswered in this study. Future studies may include similarity metrics before using the clustering method to solve the above issues.

REFERENCES

Atkinson, P. M., Jeganathan, C., Dash, J., & Atzberger, C. (2012). Inter-comparison of four models for smoothing satellite sensor time-series data to estimate vegetation phenology. *Remote Sensing of Environment*, 123, 400–417.

Bagan, H., & Yamagata, Y. (2012). Landsat analysis of urban growth: How Tokyo became the world's largest megacity during the last 40years. *Remote Sensing of Environment*, 127, 210–222.

Barsi, J. A., Schott, J. R., Palluconi, F. D., & Hook, S. J. (2005, August). Validation of a web-based atmospheric correction tool for single thermal band instruments. In *Optics & Photonics 2005* (pp. 58820E–58820E). International Society for Optics and Photonics.

Bhandari, S., Phinn, S., & Gill, T. (2012). Preparing Landsat Image Time Series (LITS) for monitoring changes in vegetation phenology in Queensland, Australia. *Remote Sensing*, 4(6), 1856–1886.

Chander, G., & Markham, B. (2003). Revised Landsat-5 TM radiometric calibration procedures and postcalibration dynamic ranges. *IEEE Transactions on Geoscience and Remote Sensing*, 41(11), 2674–2677.

Deng, C., & Wu, C. (2012). BCI: A biophysical composition index for remote sensing of urban environments. *Remote Sensing of Environment*, 127, 247–259.

Du, W., Inoue, K., & Urahama, K. (2005). Robust kernel fuzzy clustering. In *Fuzzy systems and Knowledge Discovery* (pp. 454–461). Springer Berlin Heidelberg.

Du, P., Xia, J., & Feng, L. (2015). Monitoring urban impervious surface area change using China–Brazil Earth Resources Satellites and HJ-1 remote sensing images. *Journal of Applied Remote Sensing*, 9(1), 096094–096094.

Dupuy, S., Barbe, E., & Balestrat, M. (2012). An object-based image analysis method for monitoring land conversion by artificial sprawl use of RapidEye and IRS data. *Remote Sensing*, 4(2), 404–423.

Durieux, L., Lagabrielle, E., & Nelson, A. (2008). A method for monitoring building construction in urban sprawl areas using object-based analysis of Spot 5 images and existing GIS data. *ISPRS Journal of Photogrammetry and Remote Sensing*, 63(4), 399–408.

Fu, T. C. (2011). A review on time series data mining. *Engineering Applications of Artificial Intelligence*, 24(1), 164–181.

Gao, F., de Colstoun, E. B., Ma, R., Weng, Q. et al. (2012). Mapping impervious surface expansion using medium-resolution satellite image time series: A case study in the Yangtze River Delta, China. *International Journal of Remote Sensing*, 33(24), 7609–7628.

Garcia, D. (2010). Robust smoothing of gridded data in one and higher dimensions with missing values. *Computational Statistics & Data Analysis*, 54(4), 1167–1178.

Geerken, R. A. (2009). An algorithm to classify and monitor seasonal variations in vegetation phenologies and their inter-annual change. *ISPRS Journal of Photogrammetry and Remote Sensing*, 64(4), 422–431.

Hutyra, L. R., Duren, R., Gurney, K. R., Grimm, N., Kort, E. A., Larson, E., & Shrestha, G. (2014). Urbanization and the carbon cycle: Current capabilities and research outlook from the natural sciences perspective. *Earth's Future*, 2(10), 473–495.

Jacquin, A., Misakova, L., & Gay, M. (2008). A hybrid object-based classification approach for mapping urban sprawl in periurban environment. *Landscape and Urban Planning*, 84(2), 152–165.

Li, X., Gong, P., & Liang, L. (2015). A 30-year (1984–2013) record of annual urban dynamics of Beijing City derived from Landsat data. *Remote Sensing of Environment*, 166, 78–90.

Liao, T. W. (2005). Clustering of time series data—A survey. *Pattern Recognition*, 38(11), 1857–1874.

Liao, J., Wang, T., Jiang, Z., Zhuang, B. et al. (2015). WRF/Chem modeling of the impacts of urban expansion on regional climate and air pollutants in Yangtze River Delta, China. *Atmospheric Environment*, 106, 204–214.

Liu, Z., He, C., Zhang, Q., Huang, Q., & Yang, Y. (2012). Extracting the dynamics of urban expansion in China using DMSP-OLS nighttime light data from 1992 to 2008. *Landscape and Urban Planning*, 106(1), 62–72.

Ma, T., Zhou, C., Pei, T., Haynie, S., & Fan, J. (2012). Quantitative estimation of urbanization dynamics using time series of DMSP/OLS nighttime light data: A comparative case study from China's cities. *Remote Sensing of Environment*, 124, 99–107.

Masek, J. G., Vermote, E. F., Saleous, N. E., Wolfe, R., Hall, F. G., Huemmrich, K. F., Gao, F., Kutler, J., & Lim, T. K. (2006). A Landsat surface reflectance dataset for North America, 1990–2000. *IEEE Geoscience and Remote Sensing Letters*, 3(1), 68–72.

Michishita, R., Jiang, Z., & Xu, B. (2012). Monitoring two decades of urbanization in the Poyang Lake area, China through spectral unmixing. *Remote Sensing of Environment*, 117, 3–18.

Panagopoulos, T., Duque, J. A. G., & Dan, M. B. (2015). Urban planning with respect to environmental quality and human well-being. *Environmental Pollution*. DOI:10.1016/j.envpol.2015.07.038.

Pathirana, A., Denekew, H. B., Veerbeek, W., Zevenbergen, C., & Banda, A. T. (2014). Impact of urban growth-driven landuse change on microclimate and extreme precipitation—A sensitivity study. *Atmospheric Research*, 138, 59–72.

Schneider, A. (2012). Monitoring land cover change in urban and peri-urban areas using dense time stacks of Landsat satellite data and a data mining approach. *Remote Sensing of Environment*, 124, 689–704.

Sexton, J. O., Song, X. P., Huang, C., Channan, S., Baker, M. E. et al. (2013a). Urban growth of the Washington, DC–Baltimore, MD metropolitan region from 1984 to 2010 by annual, Landsat-based estimates of impervious cover. *Remote Sensing of Environment*, 129, 42–53.

Sexton, J. O., Urban, D. L., Donohue, M. J., & Song, C. (2013b). Long-term land cover dynamics by multi-temporal classification across the Landsat-5 record. *Remote Sensing of Environment*, 128, 246–258.

Sobrino, J. A., Jiménez-Muñoz, J. C., & Paolini, L. (2004). Land surface temperature retrieval from LANDSAT TM 5. *Remote Sensing of Environment*, 90(4), 434–440.

Sun, C., Wu, Z. F., Lv, Z. Q., Yao, N., & Wei, J. B. (2013). Quantifying different types of urban growth and the change dynamic in Guangzhou using multi-temporal remote sensing data. *International Journal of Applied Earth Observation and Geoinformation*, 21, 409–417.

Verbesselt, J., Hyndman, R., Zeileis, A., & Culvenor, D. (2010). Phenological change detection while accounting for abrupt and gradual trends in satellite image time series. *Remote Sensing of Environment*, 114(12), 2970–2980.

Wang, G., Garcia, D. et al. (2012). A three-dimensional gap filling method for large geophysical datasets: Application to global satellite soil moisture observations. *Environmental Modelling & Software*, 30, 139–142.

Weng, Q., & Hu, X. (2008). Medium spatial resolution satellite imagery for estimating and mapping urban impervious surfaces using LSMA and ANN. *IEEE Transactions on Geoscience and Remote Sensing*, 46(8), 2397–2406.

Xian, G., & Crane, M. (2005). Assessments of urban growth in the Tampa Bay watershed using remote sensing data. *Remote Sensing of Environment*, 97(2), 203–215.

Yuan, F., & Bauer, M. E. (2007). Comparison of impervious surface area and normalized difference vegetation index as indicators of surface urban heat island effects in Landsat imagery. *Remote Sensing of Environment*, 106(3), 375–386.

Zhang, D. Q., & Chen, S. C. (2003, June). Kernel-based fuzzy and possibilistic c-means clustering. In *Proceedings of the International Conference Artificial Neural Network* (pp. 122–125).

Zhang, Q., & Seto, K. C. (2011). Mapping urbanization dynamics at regional and global scales using multi-temporal DMSP/OLS nighttime light data. *Remote Sensing of Environment*, 115(9), 2320–2329.

Zhao, W., Zhu, X., Sun, X., Shu, Y., & Li, Y. (2015). Water quality changes in response to urban expansion: Spatially varying relations and determinants. *Environmental Science and Pollution Research*, 1–15.

Zhu, Z., & Woodcock, C. E. (2012). Object-based cloud and cloud shadow detection in Landsat imagery. *Remote Sensing of Environment*, 118, 83–94.

4 InSAR Monitoring of Land Subsidence for Sustainable Urban Planning

Abduwasit Ghulam, Mark Grzovic,
Maitiniyazi Maimaitijiang, and Mamat Sawut

CONTENTS

4.1 INTRODUCTION

Underground mining can cause a number of environmental problems such as building collapse and road damage (Al-Rawahy 1995; Donnelly et al. 2001; Guéguen et al. 2009; Yerro et al. 2014), as well as disruptions in groundwater aquifers, underground gas, water, electricity, and sewage systems (Dong et al. 2013). With the largest reported bituminous coal reserves in the United States, Illinois has been one of the nation's major coal producers for nearly 150 years, and coal has been mined in 73 of the state's 102 counties (http://www.eia.doe.gov). Many of the modern and historic coal mining areas are located under towns or important infrastructure such as major roadways and railways. The Illinois State Geological Survey estimates that approximately 201,000 acres of urban and built-up lands may be close to under-ground mines with 333,000 housing units being exposed to possible mine subsidence.

Coal production–related ground deformation generally occurs in three stages: the initial phase, which usually corresponds to less than 15% of the total subsidence for an individual coal seam; the fast main phase, corresponding to approximately 75% of the maximum subsidence; and the residual phase, which presents a decreasing subsidence rate (Guéguen et al. 2009) that can occur over decades. Generally, subsidence occurs during active mining, and its occurrence can be directly related to the location of active mining. The amount of residual subsidence after active mining is often small (less than 10% relative to the total subsidence) and can easily be overlooked after the mine is exhausted. However, residual subsidence associated with abandoned mines can continue for years—or decades. Monitoring the long-term residual land subsidence is of great interest not only to individual homeowners, insurance companies, and regional policymakers charged with assessing risk and the development of hazard mitigation plans, but also to city planners and developers.

Conventional ground-based deformation monitoring techniques (e.g., Global Positioning Systems [GPSs] and leveling) are limited to discrete and sparse sites and are not able to provide detailed and comprehensive ground deformation over large areas. The Differential Interferometric Synthetic Aperture Radar (DInSAR) technique developed in the late 1980s has demonstrated its potential for high-density spatial mapping of ground displacement and has become an important tool for monitoring temporal and spatial ground movements (Chaussard et al. 2013; Gabriel et al. 1989; Massonnet and Feigl 1998). DInSAR can achieve an accuracy comparable to field measurements, but at a much higher spatial density, larger coverage, and with low cost (Raucoules et al. 2009; Zhang et al. 2011). However, the standard DInSAR technique is subject to uncertainties caused by temporal and spatial decorrelations as well as atmosphere artifacts (Sousa et al. 2011; Zebker and Villasenor 1992; Zebker et al. 1997). To overcome these limitations, advanced time-series InSAR techniques (timeSAR) have been proposed, utilizing multiple interferograms derived from a large set of SAR images. There are two schools of timeSAR techniques: Persistent Scatterer Interferometry (PSI) (Ferretti et al. 2000, 2001; Hooper et al. 2004; Kampes 2006) and Small Baseline Subset (SBAS) (Berardino et al. 2002; Lanari et al. 2004, 2007; Usai 2003). Millimetric precision of velocity estimation is possible with timeSAR approach.

Each of these methods has its advantages and limitations. For example, PSI is based on analysis of persistent point targets and works better in urban areas than in suburban or vegetated areas (Bell et al. 2008; Colesanti et al. 2003; Raucoules et al. 2013), providing high-resolution measurement of surface motions (Prati et al. 2010). Limitations associated with PSI processing include the following: (1) a large number of SAR images (at least 20 images) is required to obtain reliable results (Guéguen et al. 2009); (2) it assumes the temporal deformation to be linear; and (3) it is known to underestimate high deformation rates because of temporal unwrapping issues (Raucoules et al. 2009). In contrast, SBAS exploits distributed scatterers using small baseline interferogram subsets and performs better for both urban and nonurban vegetated areas, and also in areas with high deformation rates (Chaussard et al. 2014; Gourmelen et al. 2010; Hooper 2006). SBAS is able to estimate nonlinear deformation rates but may not be optimal for localized displacements that may affect, for example, small areas or single buildings.

In multitemporal InSAR processing, both PSI and SBAS approaches are optimized to obtain ground displacement rates with a nominal accuracy of millimeters per year (Yan et al. 2012). These methods have been successfully applied to detect ground movements caused by not only natural phenomena such as tectonic or volcanic activities (Chaussard et al. 2014; Yan et al. 2012), salt movement (Abir et al. 2015), glacial rebound and landslides (Lauknes et al. 2010; Liu et al. 2013), ground dissolution (Gutiérrez et al. 2011; Paine et al. 2012), and freeze–thaw of permafrost (Chen et al. 2013; Short et al. 2014), but also anthropogenic activities including underground mining (coal, oil and gas, etc.) (Abdikan et al. 2014; Grzovic and Ghulam 2015; Guéguen et al. 2009; Herrera et al. 2007; Raucoules et al. 2003; Yerro et al. 2014), groundwater extraction (Chaussard et al. 2013; Zhu et al. 2013), tunnel construction (Knospe and Busch 2009; Strozzi et al. 2011), and load increase of constructions (Abidin et al. 2011; Mazzotti et al. 2009).

The objectives of this chapter are to map risk areas of structural damage and infrastructure failure attributed to mining extraction using both PSI and SBAS techniques and to demonstrate the potential of InSAR for sustainable urban planning that avoids developing over subsiding areas.

4.2 STUDY AREA AND DATA

4.2.1 STUDY AREA

The St. Louis Metropolitan area is located just south of the confluence of the Missouri and Mississippi rivers, straddling the Missouri–Illinois border (Figure 4.1). The study area encompasses all of St. Louis City and St. Louis County as well as portions of Jefferson and St. Charles Counties in Missouri, and Madison, Monroe, and St. Clair Counties in Illinois. The region is susceptible to many geologic hazards (e.g., surface subsidence, sinkholes) because of the presence of numerous sinkholes and abandoned underground mines (Louchios et al. 2013; MoDNR 2014a,b). The area presents significant problems because of the density of urban infrastructure and population. It is situated between the relatively higher topography of the Ozark Plateau in the southwest and the lower topography of the Illinois Basin. The elevation ranges from 68 to 530 m (Jordan et al. 2014b). It has diverse land cover classes in the area such as forest, agriculture, pasture, and urban (Jordan et al. 2012). The climate of this region is continental type with distinct seasons characterized by wide ranges in temperature and irregular annual and seasonal precipitation (Jordan et al. 2014b).

The surface cover in the St. Louis metropolitan area varies from glacial deposits to stream alluvium (Schultz 1993). Much of the surface of the study area is covered by loess consisting of sandy silts, silts, and clays that are 3–25 m thick (Schultz 1993). The floodplains of Mississippi and Missouri and their tributaries are covered with alluvium consisting of fluvial clays, silts, sands, and gravels and are up to 25 m thick (Schultz 1993). Underneath this surface cover are gently dipping (<1° northeast) Paleozoic limestones, dolostones, shales, and sandstones of marine origin (Harrison

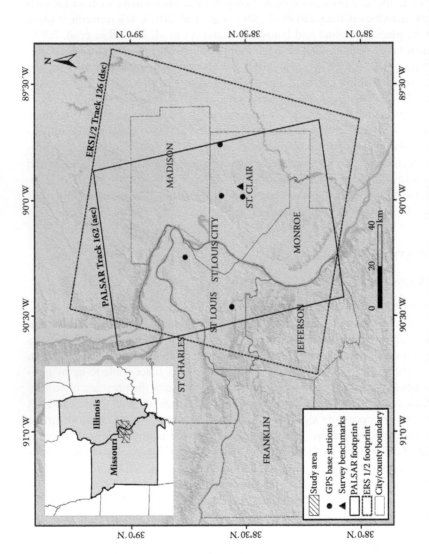

FIGURE 4.1 Location of the study area. The background is a shaded relief map from ArcGIS online map services. Solid and dashed rectangles refer to PALSAR and ERS1/2 image footprint. Black circles are the locations of GPS base stations that are referenced during InSAR processing. Long-term leveling measurements are available, and survey benchmarks are located at the solid triangle.

1997). The oldest bedrock units are in the Ozark Plateau exposed in the southwest third of the study area and consist of Cambrian, Ordovician, and Devonian dolostones, sandstones, and shales. To the northeast are younger Pennsylvania rocks in the Illinois Basin that consist of limestones, shales, and sandstones. In between the Ozark Plateau and the Illinois Basin are Mississippian limestones and sandstones topped by Pennsylvanian shales. The St. Louis Fault Zone is the only structure that significantly influences surface features. The fault zone trends north-northeast and appears to control the course of the Mississippi River, causing it to flow across the northeastern flank of the Ozark Plateau. Schultz (1993) presented evidence for right-lateral strike-slip motion on the fault, but it is *inconclusive* as most of the fault zone lies beneath the Mississippi River.

4.2.2 DATA

Thirty-seven C-band SAR images were obtained from the European Space Agency's ERS-1 and ERS-2 SAR satellites. The ERS-1/2 data were acquired on descending orbits for the period between 1992 and 2000. We also selected 16 of the Phased Array type L-band Synthetic Aperture Radar (PALSAR) images acquired from 2007 to 2011. The PALSAR sensor onboard the Advanced Land Observing Satellite (ALOS) from the Japan Aerospace Exploration Agency (JAXA) collects data with a 46-day repeat orbit cycle. The PALSAR images were acquired from ascending orbit direction in Fine Beam Single Polarization (FBS) and Fine Beam Dual Polarization (FBD) observation mode with the off-nadir angle of 34.3°. Note that both satellite radar systems only detect displacement in the line-of-sight (LOS) direction.

For validation, we used leveling data collected from 1988 to 2008. The leveling data were measured monthly as relative movement to permanent benchmarks deployed in the study area. National Elevation Dataset (NED) with a spatial resolution of 10 m was used to remove topographic effect, flattening, coregistration, and geocoding of InSAR data.

4.3 METHODS

An integrated PSI and SBAS approach can maximize the spatial density of coherent pixels, allowing the identification of persistent scatterers that dominate the scattering from the resolution cell and slowly varying filtered phase pixels representing distributed targets whose phases decorrelate little over short time intervals (Hooper 2008). We utilized both SBAS and PSI processing of ERS-1/2 data. However, only SBAS was performed on PALSAR images owing to the small number of samples. Another challenge for InSAR processing was the fact that a large portion of the area was covered by vegetation and water, which significantly decreases the coherence of InSAR pairs. To overcome this limitation, we processed the ERS-1/2 data by two smaller sections divided by the Mississippi River, St. Louis part on the west of the river and the Belleville part on the east of the river. Compared to C-band SAR images, better coherence and coregistration can be achieved with L-band data especially over vegetated terrains because of the better alignment of the reference and repeat images in the range direction (Rosen et al. 1996). We were able to process entire PALSAR data without splitting the image coverage into smaller portions. While the details of PSI and SBAS approaches can be found in other literature, for completeness, we provide a brief description of the methods with pertinent parameter settings.

4.3.1 SBAS PROCESSING

Interferograms were generated using SARscape modules of ENVI software from Exelis VIS Information Solutions. Normal baseline was set to 50% of the critical baseline and 900 days threshold was used for the temporal baseline. After coregistration and interferogram generation, NED 10 m Digital Elevation Model (DEM) resampled to 25 m was used to remove the flat earth phase component. Goldstein filtering (Goldstein and Werner 1998) with optimized parameters (Ghulam et al. 2010) were used to remove interferogram noise. A complex multilook operation was performed to produce ground resolution of approximately 25 m. Baseline-dependent phase residuals attributed to DEM inaccuracies (Berardino et al. 2002) were removed using a predefined linear displacement model that jointly estimates the DEM error and the low-pass displacement parameters. The efficacy of using a priori known displacement models in removing the DEM error has also been reported by a number of studies (Lauknes et al. 2010; Berardino et al. 2002).

Both minimum cost flow (MCF) network (Costantini 1998) and Delaunay 3D (D3D) methods (Hooper and Zebker 2007) was used to unwrap the phase values only known within $[-\pi, \pi]$. D3D unwrapping is more reliable than the standard two-dimensional approaches because it exploits the temporal information and unwraps the phase difference between neighboring pixels in the time domain to reduce atmospheric and orbital effects. However, it requires higher redundancy of SAR observations. For a given SBAS epoch, the MCF or D3D method was used depending on the redundancy of DInSAR pairs determined by a plot of time versus relative position with respect to the super master image. A spatial coherence thresholding of 0.35 was used to exclude decorrelated areas from the phase unwrapping so that a reliable phase unwrapping was feasible only with highly coherent pixels. By screening all of the wrapped and unwrapped interferograms, problems related to inaccurate orbits and uncoherent pairs were identified and further corrected or excluded.

Temporal variations in the refractive index attributed to water vapor can lead to a significant phase delay that is associated with elevation (Cavalie et al. 2007; Delacourt et al. 1998). This error component along with the phase error caused by the atmospheric signal is correlated in space and needs to be removed. To that end, we employed a two-step inversion process. First, we estimated a phase delay elevation profile for each interferogram and displacement velocity using a linear inversion model. A wavelet number of levels, which refers to the power of base 2, was used to determine the residual topography. This value was set as a function of of the reference DEM, which was used for the interferogram flattening. We processed SAR data with 25 m resolution with NED 10 m DEM resampled to 25 m; therefore, a wavelet number of level 1 was used. This means that information coarser than 50 m was removed. The higher the levels, the lower in terms of spatial frequency will be removed. Then, the first estimated residual topography and velocity phase components were subtracted from input wrapped interferograms. The difference, the remaining error component attributed to the phase delays owing to atmospheric signals, which is spatially correlated, was unwrapped and added to the first estimated displacement velocity phase. Second, the sum of the first estimated velocity and the remaining error phase was unwrapped again and reflattened, and an SBAS second inversion was conducted to jointly estimate the error source and refined velocity. Atmospheric phase screening was conducted to estimate atmospheric effects, which were removed using a low-pass spatial filter with a 1.2 km × 1.2 km window on each single acquisition followed by a high-pass filter on the time series images.

Sixty-three interferogram pairs were generated from 16 PALSAR images over the entire study area. The number of pairs generated from 37 ERS-1/2 images for the west and east of the Mississippi River was 86 and 94, respectively.

4.3.2 PSI PROCESSING

PSI is intended for the analysis of point targets characterized by high coherence behavior. A total of 37 ERS-1/2 images were processed with the PSI algorithm (Ferretti et al. 2001). The topographic reference was determined from the same DEM as for the SBAS, which was used to remove the topographic and ellipsoidal height components of the phase. Persistent scatterers were selected based on the coherence value over 0.65. Then, atmospheric phases were removed from each differential interferogram by filtering the spatial and temporal domains according to their spatial and temporal correlations using the same method utilized for SBAS. To unwrap the complex phase, the MCF method was integrated with a region growing procedure to improve the performances in areas with low signal-to-noise ratio due to predominant vegetation cover (Costantini 1998). Then, mean annual LOS velocity and displacement time series were calculated.

4.4 RESULTS AND DISCUSSION

4.4.1 SPATIAL AND TEMPORAL PATTERN OF LAND SUBSIDENCE FROM 1992 TO 2000

Figure 4.2a through d show estimated mean LOS velocity in millimeters per year from the 37 ERS-1/2 images using both SBAS (Figure 4.2a and b) and PSI (Figure 4.2c and d) methods. The mean LOS velocity image was draped on the mean intensity image. Positive LOS velocities (blue colors) represent movement toward the satellite, and negative LOS velocities (red colors) represent movement away from the satellite. Spatial patterns of ground deformation observed by both SBAS and PSI methods over St. Louis, Missouri (Figure 4.2a and c) and Belleville, Illinois (Figure 4.2b and d) are almost identical. In other words, deformation signals detected by InSAR were confirmed by both methods, demonstrating distinct subsidence or uplift patterns on several locations in the study site.

From 1992 to 2000, relative uplift, as demonstrated by the blue cool colors in Figure 4.2a and c, was observed over parks and cemeteries in south-east central areas of St. Louis city. The uplift might be attributed to an increase in precipitation in the 1990s (Jordan et al. 2014a,b) and tree growth. In Belleville, Illinois, the

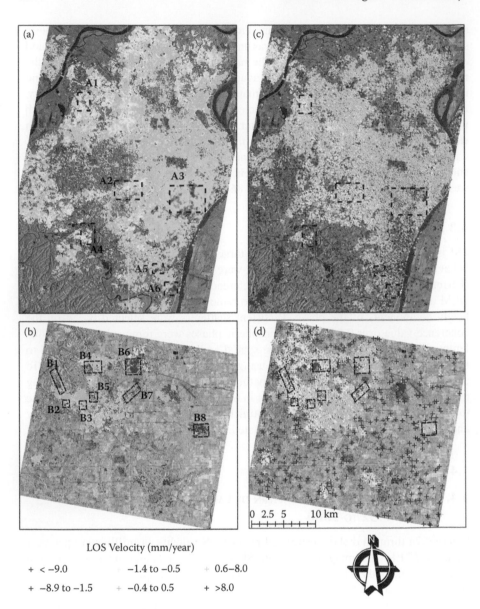

FIGURE 4.2 Mean LOS velocity (in millimeters per year) estimated by SBAS (a and b) and PSI (c and d) methods using 37 ERS-1/2 images. Background image is mean intensity. Positive LOS velocities (blue colors) represent movement toward the satellite; negative LOS velocities (red colors) represent movement away from the satellite. Dashed rectangles are typical areas with distinct subsidence or uplift patterns that will be analyzed further in succeeding sections.

deformation patterns from 1992 to 2000 were mostly subsidence, which correspond to the locations of underground mining and abandoned mine workings (Figure 4.2b and d).

Both the SBAS and PSI data sets show average annual rates of movement ranging from −9 to +8 mm/year. It is worth noting that the SBAS estimated deformation represents a pixel dimension of approximately 20 × 20 m, and the PSI estimates were retrieved at a full pixel resolution of 4 × 20 m in the azimuth and range directions, respectively. In order to compare time series, all pixels within a selected region of interest (ROI) were averaged. Figure 4.3 shows the estimated total LOS displacement

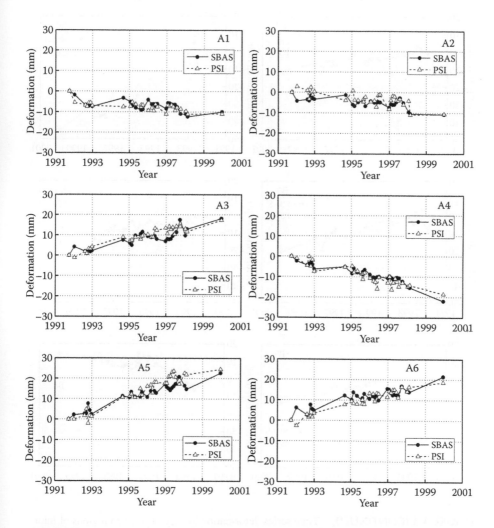

FIGURE 4.3 Time series deformation history of selected regions of interest shown in Figure 4.2. Labels A and B correspond to the St. Louis, Missouri, site and the Belleville, Illinois, site. *(Continued)*

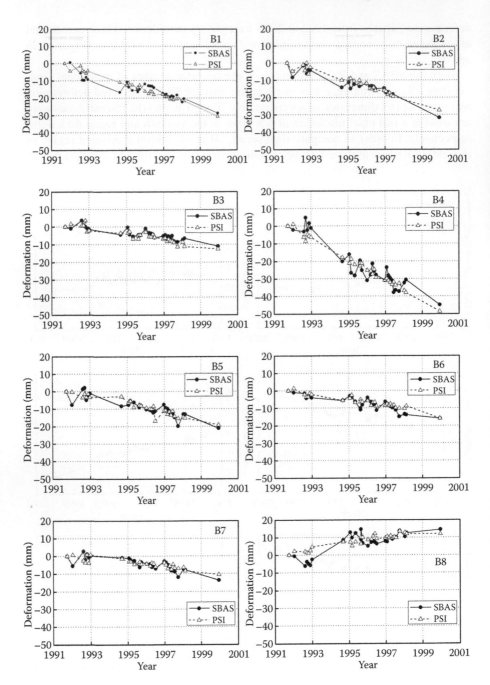

FIGURE 4.3 (CONTINUED)   Time series deformation history of selected regions of interest shown in Figure 4.2. Labels A and B correspond to the St. Louis, Missouri, site and the Belleville, Illinois, site.

time series averaged for selected ROIs using both SBAS and PSI methods. Labels A1–A6 and B1–B8 correspond to the St. Louis, Missouri, and Belleville, Illinois, sites. Main observations include the following: (1) ground deformation history from SBAS and PSI was highly consistent, demonstrating a match between highs and lows of the deformation trend line; (2) the amplitude of total deformation varied from +30 to −30 mm for most of the ROIs except for location B4, which experienced significant (up to −50 mm) subsidence from 1992 to 2000; (3) overall, SBAS-derived deformation trends were smoother than PSI results, which was possibly due to the fact that a better estimation of nonlinear deformation (note: PSI assumes a linear deformation trend) as well as residual topography was achieved with SBAS using a wavelet number approach; and (4) SBAS results seem to have a greater spatial coverage compared to PSI, especially over vegetated areas.

4.4.2 SPATIAL AND TEMPORAL PATTERN OF LAND SUBSIDENCE FROM 2007 TO 2011

Figure 4.4 shows the effects of subsidence in the St. Louis metropolitan region. The mean velocity deformation map results (Figure 4.4a) for the period from January 2, 2007 to January 13, 2011 are from the ALOS PALSAR data set using the SBAS method. Note that not all subsidence takes place over underground mines. In Figure 4.4b, an area affected by karst processes (purple outline) is shown. The karst plain covers most of Monroe county and a small portion of St. Claire County, but much of this area is agriculture or undeveloped.

Mining on the Missouri side of the Mississippi River was active up into the 1940s, but today mining activity is confined to the open pit mining of limestone (Seeger 2014). On the Illinois side, however, mining continues and consists primarily of underground coal mining. Active underground mines are required by law to have a surface subsidence mitigation plan as well as the requirement to monitor and repair mine subsidence damages that occur (Bauer 2008). Mine subsidence from these active mines causes little damage to surface infrastructure after initial collapse of the surface because the method used for mining leaves little to no open space underground (Bauer 2008). There are, however, many abandoned underground coal mines in the region (Chenoweth et al. 2004a,b, 2005). For these mines, the potential for collapse still exists after many years because the mining techniques used leave open spaces underground after mining has stopped (Bauer 2006, 2008).

Figure 4.4c shows the time series of ground subsidence, indicated by the red line, for the large southern subsiding area in Figure 4.4d, which is an area in Illinois affected by mine subsidence during the period. Groundwater depth (blue line) and the total yearly precipitation (light gray shading) for the area are also plotted. The location of the well where groundwater measurements were taken is indicated by the star on Figure 4.4a. We can see that a drastic subsidence occurred when the groundwater level drops during the drought year (2007), possibly caused by the excessive groundwater withdrawals. It suggests that the deformation pattern observed in the area is likely to be related to the drop in the groundwater level. Then, subsidence slows or reverses to uplift when groundwater levels rise. Note, however, that the surface

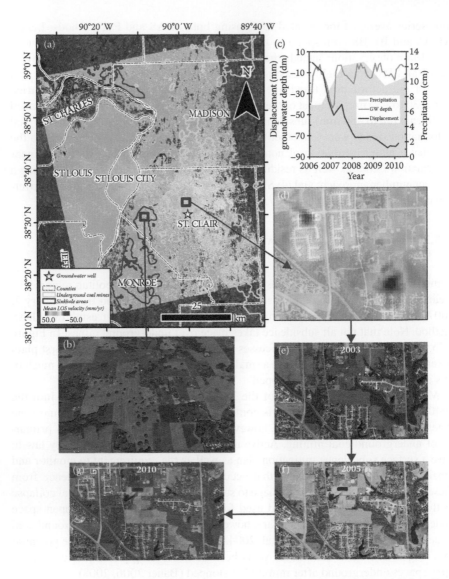

FIGURE 4.4 (a) Mean LOS velocity (in millimeters per year) using the small baseline subset (SBAS) method for the ALOS PALSAR data set, a period from January 2, 2007 to January 13, 2011. Bright colors indicate movement away from the satellite and dark colors indicate movement toward the satellite. Light gray outlines indicate underground coal mine locations. Purple outlines indicate sinkhole-prone areas. The star indicates the location of the well where groundwater depths were taken. (b) A Google Earth image from 2014 showing a typical sinkhole-prone area. (c) Time series plot of ground deformation for the southern subsiding area shown in (d). The red line indicates ground deformation, the blue line denotes groundwater depth from the well indicated in (a), and the light gray shaded area indicates yearly total precipitation. (d) Close-up of indicated area showing mean LOS velocity of an area undergoing development during the period ALOS PALSAR images were acquired. (e through g) Google Earth images showing recent urban development of the area shown in the figure.

subsidence, unlike groundwater, never returns to the level it started at. One may expect that subsidence closely follows the groundwater recharge cycle, for example, ground uplifts or subsidence corresponding to increase or decrease of groundwater tables. This is particularly true for areas where there are no underground mine pillar failures. The fact that ground subsidence did not reverse, corresponding to water level changes after a drought year, hints a high likelihood of pillar failure.

Several significant subsidence zones can be found in Figure 4.4d (see the red spots). Aided by aerial photos and Google images, we determined that these red spots have experienced recent urban developments for both residential and school building (Figure 4.4e, f, and g). It is apparent that development on the areas located on the upper-left corner of Figure 4.4d continued despite reported land subsidence incidents in a nearby school (red spot on the lower-right portion of Figure 4.4d). InSAR data measured subsidence of up to −25.6 mm/year at the two subsiding areas.

4.4.3 VALIDATION OF InSAR DEFORMATION USING LEVELING DATA

In order to validate InSAR-derived measurements, we selected two bench mark (BM) points (BM3 and BM29) where two decades of leveling measurements were available. Figure 4.5 shows the deformation history of BM3 and BM29 obtained from SBAS and leveling measurements, which confirms the accuracy of InSAR measurements. It is evident from Figure 4.5 that BM3 experienced a linear deformation from 1990 to 2011 while BM29 had an abrupt change in deformation trend in 1997. The overall trend observed in BM29 indicates that there was a possible land collapse or failure in underground mine pillars occurring in 1997, which was followed by a *no deformation* period.

The room-and-pillar mining technique, which leaves vacant space underground after abandonment of the mine, is a common type of mining technique used in the majority of the abandoned mines in our study area (Grzovic and Ghulam 2015). In this type of mining extraction, a series of pillars are left to support the roof of

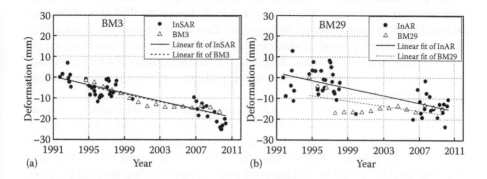

FIGURE 4.5 Comparison of InSAR-derived deformation against field measured data at BMs located in Belleville, Illinois. (a) Deformation trend at BM3. (b) Deformation trend at BM29. It indicates an abrupt change in deformation trend at BM29 in 1998. This type of land collapse might be attributed to an increase of crustal overburden due to a new development over abandoned mining areas or changes in soil properties from lawn watering or rainfall.

the mine while rooms are formed by extracting coal (Bauer 2008). Arguably, the deformation trend observed from BM29 may indicate a possible pillar failure, which might have been caused by residual land subsidence accelerated by extreme changes in precipitation, groundwater level, or a new development followed by excessive lawn watering or water use.

As shown in Figure 4.5, the SBAS time series exhibit a larger spread around the fitted linear curve than the BM survey measurements. This is due to the fact that *ground-truth* data from leveling represent measurements of a single point while InSAR-derived deformation represents a 20-m pixel average subsidence. One may expect discrepancies between absolute deformation values of leveling and SBAS, but consistency of overall deformation trends between the two measurements should be regarded as compelling.

4.4.4 IMPLICATIONS FOR SUSTAINABLE URBAN PLANNING

Commercial coal mining in Illinois started around 1810 in Jackson County, and mining for coal, fluorspar, lead, limestone, and zinc has occurred in 72 counties (Bauer 2008). The Illinois Department of Natural Resources estimates that there are approximately 5000 closed or abandoned coal mines in 53 counties across the state, about half of which have not been mapped or documented. Approximately 201,000 acres of urban and built-up lands may be close to underground mines with 333,000 housing units being exposed to possible mine subsidence. In recent decades, subsidence damage to structures has become increasingly common in Illinois as new developments expanded over abandoned, long-forgotten mines. Developing low-cost, effective tools to monitor potential risk from long-term residual land subsidence and predicting impending collapse before it actually causes economic loss is of great importance not only to individual property owners but also to insurance companies and regional policymakers charged with assessing risk related to abandoned coal mines.

Our preliminary studies indicated that sudden collapses of buildings, roads, and other infrastructure have occurred in the region. Although the thickness of the mining roof (depth to mine working) is far greater than the increasing overburden from urban developments, ground subsidence can be accelerated after a new development built on abandoned mining areas. Arguably, this might be explained by a combined effect of increasing overburden, lawn watering, abrupt changes in precipitation patterns, and groundwater table.

For example, the largest subsidence was observed at the B4 location, reaching approximately −50 mm/year over the 1992–2000 observation period. The prominent red spot on the upper-right corner of the B4 rectangle (Figure 4.2b) was right over Wolf Branch Middle School, Swansea, Illinois. The school was built around 2003 and quickly started to develop cracks and severe building damage by 2006. Significant mining subsidence at Bridle Lane, Swansea, Illinois, and surrounding areas (including the Wolf Branch Middle School) was also detected from InSAR maps of 2007–2011 (Figure 4.4d through g). These findings serve as compelling evidences that InSAR can reveal risk areas to avoid for new construction and guiding suggestions for prioritizing resources for targeted surveying, which, when combined with soil properties, groundwater withdrawal, and modern fault activities and distribution of abandoned mining areas, provides valuable information regarding impending land collapses. Doing so will (1) benefit policymakers and city planners in making informed decisions about future developments and (2) save potential loss of investments in such hazardous areas.

4.5 CONCLUSION

Coal mining can cause a number of environmental and social problems, such as building collapse, road damage, and disruptions in groundwater aquifers that are the result of mining-induced persistent land subsidence. In recent decades, subsidence damage to structures has become increasingly common as towns and cities expanded over abandoned, long-forgotten mines. Therefore, developing cost-effective methods to monitor long-term residual land subsidence is of great interest not only to individual homeowners but also to city planners, insurance companies, and regional policymakers charged with assessing risk related to abandoned coal mines. The objective of this chapter was to develop new techniques for using space-based data to detect irregular patterns of land subsidence and to design efficient strategies for identifying and mapping areas with elevated risk of structural and infrastructural damage. Thirty-seven C-band (ERS1/2) and 16 L-band (PALSAR) Synthetic Aperture Radar (SAR) images acquired from 1992 to 2000 and from 2007 to 2011 were processed with SBAS and PSI techniques to investigate land deformation. The correlation between land deformation and mining activities, regional geological background, groundwater, and precipitation were discussed. The results showed hot spots of ongoing and potential land collapses in the region with an average subsidence of −9 mm/year. Validated by conventional surveying, GPS measurements, and related geological survey, this study presented a comparative study between SBAS and PSI methods for displacement mapping, in an area made challenging by a humid climate, extensive surface waters, and dense vegetation cover.

Our results confirmed that subsidence took place mostly above the mining areas where deformation was well known and had experienced land collapses and significant infrastructure damage. Observed ground deformation, precipitation, and the variations of groundwater table show unquestionable similarities. Mechanisms that cause uplift over parks and cemeteries, however, are not clear and may be of various sources: unloading attributed to water level rise, pore pressure variations, swelling of soil or tree growth, and so on.

In most of the validation and comparison, we used vertical subsidence rates converted from the observed LOS rates. The limitation of this work is that only descending ERS-1/2 or ascending ALOS acquisitions were utilized for deformation mapping. Therefore, InSAR-derived LOS deformation represents the movement away from or toward the satellite. However, both GPS and leveling measurements conducted in the study area showed that ground displacement was occurring mainly vertically. We have projected the LOS displacement onto the vertical direction using satellite look angle and LOS displacement vector assuming that the horizontal ground displacement is negligible. However, this assumption may need comprehensive validation with additional information, for example, pairing ascending and descending

InSAR measurements or extensive GPS network. It should be noted that our work mainly focused on detecting ground deformation and identifying potential collapse hot spots. Detailed investigation on the mechanisms and structural, anthropogenic, and geological causes of mining related land collapses will be our future research.

ACKNOWLEDGMENT

This work was supported by Saint Louis University's President's Research Fund.

REFERENCES

Abdikan, S., M. Arikan, F. B. Sanli, and Z. Cakir. 2014. Monitoring of coal mining subsidence in peri-urban area of Zonguldak city (NW Turkey) with persistent scatterer interferometry using ALOS-PALSAR. *Environmental Earth Sciences* 71 (9):4081–4089.

Abidin, H. Z., H. Andreas, I. Gumilar, Y. Fukuda, Y. E. Pohan, and T. Deguchi. 2011. Land subsidence of Jakarta (Indonesia) and its relation with urban development. *Natural Hazards* 59 (3):1753–1771.

Abir, I., S. Khan, A. Ghulam, S. Tariq, and M. Shah. 2015. Active tectonics of western Potwar Plateau-Salt Range, Northern Pakistan from InSAR observations and seismic imaging. *Remote Sensing of Environment* 168:265–275.

Al-Rawahy, S. 1995. Effects of mining subsidence observed by time-lapse seismic reflection profiling, Durham University.

Bauer, R. A. 2006. Mine Subsidence in Illinois: Facts for Homeowners. Champaign, Illinois: Illinois State Geological Survey.

Bauer, R. A. 2008. Planned Coal Mine Subsidence in Illinois: A Public Information Booklet: Illinois State Geological Survey Circular, Champaign, Illinois.

Bell, J. W., F. Amelung, A. Ferretti, M. Bianchi, and F. Novali. 2008. Permanent scatterer InSAR reveals seasonal and long-term aquifer-system response to groundwater pumping and artificial recharge. *Water Resources Research* 44 (2).

Berardino, P., G. Fornaro, R. Lanari, and E. Sansosti. 2002. A new algorithm for surface deformation monitoring based on small baseline differential SAR interferograms. *IEEE Transactions on Geoscience and Remote Sensing* 40 (11):2375–2383.

Cavalie, O., M. P. Doin, C. Lasserre, and P. Briole. 2007. Ground motion measurement in the Lake Mead area, Nevada, by differential synthetic aperture radar interferometry time series analysis: Probing the lithosphere rheological structure. *Journal of Geophysical Research—Solid Earth* 112 (B3).

Chaussard, E., F. Amelung, H. Abidin, and S.-H. Hong. 2013. Sinking cities in Indonesia: ALOS PALSAR detects rapid subsidence due to groundwater and gas extraction. *Remote Sensing of Environment* 128:150–161.

Chaussard, E., S. Wdowinski, E. Cabral-Cano, and F. Amelung. 2014. Land subsidence in central Mexico detected by ALOS InSAR time-series. *Remote Sensing of Environment* 140:94–106.

Chen, F., H. Lin, W. Zhou, T. Hong, and G. Wang. 2013. Surface deformation detected by ALOS PALSAR small baseline SAR interferometry over permafrost environment of Beiluhe section, Tibet Plateau, China. *Remote Sensing of Environment* 138:10–18.

Chenoweth, C. A., M. E. Barrett, and S. D. Elrick. 2004a. Directory of Coal Mines in Illinois 7.5-minute Quadrangle Series: O'Fallon Quadrangle St. Clair County. Champaign, Illinois: Illinois State Geological Survey.

Chenoweth, C. A., S. D. Elrick, and M. E. Barrett. 2004b. Directory of Coal Mines in Illinois 7.5-minute Quadrangle Series: French Village Quadrangle St. Clair County. Champaign, Illinois: Illinois State Geological Survey.

Chenoweth, C. A., S. D. Elrick, and M. E. Barrett. 2005. Directory of Coal Mines in Illinois 7.5-minute Quadrangle Series: Collinsville Quadrangle Madison and St. Clair Counties. Champaign, Illinois: Illinois State Geological Survey.

Colesanti, C., A. Ferretti, F. Novali, C. Prati, and F. Rocca. 2003. SAR monitoring of progressive and seasonal ground deformation using the permanent scatterers technique. *IEEE Transactions on Geoscience and Remote Sensing* 41 (7):1685–1701.

Costantini, M. 1998. A novel phase unwrapping method based on network programming. *IEEE Transactions on Geoscience and Remote Sensing* 36 (3):813–821.

Delacourt, C., P. Briole, and J. Achache. 1998. Tropospheric corrections of SAR interferograms with strong topography. Application to Etna. *Geophysical Research Letters* 25 (15):2849–2852.

Dong, S., S. Samsonov, H. Yin, S. Ye, and Y. Cao. 2013. Time-series analysis of subsidence associated with rapid urbanization in Shanghai, China measured with SBAS InSAR method. *Environmental Earth Sciences*:1–15.

Donnelly, L. J., H. De La Cruz, I. Asmar, O. Zapata, and J. D. Perez. 2001. The monitoring and prediction of mining subsidence in the Amaga, Angelopolis, Venecia and Bolombolo Regions, Antioquia, Colombia. *Engineering Geology* 59 (1–2):103–114.

Ferretti, A., C. Prati, and F. Rocca. 2000. Nonlinear subsidence rate estimation using permanent scatterers in differential SAR interferometry. *IEEE Transactions on Geoscience and Remote Sensing* 38 (5):2202–2212.

Ferretti, A., C. Prati, and F. Rocca. 2001. Permanent scatterers in SAR interferometry. *IEEE Transactions on Geoscience and Remote Sensing* 39 (1):8–20.

Gabriel, A. K., R. M. Goldstein, and H. A. Zebker. 1989. Mapping small elevation changes over large areas: Differential radar interferometry. *Journal of Geophysical Research: Solid Earth (1978–2012)* 94 (B7):9183–9191.

Ghulam, A., R. Amer, and R. Ripperdan. 2010. A filtering approach to improve deformation accuracy using large baseline, low coherence DInSAR phase images. In *2010 IEEE International Geoscience and Remote Sensing Symposium*. Honolulu, Hawaii: IEEE.

Goldstein, R. M., and C. L. Werner. 1998. Radar interferogram filtering for geophysical applications. *Geophysical Research Letters* 25 (21):4035–4038.

Gourmelen, N., F. Amelung, and R. Lanari. 2010. Interferometric synthetic aperture radar–GPS integration: Interseismic strain accumulation across the Hunter Mountain fault in the eastern California shear zone. *Journal of Geophysical Research: Solid Earth (1978–2012)* 115 (B9).

Grzovic, M., and A. Ghulam. 2015. Evaluation of land subsidence from underground coal mining using TimeSAR (SBAS and PSI) in Springfield, Illinois. *Natural Hazards* 79:1739–1751.

Guéguen, Y., B. Deffontaines, B. Fruneau et al. 2009. Monitoring residual mining subsidence of Nord/Pas-de-Calais coal basin from differential and Persistent Scatterer Interferometry (Northern France). *Journal of Applied Geophysics* 69 (1):24–34.

Gutiérrez, F., J. P. Galve, P. Lucha, C. Castañeda, J. Bonachea, and J. Guerrero. 2011. Integrating geomorphological mapping, trenching, InSAR and GPR for the identification and characterization of sinkholes: A review and application in the mantled evaporite karst of the Ebro Valley (NE Spain). *Geomorphology* 134 (1):144–156.

Harrison, R. W. 1997. Bedrock geologic map of the St. Louis 30′ × 60′ quadrangle, Missouri and Illinois. U.S. Department of the Interior, U.S. Geological Survey.

Herrera, G., R. Tomas, J. M. Lopez-Sanchez et al. 2007. Advanced DInSAR analysis on mining areas: La Union case study (Murcia, SE Spain). *Engineering Geology* 90 (3–4):148–159.

Hooper, A. 2008. A multi-temporal InSAR method incorporating both persistent scatterer and small baseline approaches. *Geophysical Research Letters* 35:L16302.

Hooper, A. 2006. Persistent scatter radar interferometry for crustal deformation studies and modeling of volcanic deformation. Ph.D. thesis, Stanford University.

Hooper, A., and H. A. Zebker. 2007. Phase unwrapping in three dimensions with application to InSAR time series. *Journal of the Optical Society of America A—Optics Image Science and Vision* 24 (9):2737–2747.

Hooper, A., H. Zebker, P. Segall, and B. Kampes. 2004. A new method for measuring deformation on volcanoes and other natural terrains using InSAR persistent scatterers. *Geophysical Research Letters* 31 (23).

Jordan, Y. C., A. Ghulam, and M. L. Chu. 2014a. Assessing the impacts of future urban development patterns and climate changes on total suspended sediment loading in surface waters using geoinformatics. *Journal of Environmental Informatics* 24 (2):65–79.

Jordan, Y. C., A. Ghulam, and S. Hartling. 2014b. Traits of surface water pollution under climate and land use changes: A remote sensing and hydrological modeling approach. *Earth-Science Reviews* 128:181–195.

Jordan, Y. C., A. Ghulam, and R. B. Herrmann. 2012. Floodplain ecosystem response to climate variability and land-cover and land-use change in Lower Missouri River basin. *Landscape Ecology* 27 (6):843–857.

Kampes, B. M. 2006. *Radar Interferometry: Persistent Scatterer Technique.* Vol. 12: Springer.

Knospe, S., and W. Busch. 2009. Monitoring a tunneling in an urbanized area with Terrasar-X interferometry—Surface deformation measurements and atmospheric error treatment. Paper read at Geoscience and Remote Sensing Symposium, 2009 IEEE International, IGARSS 2009.

Lanari, R., F. Casu, M. Manzo et al. 2007. An overview of the small baseline subset algorithm: A DInSAR technique for surface deformation analysis. *Pure and Applied Geophysics* 164 (4):637–661.

Lanari, R., O. Mora, M. Manunta, J. J. Mallorquí, P. Berardino, and E. Sansosti. 2004. A small-baseline approach for investigating deformations on full-resolution differential SAR interferograms. *IEEE Transactions on Geoscience and Remote Sensing* 42 (7):1377–1386.

Lauknes, T. R., A. Piyush Shanker, J. F. Dehls, H. A. Zebker, I. H. C. Henderson, and Y. Larsen. 2010. Detailed rockslide mapping in northern Norway with small baseline and persistent scatterer interferometric SAR time series methods. *Remote Sensing of Environment* 114 (9):2097–2109.

Liu, P., Z. Li, T. Hoey et al. 2013. Using advanced InSAR time series techniques to monitor landslide movements in Badong of the Three Gorges region, China. *International Journal of Applied Earth Observation and Geoinformation* 21:253–264.

Louchios, A. G., S. D. Elrick, C. P. Korose, and P. Wise. 2013. Coal Industry in Illinois. Champaign, Illinois: Illinois State Geological Survey.

Massonnet, D., and K. L. Feigl. 1998. Radar interferometry and its application to changes in the Earth's surface. *Reviews of Geophysics* 36 (4):441–500.

Mazzotti, S., A. Lambert, M. Van der Kooij, and A. Mainville. 2009. Impact of anthropogenic subsidence on relative sea-level rise in the Fraser River delta. *Geology* 37 (9):771–774.

MoDNR. 2014a. MO 2014 Inventory of Mines, Occurences, and Prospects (IMOP). Missouri Department of Natural Resources (MoDNR), Missouri Geological Survey (MGS), Geological Survey Program (GSP).

MoDNR. 2014b. MO 2014 Sinkholes. Missouri Department of Natural Resources (MoDNR), Missouri Geological Survey (MGS), Geological Survey Program (GSP), Environmental Geology Section (EGS).

Paine, J. G., S. M. Buckley, E. W. Collins, and C. R. Wilson. 2012. Assessing collapse risk in evaporite sinkhole-prone areas using microgravimetry and radar interferometry. *Journal of Environmental and Engineering Geophysics* 17 (2):75–87.

Prati, C., A. Ferretti, and D. Perissin. 2010. Recent advances on surface ground deformation measurement by means of repeated space-borne SAR observations. *Journal of Geodynamics* 49 (3–4):161–170.

Raucoules, D., B. Bourgine, M. De Michele et al. 2009. Validation and intercomparison of persistent scatterers interferometry: PSIC4 project results. *Journal of Applied Geophysics* 68 (3):335–347.

Raucoules, D., C. Cartannaz, F. Mathieu, and D. Midot. 2013. Combined use of space-borne SAR interferometric techniques and ground-based measurements on a 0.3 km² subsidence phenomenon. *Remote Sensing of Environment* 139:331–339.

Raucoules, D., C. Maisons, C. Camec, S. Le Mouelic, C. King, and S. Hosford. 2003. Monitoring of slow ground deformation by ERS radar interferometry on the Vauvert salt mine (France)—Comparison with ground-based measurement. *Remote Sensing of Environment* 88 (4):468–478.

Rosen, P. A., S. Hensley, H. A. Zebker, F. H. Webb, and E. J. Fielding. 1996. Surface deformation and coherence measurements of Kilauea volcano, Hawaii, from SIR-C radar interferometry. *Journal of Geophysical Research—Planets* 101 (E10):23109–23125.

Schultz, A. P. 1993. Map showing surficial geology of the St. Louis 30 × 60 minute quadrangle.

Seeger, C. 2014. MO 2014 Inventory of Mines, Occurences, and Prospects (SHP). Rolla, Missouri: Missouri Geological Survey.

Short, N., A.-M. LeBlanc, W. Sladen, G. Oldenborger, V. Mathon-Dufour, and B. Brisco. 2014. RADARSAT-2 D-InSAR for ground displacement in permafrost terrain, validation from Iqaluit Airport, Baffin Island, Canada. *Remote Sensing of Environment* 141:40–51.

Sousa, J. J., A. J. Hooper, R. F. Hanssen, L. C. Bastos, and A. M. Ruiz. 2011. Persistent Scatterer InSAR: A comparison of methodologies based on a model of temporal deformation vs. spatial correlation selection criteria. *Remote Sensing of Environment* 115 (10):2652–2663.

Strozzi, T., R. Delaloye, D. Poffet, J. Hansmann, and S. Loew. 2011. Surface subsidence and uplift above a headrace tunnel in metamorphic basement rocks of the Swiss Alps as detected by satellite SAR interferometry. *Remote Sensing of Environment* 115 (6):1353–1360.

Usai, S. 2003. A least squares database approach for SAR interferometric data. *IEEE Transactions on Geoscience and Remote Sensing* 41 (4):753–760.

Yan, Y., M.-P. Doin, P. Lopez-Quiroz et al. 2012. Mexico City subsidence measured by InSAR time series: Joint analysis using PS and SBAS approaches. *IEEE Journal of Selected Topics in Applied Earth Observations and Remote Sensing* 5 (4):1312–1326.

Yerro, A., J. Corominas, D. Monells, and J. J. Mallorqui. 2014. Analysis of the evolution of ground movements in a low densely urban area by means of DInSAR technique. *Engineering Geology* 170:52–65.

Zebker, H. A., P. A. Rosen, and S. Hensley. 1997. Atmospheric effects in interferometric synthetic aperture radar surface deformation and topographic maps. *Journal of Geophysical Research—Solid Earth* 102 (B4):7547–7563.

Zebker, H. A., and J. Villasenor. 1992. Decorrelation in interferometric radar echoes. *IEEE Transactions on Geoscience and Remote Sensing* 30 (5):950–959.

Zhang, Y. H., J. X. Zhang, H. A. Wu, Z. Lu, and G. T. Sun. 2011. Monitoring of urban subsidence with SAR interferometric point target analysis: A case study in Suzhou, China. *International Journal of Applied Earth Observation and Geoinformation* 13 (5):812–818.

Zhu, W., Q. Zhang, X. L. Ding, C. Y. Zhao, C. S. Yang, and W. Qu. 2013. Recent ground deformation of Taiyuan basin (China) investigated with C-, L-, and X-bands SAR images. *Journal of Geodynamics* 70:28–35.

5 A Tale of Two Cities

Urbanization in Greensboro, North Carolina, and Guiyang, Guizhou, China

Honglin Xiao and Qihao Weng

CONTENTS

5.1 INTRODUCTION

Industrialization and urbanization are the primary revolutionary forces that remake and reshape the world (Carpenter 1966). Urbanization is the process when other types of land cover, such as agriculture and vegetation, are largely replaced by paved concrete surfaces. In the past several decades, significant population increases, migration, and accelerated socioeconomic development have intensified the urbanization process in both China and the United States. Not surprisingly, this rapid urbanization process, along with the ongoing social and economic transitions, has been creating many environmental problems. However, there are important differences in development stages, cultures and societies, urban structures, construction materials, and physical settings that result in differences of magnitudes, patterns, spatial extents, and natures of urban growth between the two countries. These differences have then further resulted in variations in the urban heat budget, surface runoff, water quality, local climate, vegetation change, and other environmental processes between the two (Boggs and Sun 2011; Cervero 2000; Gibbard et al. 2005; Jones et al. 1990). In China, land use and land cover patterns have undergone a fundamental change since the inception of the economic reform and open-door policies in 1978. Urban growth has been sped up, and extreme stress on the environment has occurred. Massive amounts of agricultural and forest land are disappearing each year, converted to urban or related uses. In recent years, urbanization in China has taken place at an unprecedented pace and will continue over the next decades. The level of urbanization in China has risen from 18% in 1978 to 30% in 1995, and to 42% in 2004. According to the latest sixth national census data, as of November 1, 2010, China's urban population stood at 665.57 million, with an urbanization level of 49.68%. At the rate of 35 million people per year moving to the cities, China will have nearly 1 billion in the cities and towns by 2030 or a 70% urban population (*The Economist* 2014). Although the United States is a far more developed country than China, it also loses a large amount of agricultural and forest land each year to urbanization as a result of the increasing population, desires for larger and better living spaces, and economic development. North Carolina in particular has changed significantly since 1990. In 1990, the state was home to 6.6 million people; by 2000, its population surpassed the 8 million mark, an increase of 21.4% from 1990. From 2000 to 2010, the state grew 18.5% compared with the 9.7% overall growth in the United States. This has resulted in a population of 9.5 million people and led North Carolina to become the 10th most populous state in the United States (Bauerlein 2011). By 2030, North Carolina is predicted to have a population of more than 13 million—an increase of more than 30% from 2010—making it the seventh most populated state in the country (US Bureau of the Census).

Understanding urbanization characteristics, patterns, and processes is an effective way to reduce the risk level that the rapid urbanization has brought about to the environment. Measuring the degrees of the impact from urbanizations in countries with different cultures, societies, technology backgrounds, and historical traditions has long been an interest of scientific inquiry (Carpenter 1966; Wirth 1938). Although no two places experienced identical urban transformation, it is obvious that there would be similarities in the processes. It is particularly important to compare the urbanization processes in the United States and China. These two countries account for approximately a quarter of the world population and have the two largest economies globally. Urbanization from the two countries has had and will have an even bigger impact on the surface of the Earth.

In satisfying the need, this study compared the urbanization processes in the North Carolina urban crescent from Raleigh–Durham–Chapel Hill (Research Triangle Park, RTP) to Greensboro (TRIAD) and Guiyang–Anshun in southwestern Guizhou Province, China, from the 1980s to the 2010s. Specifically, this study will first use Landsat TM images from 1991, 2001, and 2007 for Guizhou and TM images from 1989, 2001, and 2010 for North Carolina to extract land use/land cover (LULC) information. It will then examine how the urban areas had developed and what the similarities and differences in the urbanization processes were.

5.2   STUDY AREAS

In order to make the study results comparable, two study sites were carefully selected. Guizhou Province in the southwestern China (Figure 5.1a) and North Carolina in the southeastern United States (Figure 5.1b) were ideal for this study for the following reasons: First, both provinces/states are relatively rural and rely mostly on agricultural products despite having experienced tremendous urban growth in the past several decades. Second, North Carolina is located in the Piedmont area of the Appalachian Mountain leaning toward the Ocean, while Guizhou Province is on the slope of the Himalaya Mountains. Third, their statuses and degrees of development in comparison to their respective countries are similar. Combined together, Raleigh–Durham–Chapel Hill is ranked the 25th largest urban cluster in the United States. Guiyang, the capital of Guizhou Province, is ranked the 27th largest city in China. Additionally, Greensboro, the third largest city in North Carolina and part of the Piedmont Triad that consists of the area within and surrounding the three major cities of Greensboro, Winston-Salem, and High Point, has approximately 225,000 residents, while Anshun (the third largest city in Guizhou Province) has a population size of 220,000 in its metro area. Furthermore, the TRIAD–RTP area is one of the most developed areas in North Carolina, while the area that includes Guiyang, Anshun, Qingzheng, and Pingba County is among the most developed and flattest areas in Guizhou.

The physical settings are very comparable as well. Guiyang–Anshun in Guizhou Province is located in the east side of the Yunnan–Guizhou Plateau. It covers 6497 km² and has a population of 10 million. It has a subtropical wet monsoon climate. The mean annual temperature is 20°C, with the hottest month being July and the coldest being January. Annual average precipitation is 1140 mm, with a distinct summer wet season and a winter dry season. Average monthly humidity ranges from 74% to 78%. LULC varies from agricultural fields (with crops such as rice, corn, soybeans, wheat, rapeseed, oats, barley, and sweet potatoes) to natural

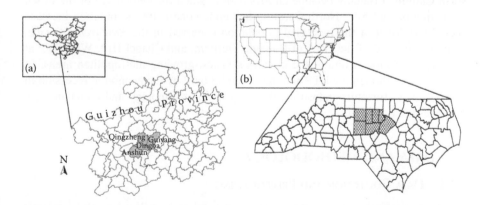

FIGURE 5.1   The study areas. (a) Guiyang–Anshun, Guizhou, China. (b) RTP–Greensboro, North Carolina.

forests in remote areas and local Feng-Shui (literally *wind and water*) preserves. In addition, there are barren land, water, and urban/built-up areas. Vegetation is subtropical with largely evergreen trees, although deciduous trees can also be found in some areas.

RTP-Greensboro in North Carolina is characterized by hilly, rolling land. It covers an area of 10,756 km². According to the recent census data, this area has become one of the most populous areas in the United States with a population size of approximately 1.4 million (US Census Bureau 2010). It has a humid, subtropical climate. Winters are short and mild, while summers are usually very hot and humid. Its average January temperatures range from −1°C to 10°C and July temperatures range from 20°C to 27°C. The average annual precipitation in the Piedmont is approximately 1200 mm. This region, located at the slope area of the Smoky Mountains, contains many of the state's largest cities. Elevations in the area vary from 100 to 400 m above the mean sea level. The major agricultural products include tobacco, corn, cotton, hay, peanuts, and vegetable crops. Forests are mainly mixed deciduous and evergreen trees. The area houses the nation's largest furniture, tobacco, brick, and textile producers. Metalworking, chemicals, mining, and paper are also important industries.

Since the beginning of China's economic reforms in 1978, China has experienced dramatic changes in its environment. Because it is dominated by Karst topography, Guiyang–Anshun has experienced more significant environmental problems than most other parts of the nation (Xiao et al. 2003). In recent years, especially since 1990, many of the high-quality agricultural lands are no longer used to produce crops. Accelerated economic growth has led to the encroachment of urban areas into nearby agricultural lands. New industrial districts have occupied much of the available flat land near major cities, while in the countryside, farmers use flat land to build new houses. In comparison, North Carolina has experienced dramatic change in the past three decades, with impressive growth and remarkable economic and educational development. This once rural state has been rapidly urbanized, and in 1992, North Carolina officially became an *urban state* when the percentage of the urban population passed 51%. North Carolina has undergone a 10% population increase since 2000. Most of the increased population occurred in the *Research Triangle*, an area composed of the cities of Raleigh, Durham, and Chapel Hill. With RTP at its core, the region is one of the country's fastest-growing metropolitan areas. In recent years, although the RTP area has been hit hard by the economic recession in the textile and furniture industries, the information technology and biotechnology industries are booming.

5.3   DATA AND METHODOLOGY

5.3.1   Data Collection and Preprocessing

In order to make the comparison more meaningful and, at the same time, cover a sufficiently long period, Landsat TM scenes were carefully chosen for each study area. Although obtaining images of exactly matching times in the same years over the same period proved impossible because of the limitation in data availability, we had managed to obtain three cloud-free images for each area that were acquired in approximately the same seasons and covered a similar time frame. The images for Guizhou were acquired in November or December 1991, 2000, and 2007 and the images for North Carolina were obtained in July or August 1989, 2001, and 2010. Because these images in each area were taken at a similar time of year, the phonological effect was minimized. Other ancillary data, such as county boundary files, topographic maps, vegetation maps of Guizhou, DEMs, historical aerial photographs, and LULC records from the two study areas, were also collected for the study.

The satellite images were corrected to remove atmospheric effects by using Erdas Imagine's Atmospheric Correct Extension (ATCOR 3), which used an algorithm based on the methods of Richter (1996) and Zhang et al. (2002). The images were then georectified using ground control points collected by GPS. The images were resampled to 30 m pixel size for all bands using the nearest-neighbor method. All the data were projected to a common UTM coordinate system for the purpose of analysis.

5.3.2 Image Classification

On the basis of the spatial resolution of the satellite images used and the study objective, LULC types for Guizhou were classified in accordance with the classification system developed by the Resource and Environment Information Center of Chinese Academy of Sciences (CAS) (Peng et al. 2011) (Table 5.1). For North Carolina, LULC was classified with a modified Anderson's system (Anderson et al. 1976) in reference to the US Geological Survey NLCD system (Homer et al. 2007) (Table 5.2). A hybrid image classification of unsupervised and supervised classifiers was applied. Studies have found that hybrid classifiers are particularly valuable in the analysis of LULC change, particularly in the area where there is complex variability in the spectral patterns or individual cover types (Lillesand et al. 2008; Mas 1999). Hybrid unsupervised/supervised classifications have been conducted in many LULC studies, and in most cases, the results are better than those from supervised or unsupervised classifications alone (García and Álvarez 1994; Rutchey and Vilcheck 1994). The unsupervised classification was carried out using the Iterative Self-Organizing Data Analysis algorithm to identify spectral clusters in the images. On the basis of the results of the unsupervised classifications, training sites were chosen from the images. For each image, spectral signatures for the training sites were carefully chosen and examined. A maximum likelihood classifier was then employed for the image classification. Land cover types for Guizhou were first classified into 11 types, including water, urban construction land, rural settlement, transportation and mining land, dense forest land, shrub land, fruit land, tea gardens, grass land, rice paddy land, arid agricultural land, vegetable land, and rocky desertification land. Land covers for North Carolina were first classified into 12 types of second-order land covers, including water, open space, low intensity, medium intensity, high intensity, deciduous forest, evergreen forest, mixed forest, shrub/scrub, grassland/herbaceous, pasture/hay, and cultivated crops. They were then combined to form into the first-order lands (Table 5.2).

TABLE 5.1
Land Cover Classification Scheme for Guizhou, China

Level I Types		Level II Types		
Order	Type	Order	Type	Description
1	Water area	40	Water	Mainly rivers and lakes and looking nearly black and homogeneous with clear boundary.
2	Urban or built-up	21	Urban construction land	Built-up areas in cities and towns. Looking bright green on RS image.
		22	Rural settlements	Mainly big villages, and usually scattered into arable lands. Look like urban built-up areas but much smaller on RS image.
		23	Transportation and mining land	Mainly mining areas and transportation facilities, such as motorways. Looking bright green and discernable by its line shape.
3	Wood land	31	Dense forest land	Patches of secondary coniferous forest, mainly dense redpines. Looking dark red and being discernable on RS image with clear boundary.
		32	Shrub land	Mainly growing dense broad-leaf shrubbery usually less than 3 m tall, usually distributing on mountainous areas with bright red and veined patches on RS image.
		33	Fruit tree land	Mainly different types of fruit trees.
		34	Tea garden land	Mainly tea gardens. Looking shiny red but homogeneous with clear boundary on RS image.
4	Grass land	40	Grass land	Covered by grasses, usually containing lots of thorns and scattered shrubbery with crown density<10%. Looking light pink with veins.
5	Arable land	51	Paddy land	Mainly used for growing rice. Usually distributing in bottom of valleys and basins and lying near rivers and lakes. Being homogeneous patches with clear boundary.
		52	Arid land	Mainly used for growing corn and wheat with poor irrigation facility. Usually lying in hilly areas and looking bright gray on false-colored RS image.
		53	Vegetable land	Mainly used for growing different kinds of vegetables. Usually distributed around cities and towns as patches with clear boundaries. Normally looking pinkish on RS image.
6	Barren land	60	Rocky desertification land	Rocky karst hills with lot of rocks exposed and bare, and usually covered by limited grasses and unused. Looking dark gray on RS image.

TABLE 5.2

Land Cover Classification Scheme for North Carolina

Level I Types		Level II Types		
Order	Type	Order	Type	Description
1	Water area	40	Water	All areas of open water, generally with less than 25% cover of vegetation or soil, looking nearly black and homogeneous with clear boundary.
2	Urban or built-up	21	Open space	Includes areas with a mixture of some constructed materials, but mostly vegetation in the form of lawn grasses. Impervious surfaces account for less than 20% of total cover. These areas most commonly include large-lot single-family housing units, parks, golf courses, and vegetation planted in developed settings for recreation, erosion control, or aesthetic purposes. Looking bright green on RS image.
		22	Low intensity	Includes areas with a mixture of constructed materials and vegetation. Impervious surfaces account for 20%–49% of total cover. These areas most commonly include single-family housing units. Looking like urban built-up areas but much smaller on RS image.
		23	Medium intensity	Includes areas with a mixture of constructed materials and vegetation. Impervious surfaces account for 50%–79% of the total cover. These areas most commonly include single-family housing units. Looking bright green and discernable by its line shape.
		24	High intensity	Includes highly developed areas where people reside or work in high numbers. Examples include apartment complexes, row houses, and commercial/industrial. Impervious surfaces account for 80%–100% of the total cover.
3	Wood land	31	Deciduous forest	Areas dominated by trees generally greater than 5 m tall, and greater than 20% of total vegetation cover. More than 75% of the tree species shed foliage simultaneously in response to seasonal change. Looking dark red and being discernable on RS image with clear boundary.

(Continued)

TABLE 5.2 (CONTINUED)
Land Cover Classification Scheme for North Carolina

Level I Types		Level II Types		
Order	Type	Order	Type	Description
		32	Evergreen forest	Areas dominated by trees generally greater than 5 m tall, and greater than 20% of total vegetation cover. More than 75% of the tree species maintain their leaves all year. Canopy is never without green foliage.
		33	Mixed forest	Areas dominated by trees generally greater than 5 m tall, and greater than 20% of total vegetation cover. Neither deciduous nor evergreen species are greater than 75% of total tree cover.
		34	Shrub/scrub	Areas dominated by shrubs; less than 5 m tall with shrub canopy typically greater than 20% of total vegetation. This class includes true shrubs, young trees in an early successional stage, or trees stunted from environmental conditions with bright red and veined patches on RS image.
4	Grassland herbaceous	40	Grassland herbaceous	Areas dominated by grammanoid or herbaceous vegetation, generally greater than 80% of total vegetation. These areas are not subject to intensive management such as tilling, but can be utilized for grazing. Looking light pink with veins.
5	Agricultural land	51	Pasture/hay	Areas of grasses, legumes, or grass-legume mixtures planted for livestock grazing or the production of seed or hay crops, typically on a perennial cycle. Pasture/hay vegetation accounts for greater than 20% of total vegetation. Being homogeneous patches with clear boundary.
		52	Cultivated crops	Areas used for the production of annual crops, such as corn, soybeans, vegetables, tobacco, and cotton, and also perennial woody crops such as orchards and vineyards. Crop vegetation accounts for greater than 20% of total vegetation. This class also includes all land being actively tilled and looking bright gray on false-colored RS image.
		53	Vegetable land	Mainly used for growing different kinds of vegetables. Usually distributed around cities and towns as patches with clear boundaries. Normally looking pinkish on RS image.

(Continued)

TABLE 5.2 (CONTINUED)
Land Cover Classification Scheme for North Carolina

Level I Types		Level II Types		
Order	Type	Order	Type	Description
6	Woody wetlands	69	Woody wetlands	Areas where forest or shrub land vegetation accounts for greater than 20% of vegetative cover and the soil or substrate is periodically saturated with or covered with water.
7	Barren land	70	Rock/sand/ clay	Barren areas of bedrock, desert pavement, scarps, talus, slides, volcanic material, glacial debris, sand dunes, strip mines, gravel pits, and other accumulations of earthen material. Generally, vegetation accounts for less than 15% of total cover. Looking dark gray on RS image.

It is difficult to classify land cover types by remote sensing interpretation alone, especially in the mountainous karst areas in Guizhou, because of the influence of rugged terrain and fragmented land covers. The authors' knowledge of and familiarity to this study areas played an important role in the LULC classification process. Postclassification sorting was performed to enhance the classification results. Occasionally, Google Earth was used to help correct the misclassifications. The overall accuracy for the three images in Guizhou was 89.3%, 90.5%, and 91.5%, respectively, while the overall accuracy for the three maps in North Carolina was 91.3%, 93.4%, and 90.6%, respectively. To identify the spatial–temporal patterns of LULC changes, and to examine the conversion from one class to another, a thematic change detection analysis was conducted using the Image Analyst Extension from ArcGIS software. LULC maps were superimposed to create change maps between different periods. To emphasize changes in urban and built-up, other types of changes were dropped from the analysis of change detection.

5.4 URBAN LAND CHANGES

Both study areas have experienced significant changes on urban land during the past two decades (Figures 5.2 and 5.3). In Guizhou, urban land use only accounted for 2.5% of the study area in 1989. It increased to 4.75% in 2000, and then further jumped to 16.79% in 2007. In less than 20 years, the urban land increased sixfold, from 40,322 to 269,530 acres. Of the lands changed to urban, rice paddy land contributed the most at 88,310 acres (Figure 5.4a). Substantial amounts of dry agricultural and forest lands had also been converted into urban use. The loss of dry agricultural land was slightly less than rice paddy at 77,330 acres, which was followed by forest at 53,160 acres and then grassland at 17,460 acres. The amounts of shrub land, non-rice crop land, and barren land that were converted into urban were 13,010, 12,552, and 14,790 acres, respectively.

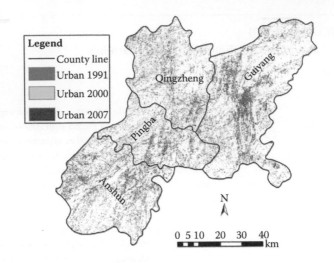

FIGURE 5.2   Urban expansion map in Guiyang–Anshun, Guizhou, China, from 1991 to 2007.

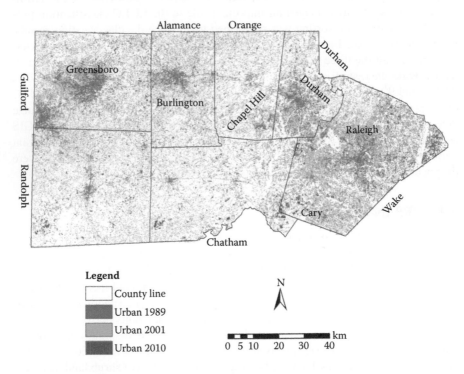

FIGURE 5.3   Urban expansion map for RTP–Greensboro, North Carolina, from 1989 to 2010.

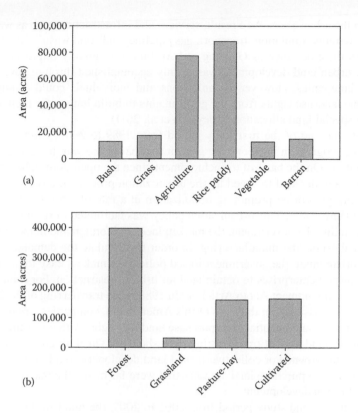

FIGURE 5.4 The amount of land converted into urban area. (a) Guiyang–Anshun from 1989 to 2007. (b) RTP–TRIAD in North Carolina from 1989 to 2010.

In the RTP–TRIAD area, urban and built-up land accounted for 5.18% of the total study area in 1989 (Figure 5.3). It more than tripled to 18.01% in 2001, and then almost doubled to 31.37% in 2010. In the past two decades, 397,790 acres of forest lands had been converted into urban or built-up land (Figure 5.4b). Almost equal amounts of pasture/hay land and cultivated lands were transformed into houses, offices, impervious surfaces, golf courses, and so on. Their loss stood at 159,109 and 162,385 acres, respectively. In addition, approximately 30,000 acres of grassland was converted into urban and built-up from 1989 to 2010.

5.5 URBANIZATION PROCESS ANALYSIS

5.5.1 Urbanization in Guizhou, China

Guizhou has experienced tremendous change in the urban area since the induction of China's open-door policy in the late 1970s. It has gone through two stages of urban development during the study period. During the first phase from 1989 to 2001, urban development was focused on the renewal of the urban space as a result of gentrification. Much of the effort was put on improving the quality of life and human welfare in the urban areas. As a result, services and infrastructure such as water and sewerage systems, sanitation, transport, gas pipeline, and solid waste collection and disposal in the cities such as Guiyang and Anshun were greatly improved. During that time, urban land development was mainly accomplished through government-affiliated companies. However, organizations and individuals could acquire non-transferable land-use rights from the governments to build houses for their workers through a special land allocation policy (Du et al. 2011).

Close to the end of the first study period from 1989 to 2000, China had made tremendous progress in its economic growth that paved the way for China's future modernization. On the basis of the achievement since the open-door policy, the central government decided to speed up the urbanization process in order to stimulate the economy growth by promoting urbanization at a rate of 1.5% to 2% annually (Song and Ding 2007). This urbanization policy was facilitated in several combined factors including the government, the marketplace, and foreign and domestic investment in real estate and manufacturing. In order to stimulate the demand for urban land and urban labor, the government issued policies to make it easy for companies and state-owned enterprises to obtain land for urban construction. For example, the Constitution Amendment Act to Article 10 in 1988 made transferring of the land use right to others possible. On the basis of this Amendment, companies, organizations, and individuals could negotiate and purchase land-use rights with governments or in the secondary land market from existing rights-holders (Du et al. 2011). During this period, the state-owned or collective-owned land developers could often work with the government to purchase land at prices that were far below the market prices for the sake of urban development.

During the second study period from 2001 to 2007, the main strategies were to expand the established cities by creating new urban districts to absorb the hundreds of thousands of migrants from the countryside and reduce the amount of peasants by merging rural townships and villages (Che xiang bing cun). A large percentage of the urban sprawl was a result of the establishment of several suburban Economic and Technological Development Zones (EDZs), such as "Jingyang New Development District" in Guiyang City, where the local government attracted companies to establish business or to build factories and real estate companies to build thousands of residential building complexes by providing low-priced land, incentives, and free infrastructure. To boost this new urban development, the city moved all of its government developments from the old core to this newly established city. Approximately half of the urban land expansion in Guiyang was caused by government-led development of EDZs (Ding 2003; Song and Ding 2007).

5.5.2 Urbanization in North Carolina

During the two study periods, the urban growth in the RTP–TRIAD area had unique expansion patterns as well. In the first phase, urban expansion permeated almost everywhere throughout the study area (Figure 5.3). However, the expansions were not distributed evenly among the various types of areas because location was a critical determinant on the types of growth each area might have. For example, the growth in the RTP and Greensboro area were mostly in highly developed areas for commercial/industrial buildings and apartment complexes; in the rest of the area, the growth was mainly sporadic single-family housing units. In the early years of the first study period, because of easy access, relatively cheap land, and convenience of doing business, the new development expanded around the previously established urban/business core. The urban growth patterns depended heavily on the prior density: growth went where there was room for it, filling in the urban tracts and lower-density edge of the city tracts. However, as the open lands were filled up toward their practical maximum density, their growth slowed down (Hartgen 2003).

As the land parcels became harder to assemble, prices rose, and the remaining lots were too expensive. Many companies or real estate realtors started to build their new businesses or homes in the areas that were not too far away from the urban core, much cheaper with low density and still close enough for daily commuters. This urban development phenomenon has been called *counter-urbanization*, a process of population de-concentration characterized by urban areas or communities with smaller sizes, lower densities, and more local homogeneity (Brian 1977, 1980). Most of the growth has become housing and entertainment communities for the burgeoning RTP and Greensboro areas. It is estimated that approximately half of the working population were living in places at least 15 min away from their workplaces. As more and more land went toward low-density housing subdivisions, streets, golf courses, strip malls, and other related developments, this came at the cost of massive tracts of forest land and farmland, sometimes wasted because of the low land use efficiency. The sprawling growth of suburban areas was largely responsible for the loss of almost half a million acres of farmland between 1997 and 2002 in North Carolina (Stuart 2010). During the time, land consumption was increasing four times faster than the rate of population growth. When government policies were issued to make it more difficult for new development to get approval, it pushed the additional growth to the region's remote areas.

The urbanization process accelerated in the beginning of the second study period at around 2001 when more land was converted into residential houses to fulfill housing demand. The financial institutions artificially created an easier environment for people to access home mortgages, which boosted a housing bubble in this area. On top of a primary residence home, many households also purchased additional houses for investment/speculative purposes. This development pattern, however, took a sharp turn in the second study phase. Since approximately 2006, the area was hit hard by the economic meltdown and when the housing bubble burst. The decreased house values and the sluggish economy had caused a serious budget crisis for local governments in maintaining normal operations. In order to find ways to fill the budget deficit, some cities in the study area had launched projects targeted at *urban renewal or development*, although much different from the earlier versions in the 1950s to the 1960s. For example, the town of Burlington in Alamance County was once a hub for the textile industry. In recent years, because of the challenges from foreign countries, most of the factories were closed down. This had led to mass unemployment and residential houses foreclosed or abandoned. To deal with the budget crisis and find a new economic engine, the city initiated a massive project by relocating its downtown shopping center to a location right off the nearby major highway, interstate highway I-85/40, aimed at drawing business from the travelers or nearby towns in order to revitalize its troubled economy. This new trend of renewal has resulted in urban sprawl into the less congested areas while creating easier access to the freeways and expressways.

5.6   COMPARISON OF THE URBANIZATION PROCESS

Although both of the two study areas experienced tremendous increases in urban area and shared many similarities in the urbanization process, the two study areas had profoundly different cultures and economic and political systems. Consequently, they followed differing ways of development.

5.6.1   VERTICAL VERSUS HORIZONTAL DEVELOPMENT

The urbanization in China can be characterized as *vertically expanded*, in contrast to *horizontally expanded* in the United States. Guiyang–Anshun had 269,530 acres of urban land with a population of 8 million, while RTP–TRIAD only had a population of approximately 2 million but with 833,897 acres of urban area. In order to use the limited land efficiently and effectively, the Chinese cities gravitated toward building concentrated housing in tall complexes. To maximize land use efficiency, city governments issued regulations ensuring that newly built commercial and residential buildings reached a minimum number of floors. In 1991, the average number of floors for the residential buildings was approximately six, which increased to approximately 15 in 2008. The tallest building in Guiyang was 28 stories in 1990. In 2008, buildings of 30 to 40 stories became a norm in Guiyang and Anshun. As a result, the cities were congested with clusters of so-called building forests (Figure 5.5a) with not much open space between the buildings. On the other hand, the urban area in the United States was more spread out with low density (Figure 5.5b). Residential buildings in the United States were mostly one to two stories with front yards and backyards. Houses had ample space between with lawns and fences/shrubs separating them. This difference in the dimension of the urban development has resulted in the observed difference in transformations of the urban layout between the two study areas. The Chinese cities were much more congested, with more homogeneous construction materials and roof and surface types. The US cities, on the other hand, were diverse with large portions of land allocated to nonurban use.

The vertical development pattern faced a change over the past few years in Guiyang. As the Chinese economy continued to boom, the relatively affluent citizens wanted to have a US standard of living. Consequently, many luxurious single-house villas were built to meet this demand. The scale of horizontal development in Guiyang was small, but had a great impact on its environment. The majority of the development was at the expense of the limited arable land in this karst region. The US cities, in contrast, had changed to add more vertical development. As the land became scarcer and more expensive, large backyards and the unused lands between communities were consequently more expensive owing to opportunity cost. For example, in the city of Chapel Hill, the lot size of the houses built in the years of

(a)   (b)

FIGURE 5.5 Vertical versus horizontal development in Guizhou and North Carolina. Both photos were captured from Google Earth. (a) A small part of Jingyang New Development Zone in Guiyang City. (b) A typical residential area in the RTP–Greensboro area.

the study period had been shrinking and there were more high-rise complexes being built than before. In addition, these new construction projects were inserted into the forest or open space between the established communities as *in-fill* developments instead of *leap-frog* clusters into new open areas.

5.6.2 CENTRALIZATION VERSUS DECENTRALIZATION

During the study periods, most urban growth in Guizhou Province had been centered on the old established cores, before gradually radiating to the surrounding areas. Cities expanded outward in the form of rings through different development stages. The first phase started in the early 1990s from the renewal or redevelopment of the old city cores, where aging buildings were demolished and replaced by modern skyscrapers and residential high-rise complexes. After 1995, the city started its second phase of development by encroaching from different directions into the closest available lands. Once the new expansion could not keep up with the ever-increasing demand, it would then continue its new phase of expansion. In China, there was a big difference between the quality of life in the city and the surrounding rural area in terms of public facilities, services, job opportunities, and so on, and thus it had become a universal goal to own an apartment in the urban area for people both living inside the cities and in the rural villages. When possible, they would spend their life savings to purchase a house in the city. Because of the massive influx of the rural people into the cities, the city boundaries had been pushed further away into the neighboring agricultural or other nonurban lands.

The urbanization process in North Carolina was more marked in contrast with that of Guiyang, especially during the first study period. Instead of agglomerating around the old city cores, urban sprawl was evident throughout the study area. Most urban areas experienced a population shift from the central city to suburban regions. Directly opposite to the situation in China, the quality of life in the suburban areas was much better than the condensed urban area. This decentralized urbanization phenomenon was commonly visible around the US cities, where the urban expansions were scattered and far away from the central urban cores (Schneider and Woodcock 2008). Many new shopping malls, public facilities, parking lots, and highway extended into those once rural areas to accompany the newly built residential communities. Furthermore, the materials used for urban construction were very different between the urban core and the local communities. In order to sustain the weight of the high-rises in the downtown area, brick, cement, and steel were the main materials for the construction. For the suburban cities, most buildings (especially residential ones) had wooden frames. This construction design in the suburbs had a more *natural* environment than the city core.

5.6.3 Centrally Planned versus Market Driven

The urbanization process in China was closely associated with the government's policies and plans. In China, all the land belonged either to the state or collectives. Individuals by law are unable to own land, but could obtain land use rights for specific purposes. The rights could not be transferred or sold in the market. When necessary, the government could take away the land from individuals for use in other government planned purposes. As a result, the degree, extent, and pattern of urban development were largely controlled by the government. In the early 1990s, the urban development was limited within the established old urban cores because of the government's policy to protect the farmland (Xiao and Weng 2007). In the 1990s, because of worries about food shortages as a result of farmland loss and population increases, the central government issued a series of strict regulations preventing land conversion. Starting around 2001, the urbanization process accelerated because of the government's policy shift to stimulate economy growth by promoting urbanization (Song and Ding 2007). For the first time, private real estate developers were allowed to share the market with the government, and the urbanization process accelerated at an unprecedented speed as the result. During this period, most of the urban expansion was a result of the establishment of New Development Zones (NDZs) by the government. For example, the site of Jingyang New District in Guiyang was a prisoner training camp surrounded by farmland before 2000. After the government decided to turn this area into another new city center by providing low-priced land and free infrastructure to factories and real estate developers, this once remote area had grown into a metro area as large as the old urban area. New urban districts expanded from the existing cities into the adjacent rural areas by taking over large areas of rice paddies or other lands (Ding 2003).

The urbanization process in the United States was mainly driven by market forces, since it was the developers and individuals and not government policies that drove most of the urban development. The housing markets as a whole functioned systematically and orderly from many seemingly random decisions between the consumer preferences and the products and services from the entrepreneurs and producers (Blair 1995). Decisions about what land to develop, and whether to build single houses or condominiums, were based on the developers' judgments about the conditions of the market. The land developers would completely take on the risks and rewards associated with their own decisions. When developers saw opportunities to make profits from developing certain areas, they would first buy the lands from the owners, and then improve the properties by turning them into homes, office buildings, or shopping malls, and later sell or lease them to businesses and families willing to pay their prices (McDonald 1997; Staley 1999). In the land market, the developers and consumers made decisions on the land and building prices from different sources including the home supply, office availability, and factory inventories (Henderson 1988). In summation, the urban area expanded in accordance with the developers and the consumers' interests. Over the past two decades, because there was a strong demand for homes and office buildings, large areas of farmland and forest land had been turned into urban and built-up areas in the RTP area. In Burlington, while one side of the city was being abandoned, the area closest to interstate I-40/85 boomed because of the new modern shopping malls and cheap but high-quality residential homes.

5.6.4 URBANIZED VERSUS URBANIZING

North Carolina was much more urbanized than Guizhou before the study period. It had undergone several decades of speedy urbanization after World War II. Its infrastructure, transportation system, and public facilities were very well developed at the beginning of the study period. The creation of the RTP in the 1950s had helped greatly in elevating this area into one of the most urbanized areas in the United States. Guizhou Province, however, was at the early stage of its development with only 2.5% urban land use. Back then, China's urbanization rate was generally among some of the lowest in the world, and Guizhou was one of the least developed areas in the nation. Because of the difference in development stages, the characteristics of urbanization were quite different between the two places. The type of development in North Carolina was much more patched, in-filling, and sporadic. Most of the construction was to improve the already established ones and was fairly spread throughout the study area. A large portion of the urban expansion was building golf courses, baseball fields, or parks to improve the quality of life. In Guizhou, on the other hand, the development was geared more toward fundamental infrastructure. Most of the urban increase was to build basic infrastructures such as airports, New Development Districts, new residential centers and neighborhoods, or modern highways to loop around the cities and connect them to other cities.

5.7 CONCLUSION

Our study result indicates that the United States and China share a lot of similarities and differences in the processes of urbanization. In the past two decades, urban areas have grown significantly at an alarming speed of approximately 500% in the

Remote Sensing for Sustainability

two study sites. In Guiyang–Anshun, Guizhou Province, China, the urban area grew
from 2.75% in 1991 to 16.78% in 2007, while in RTP–TRIAD, North Carolina, the
increase was from 5.18% in 1989 to 31.37% in 2010. As a result of the urban growth, a
great deal of nonurban land was lost. In Guizhou, the urban land was converted from
vegetable land, shrub land, grass land, dry agriculture, and rice paddies. In North
Carolina, the losses were mostly from areas of forest, grassland/shrub, pasture/hay,
and cultivated lands.

   Because of the differences in political systems, economic development stages,
cultures, and physical settings, the two study areas followed different paths of urbanization. In Guizhou, because of the shortage of land resources, the cities were more
congested and built with skyscrapers. Cities in North Carolina, on the other hand,
were developed with low density. Because of the common pursuit of better living
environments and amenities, the urban areas were more spread out horizontally.
In Guizhou, urban areas grew out from the old established cores and radiated into
the surrounding areas. These areas were developed according to the central government's plans. In North Carolina, the developers and the landowners chose where and
what type of growth urban expansion should be. In addition, because the two study
sites were in different development stages, the nature of the urban expansion was
quite different. Guizhou's urban growth focused more on infrastructure and basic
facilities, while the growth in North Carolina was to build parks, golf courses, and
athletic fields and other amenities to improve the quality of life.

REFERENCES

Anderson, J. R., Hardy, E. E., Roach, J. T., and Witmer, R. E., 1976. A Land Use and Land
    Cover Classification System for Use with Remote Sensor Data: U.S. Geological Survey
    Professional Paper 964, 28 pp.
Bauerlein, V., 2011. Population Jumps in North Carolina. *The Wall Street Journal*, March 3,
    2011.
Blair, J. P., 1995. *Local Economic Development*. Beverly Hills, California: Sage Publications.
Boggs, J. L., and Sun, G., 2011. Urbanization alters watershed hydrology in the Piedmont of
    North Carolina. *Ecohydrology*, 4: 256–264.
Brian J. L. Berry (ed.), 1977. *Urbanization and Counter Urbanization*. London: Sage
    Publications.
Brian, B. J. L., 1980. Urbanization and counterurbanization in the United States. *Annals of the
    American Academy of Political and Social Science, Changing Cities: A Challenge to
    Planning*, 451: 13–20.
Carpenter, D., 1966. Urbanization in the United States and Japan. *Studies in Comparative
    International Development*, 2(3): 38–49.
Cervero, R., 2000. Efficient Urbanization: Economic Performance and the Shape of the
    Metropolis. Policy Working Paper. Lincoln Institute of Land Policy. Cambridge, MA.
Ding, C., 2003. Land policy reform in China: Assessment and prospects. *Land Use Policy*,
    20: 109–120.
Du, H., Ma, Y., and An, Y., 2011. The impact of land policy on the relation between hous-
    ing and land prices: Evidence from China. *The Quarterly Review of Economics and
    Finance*, 51(1): 19–27.
García, M. C., and Álvarez, R., 1994. TM digital processing of a tropical forest region in
    southeastern Mexico. *International Journal of Remote Sensing*, 15, 1611–1632.

Gibbard, S. G., Caldeira, K., Bala, G., Phillips, T.J., Wickett, M. 2005. Climate effects of global land cover change. *Geophysical Research Letters*, 32(L23705): 4.

Hartgen, D. T., 2003. Highways and Sprawl in North Carolina. Policy Summary.

Henderson, J. V., 1988. *Urban Development: Theory, Fact, and Illusion*. Oxford University Press, New York.

Homer, C., Dewitz, J., Fry, J., Coan, M., Hossain, N., Larson, C., Herold, N., McKerrow, A., VanDriel, J. N., and Wickham, J., 2007. Completion of the 2001 National Land Cover Database for the Conterminous United States. *Photogrammetric Engineering and Remote Sensing*, 73(4): 337–341.

Jones, P. D., Groisman, P. Ya, Coughlan, M., Plummer, N., Wang, W-C., and Karl, T. R., 1990. Assessment of urbanization effects in time series of surface air temperature over land. *Nature* 347, 169–172.

Lillesand, T., Kiefer, R. W., and Chipman, J., 2008. *Remote Sensing and Image Interpretation*, 6th Edition, Wiley. 968 pp.

Mas, J., 1999. Monitoring land-cover changes: A comparison of change detection techniques. *International Journal of Remote Sensing*, 20: 139–152.

McDonald, J. F., 1997. *Fundamentals of Urban Economics*. Prentice Hall, Inc., Englewood Cliffs, New Jersey.

Peng, J., Xu, Y. Q., Cai, Y. L., and Xiao, H. L., 2011. The role of policies in land use/cover change since the 1970s in ecologically fragile karst areas of Southwest China: A case study on the Maotiaohe watershed. *Environmental Science & Policy*. 14(4): 408–418.

Richter, R., 1996. A spatially adaptive fast atmospheric correction algorithm. *International Journal of Remote Sensing*, 17: 1201–1214.

Rutchey, K., and Vilcheck, L., 1994. Development of an everglades vegetation map using a SPOT image and the global positioning system. *Photogrammetric Engineering and Remote Sensing*, 60(6): 767–775.

Schneider, A., and Woodcock, C. E., 2008. Compact, dispersed, fragmented, extensive? A comparison of urban growth in twenty-five global cities using remotely sensed data, pattern metrics and census information. *Urban Studies*, 45: 659.

Song, Y., and Ding, C., 2007. Urbanization in china: Critical issues in an era of rapid growth. Lincoln Institute of Land Policy, Cambridge, MA. 302 pp.

Staley, S. R., 1999. "Urban Sprawl" and the Michigan Landscape: A Market-Oriented Approach. A joint project for Michigan from the Mackinac Center for Public Policy and the Reason Public Policy Institute.

Stuart, A. W., 2010. Recent Population Change in North Carolina. North Carolina Museum of History Office of Archives and History, N.C. Department of Cultural Resources.

The Economist, 2014. Building the dream, Special Report, The Economist Newspaper Limited. Apr 19, 2014.

U.S. Census Bureau. 2010. Population and housing unit counts for the United States, regions, divisions, and American Indian, Alaska Native, and Native Hawaiian Areas, from http://www.census.gov/2010census/.

Wirth, L., 1938. Urbanism as a way of life. *The American Journal of Sociology*, 44(1) 1–24.

Xiao, H., Brook, G., and Lo, C., 2003. Evidence from satellite images of human-induced environment change in Guizhou Province, China 1991–1998. *Asian Geographer*, 22 (1 and 2).

Xiao, H., and Weng, Q., 2007. The impact of land use and land cover changes on land surface temperature in a karst area of China. *Journal of Environmental Management*, 85: 245–257.

Zhang, Y., Guindon, B., and Cihlar, J., 2002. An image transform to characterize and compensate for spatial variations in thin cloud contamination of Landsat images. *Remote Sensing of Environment*, 82: 173–187.

Section II

Remote Sensing for Sustainable
Natural Resources

6 Role of Remote Sensing in Sustainable Grassland Management
A Review and Case Studies for a Mixed-Grass Prairie Ecosystem

Alexander Tong, Bing Lu,
Yuhong He, and Xulin Guo

CONTENTS

6.1 INTRODUCTION

Grasslands are critically important ecosystems that serve as habitats for a variety of species of perennial grasses and other herbaceous vegetation, birds, animals, and insects. However, a majority of grassland biomes have been transformed, converted, or altered by human intervention on a global scale, with very few natural intact areas remaining. In North America, the grassland biome was once the most extensive, but has become one of the most threatened ecosystems (Samson and Knopf 1994). Since the remaining grassland ecosystems are inherently fragile, effective management strategies are important to address impacts driven by anthropogenic activities such as overgrazing, urban expansion, agricultural intensification, invasive species, and wildfire suppression under a changing climate. In fact, grassland management strategies have been moving toward integrated ecosystem and landscape-based approaches to address the impacts by focusing on areas such as biodiversity conservation, habitat restoration, and sustainable resource management (Bizikova 2009; Estrada-Carmona et al. 2014; Jun 2006). Such broad-scale plans bring partners and stakeholders together to realize common and shared objectives that consider both local and landscape-wide needs.

Monitoring grassland changes across space and time is the first step leading to effective management plans. Since grasslands are central to the livelihoods of more than a billion low-income people, managing grasslands has been a balance between competing demands, especially between economic returns and ecosystem services. Traditional ways of grassland monitoring that rely on field surveys are typically expensive and labor-intensive. In addition, field-based methods only provide localized information that presents a challenge when extrapolating it over large areas, and the information is thus often insufficient for land managers. Alternatively, remote sensing has become increasingly important for grassland ecosystem monitoring and guiding sustainable management practices (Boval and Dixon 2012). Once remote sensing technologies and techniques are validated at a local level, they can be easily generalized and used for long-term monitoring at a range of spatial and temporal scales.

Remote sensing imagery with a variety of spatial and temporal resolutions can be utilized for different management purposes with modest budgets. For practical and economic reasons, multispectral image data including Système Probatoire d'Observation de la Terre (SPOT), Landsat, Moderate Resolution Imaging Spectroradiometer (MODIS), and Advanced Very High Resolution Radiometer (AVHRR) images acquired from spaceborne sensors are commonly used for studying large geographical areas. Whereas SPOT and Landsat images offer high and medium spatial resolution data, respectively, they also have correspondingly lower temporal resolutions in comparison to the coarse spatial, but high temporal resolutions of MODIS and AVHRR data. Still, the most extensively used satellite imagery for research and application has been Landsat, as it offers a long historic archive that can be used to map long-term spatiotemporal vegetation changes. More capable sensors that could resolve ground features more accurately are spaceborne sensors offering very high spatial resolutions such as GeoEye-1, WorldView-2/3, or Pleiades-1A with spatial resolutions of 0.41, 0.46, and 0.5 m, respectively. Less commonly used for study because of the high cost of acquisition is airborne hyperspectral imagery such as AVIRIS or CASI, which provides very high spectral and spatial resolution data, but is consequently not practical for long-term studies/monitoring application. Recently, lightweight digital cameras and hyperspectral sensors that can be mounted on unmanned aerial vehicles (UAVs) have been utilized for grassland surveys (Laliberte and Rango 2011; Laliberte et al. 2011; Rango et al. 2006; Von Bueren et al. 2015). UAVs as an emerging remote sensing platform can provide imagery with spatial resolution higher than 0.1 m, which is an unprecedented data source for species-level research. Such platforms are capable of providing very high spatial resolution images at required temporal resolutions that can be used for finer-scale grassland investigation. However, the current acquisition cost and processing time for such high spatial resolution imagery may be only appropriate for hot spot monitoring, but untenable for long-term large-area monitoring.

Over the past decade, studies using remote sensing for grassland management have been conducted all over the world, including North America (Listopad et al. 2015; Mirik and Ansley 2012), South America (Bradley and Millington 2006; Di Bella et al. 2011), Europe (Psomas et al. 2011; Redhead et al. 2012), Asia (Cui et al. 2012; Leisher et al. 2012; Zhang et al. 2008), Africa (Olsen et al. 2015), and Australia (Guerschman et al. 2009; Lawes and Wallace 2008). Remote sensing of spatiotemporal changes in vegetation attributes such as biochemical (e.g., chlorophyll pigment) and biophysical (e.g., leaf area index [LAI]) properties, plant biomass, canopy cover, and vegetation height can inform ecosystem status. Additionally, these attributes can be applied to monitor the effects of wildfire disturbances, habitat loss, or climate change on grasslands. In these studies, the most commonly used remote sensing technique for ecosystem monitoring applications has been the use of empirical–statistical models. These models involve establishing a relationship between in situ biochemical or biophysical measurements with spectral vegetation indices calculated using ground-level spectral reflectance measurements or optical remote sensing imagery.

Intensive research using remote sensing within the past decade has been conducted in an endangered mixed-grass prairie ecosystem to evaluate vegetation conditions across space and time in relation to local environmental factors, climate conditions, and disturbance events. The implications of the research for validating remote sensing technologies and techniques for the mixed-grass prairies are providing valuable information and insight to land managers for guiding future management practices; several bodies of this research are presented in the following section.

6.2 CASE STUDIES

Of the three North American prairie types—tallgrass, mixed-grass, and shortgrass, the mixed-grass prairies have seen some of the worst decline, with remnants still found in parts of southern Alberta and Saskatchewan, Canada. The Government of Canada recognized the ecological importance of preserving an intact area of the endangered mixed-grass prairie ecosystem in 1981 and the Grasslands National Park (GNP, N 49°12′, W 107°24′) was soon established in 1988 (Figure 6.1). However, portions of the GNP are fragmented into small parcels as a consequence of land within and neighboring the park boundaries being privately held and used for agricultural or grazing purposes. Nevertheless, Parks Canada, an agency of the Government of Canada, has been tasked with operating and protecting the GNP in efforts to conserve

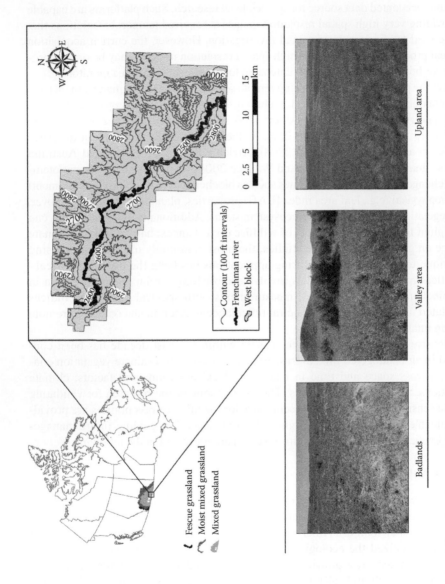

FIGURE 6.1 Map delineating the extent of the mixed grass prairies in Canada, the GNP, and three typical landscape units including badlands, valley area, and upland area in the GNP.

and restore the rich diversity of species and highly specialized communities of plants and animals in their native state that have evolved in response to a variety of stresses, such as drought, grazing, and fire (Anderson 2006; Shorthouse and Larson 2010).

Here, we introduce a select number of studies demonstrating the application of remote sensing technologies and techniques for the management of the mixed-grass prairies at the GNP. From a remote sensing perspective, the GNP has offered a unique challenge of assessing heterogeneous canopies composed of several dominant species, often with dead material or exposed soil. We focus on the role of remote sensing for addressing several key challenges currently affecting the ecological integrity of the GNP and indicate the utility of these studies for sustainable management needs. The importance of deriving vegetation attributes (e.g., LAI and chlorophyll content) for evaluating grassland health is first discussed, followed by the introduction of how remote sensing can be used for assessing the effect of wildfire disturbances and climate conditions, and finally we outline how remote sensing–derived information can aid in habitat mapping of an endangered grassland species.

6.2.1 Assessing Grassland Health Using Remote Sensing–Derived Biophysical Properties

Biophysical properties derived from remote sensing are direct indicators of grassland ecological status and can help land managers to assess the health of the ecosystem across space and aid in sustainable management. Many studies for the GNP (e.g., Banerjee et al. 2011; He et al. 2006a,b, 2007a,b, 2009; Tong and He 2013) have focused on using remote sensing–derived LAI to study vegetation spatial patterns of the ecosystem. He et al. (2009) linked LAI to other vegetation biophysical properties such as dead biomass, which solved the difficulty of mapping biophysical information owing to insufficient sampling coverage for the GNP, and many of these biophysical measures have been further incorporated in aid of spatial fire fuel modeling, habitat modeling for species at risk, and biomass monitoring for a reintroduced herd of plains bison (Figure 6.2). Fire and habitat modeling and biomass monitoring are all important remote sensing applications to address some of the current ecological challenges facing Parks Canada. For instance, fire modeling can assist in the implementation of appropriate areas for prescribed burning, habitat modeling can aid in species reintroduction and protection of critical habitats for species at risk, while biomass monitoring is essential for determining the effects of grazing regimes in the GNP.

6.2.2 Mapping Chlorophyll Content as an Indicator of Grassland Health

Leaf chlorophylls are inherently related to the photosynthetic capacity of plants and thus provide a measure of productivity to guide sustainable management practices. A study conducted by Tong (2014) used empirical–statistical models to estimate chlorophylls at a range of spatial scales. In specific, the leaf-level relationships between chlorophylls and corresponding spectral reflectance were scaled up to generate canopy and landscape-level chlorophyll maps through a scaling-up procedure proposed by Wong and He (2013) using SPOT-5 and CASI-550 images (Figure 6.3). Such

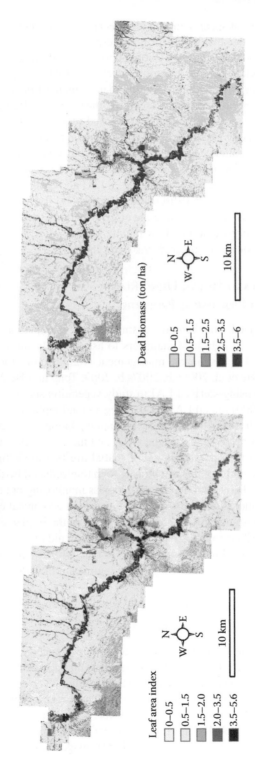

FIGURE 6.2 LAI map with a map accuracy of 66.7% obtained from SPOT-4 image using the regression model between Adjusted Transformed Soil Adjusted Vegetation Index (ATSAVI) and LAI. Dead biomass map with an accuracy of 51.2% obtained from LAI map using the regression model between dead biomass and LAI. (Adapted from He, Y., X. Guo, and J. F. Wilmshurst. 2009. Reflectance measures of grassland biophysical structure. *International Journal of Remote Sensing* 30 (10): 2509–2521. doi:10.1080/01431160802552751.)

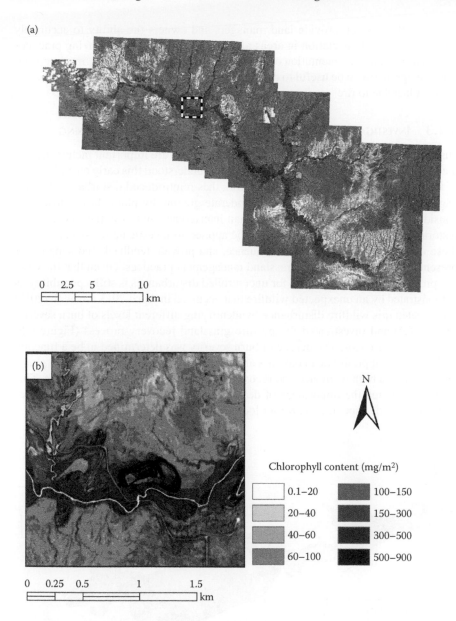

FIGURE 6.3 Chlorophyll maps derived from (a) SPOT-5 image acquired June 8, 2012, with a map accuracy of 63.52%; (b) CASI-550 image acquired June 23, 2012, with a map accuracy of 72.88%. (Adapted from Tong, A. 2014. Estimating Grassland Chlorophyll Content for a Mixed Grassland: Exploring the Performance of the Empirical–Statistical and the Physical Modelling Approach. MSc dissertation, University of Toronto.)

chlorophyll maps can provide land managers and owners the ability to accurately assess the vigor of vegetation in areas affected by agricultural or grazing practices and facilitate the implementation of strategies for sustainable resource management. These maps may also be useful to detect areas that are water stressed and help assess areas vulnerable to fire.

6.2.3 Investigating Grassland Disturbance Using Remote Sensing

Disturbances on the grasslands (e.g., wildfire, grazing) are important factors driving the evolution of grasslands. Parks Canada (2010) understood this early on and, as part of the current GNP management framework, has reintroduced disturbance regimes (e.g., prescribed burning and light to moderate grazing by plains bison) that were considered natural processes before human interference and essential to ecological restoration efforts. Remote sensing can be applied to investigate the occurrence of these disturbances, evaluate their influence, and provide feedback and support for present and future sustainable grassland management practices. Given that the GNP is a protected area, the potential for uncontrolled disturbances is still a possibility as demonstrated by an unexpected wildfire that occurred in April 2013. Lu et al. (2015) investigated this wildfire disturbance by identifying different levels of burn severity (Figure 6.4) and investigated the postfire grassland recovery process (Figure 6.5) using Landsat images. The degree of burn severity was determined to be a function of the amount of prefire dead biomass and the elevation of the landscape. The grasslands showed a high resilience and recovered quickly after the fire disturbance. This study highlighted the importance of dead biomass estimation for fire risk management and invasive species control for local ecosystem balance.

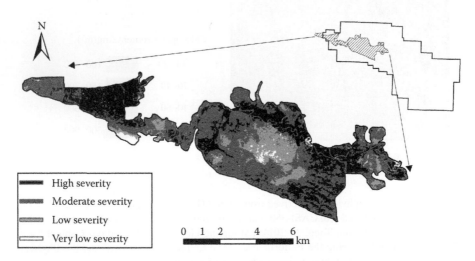

FIGURE 6.4 Burned area in GNP and burn severity estimated using Landsat imagery. (Adapted from Lu, B., Y. He, and A. Tong. 2015. Evaluation of spectral indices for estimating burn severity in semi-arid grasslands. *International Journal of Wildland Fire* 25 (2): 147–157.)

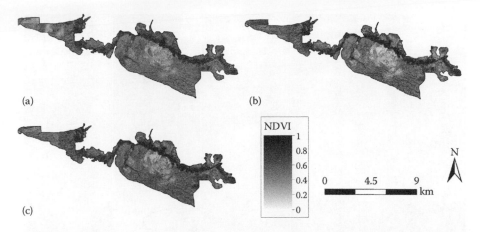

FIGURE 6.5 Normalized Difference Vegetation Index (NDVI) map derived from Landsat images showing the postfire vegetation recovery. (a) Six weeks after fire; (b) 10 weeks after fire; (c) 3 months after fire. (Adapted from Lu, B., Y. He, and A. Tong. 2015. Evaluation of spectral indices for estimating burn severity in semi-arid grasslands. *International Journal of Wildland Fire* 25 (2): 147–157.)

6.2.4 Determining the Effects of Climatic Factors on Grassland Health Using Remote Sensing

Environmental changes driven by climatic factors, primarily temperature and precipitation regimes, are important for determining ecosystem health. At the same time, understanding the spatial and temporal variability of evapotranspiration (ET) is also important for identifying soil and plant water stress that affects the health of vegetation (Girolimetto and Venturini 2013; Ritchie 1998; Yang et al. 2012). Remote sensing can provide important information on water availability and guide human activities such as irrigation or other management practices to help alleviate plant drought stress. For semiarid grasslands such as the GNP, vegetation greenness varies dramatically across space over years, as indicated by AVHRR Normalized Difference Vegetation Index (NDVI) values shown in Figure 6.6. For example, in a typical dry year, 2001, AVHRR NDVI values were generally low with an NDVI value less than 0.2 across the GNP, while in a wet year, 2002, AVHRR NDVI values are higher than 0.35. He (2014) thus investigated the effect of precipitation on vegetation cover for the GNP using climate data and SPOT-4 and AVHRR images. The amount and timing of precipitation were found to effectively control the ecosystem dynamics (Figure 6.7). Years with decreased vegetation coverage were associated with increased ET as a result of higher temperature, which stressed vegetation and was exacerbated by water deficiency in the soil. Although He (2014) did not directly estimate ET using remote sensing, studies have established its utility for a grasslands environment (e.g., Nosetto et al. 2005; Yang et al. 2012). Precipitation affected upland grass communities the highest, followed by valley grass communities, while the effect on riverside shrubs was not significant. The highest relationships occurred between percent vegetation cover and precipitation from the previous 80-day period,

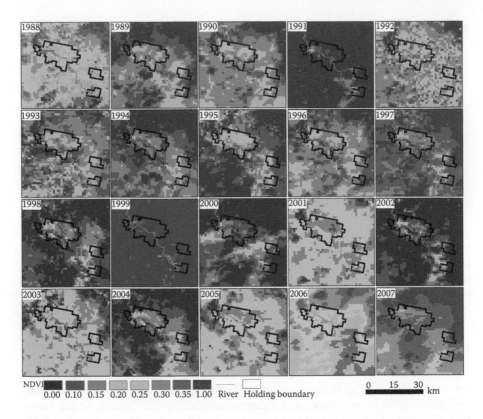

FIGURE 6.6 Vegetation greenness (i.e., NDVI derived from July AVHRR imagery) varies from 1988 to 2007. (Adapted from He, Y. 2014. The effect of precipitation on vegetation cover over three landscape units in a protected semi-arid grassland: Temporal dynamics and suitable climatic index. *Journal of Arid Environments* 109: 74–82. doi:10.1016/j.jaridenv.2014.05.022.)

suggesting that the lag effect of precipitation and the precipitation accumulated from the previous 80 days contributed the most to vegetation growth. This result could be used to aid land managers in predicting plant stress and implement water management strategies.

6.2.5 MAPPING HABITAT FOR ENDANGERED SPECIES USING REMOTE SENSING

In recent decades, wildlife populations have been on a drastic decline owing to human activities involving land conversion, pesticide use, or overhunting. In order to effectively guide land management practices to conserve, protect, and prevent further declines in wildlife populations, it has been necessary to investigate metrics that can quantify the biological activity of threatened and endangered species (e.g., foraging or reproduction behaviour), the biophysical and biochemical features of habitats, and the spatiotemporal changes of habitats as a result of human intervention. Such data need to be acquired at a broad spatial scale, which cannot

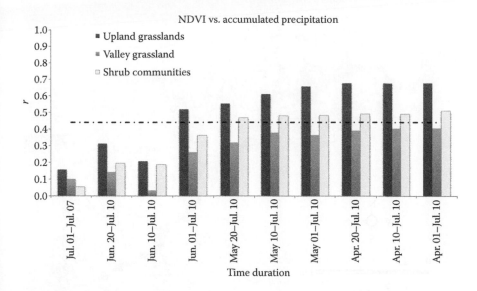

FIGURE 6.7 The correlation coefficients calculated between NDVI during the peak growing season (July 10) and accumulated precipitation from previous 10–100 days with a 10-day interval over three vegetation communities—upland/sloped grass, valley grass, and riparian shrub from 1988 to 2007. The dashed line is the significant value at the 0.05 level. (Adapted from He, Y. 2014. The effect of precipitation on vegetation cover over three landscape units in a protected semi-arid grassland: Temporal dynamics and suitable climatic index. *Journal of Arid Environments* 109: 74–82. doi:10.1016/j.jaridenv.2014.05.022.)

be easily achieved using traditional field survey methods. Remote sensing provides the capability to investigate the aforementioned metrics by quantifying the spatiotemporal changes in land use, vegetation type and cover, ecological processes and conditions, and wildlife corridors (Flaherty et al. 2014; Neumann et al. 2015; Olsen et al. 2007).

The habitats for various species in the North American prairie have experienced some of the worst decline in the past few decades (Klimek et al. 2007), and protected areas such as the GNP have established themselves as a haven for endemic species that have seen their habitats destroyed elsewhere, including the loggerhead shrike, an open country bird that was once widely distributed throughout North America. Remote sensing has also been applied in this area to support endangered species protection. Using remote sensing imagery, Shen et al. (2013) applied SPOT-4 imagery to map loggerhead shrike nesting sites in the GNP, which offers the opportunity to monitor the spatial distribution of the species for future conservation strategies (Figure 6.8). Findings indicated that nests in highly elevated open areas away from roads with scattered shrubs, particularly thorny species, were important for active shrike nesting sites. The study indicated that future park management goals should focus on preserving native grass and thorny shrub species within the shrike's breeding range.

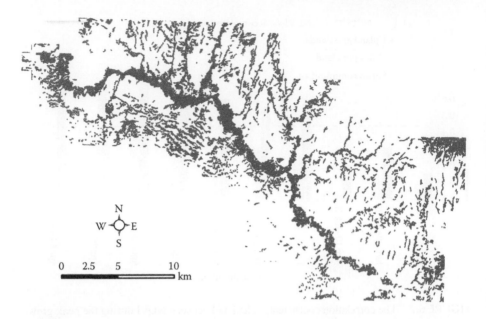

FIGURE 6.8 The suitable shrike habitat mapping based on a logistic model for loggerhead shrike in the West Block of GNP in 2006. (Adapted from Shen, L., Y. He, and X. Guo. 2013. Exploration of loggerhead shrike habitats in Grassland National Park of Canada based on in situ measurements and satellite-derived Adjusted Transformed Soil-Adjusted Vegetation Index (ATSAVI). *Remote Sensing* 5 (1): 432–453. doi:10.3390/rs5010432.)

6.3 CONCLUSIONS, CHALLENGES, AND OPPORTUNITIES

6.3.1 CONCLUSION

In the research capacity, remote sensing techniques supporting grassland management have been validated for many grassland biomes, such as the mixed-grass prairies at the GNP. By quantifying vegetation biophysical and biochemical properties and establishing empirical relationships at the local scale, upscaling procedures can extrapolate these relationships to produce vegetation maps for the entire study area. Traditional field-based methods have only been able to measure conditions at the local scale, whereas remote sensing has provided the ability to generate cost-effective vegetation maps that are a time-saving solution to accurately assess the entire extent of an area; subsequently, the application of remote sensing has not only provided vegetation health information but also detected and assessed grassland disturbances such as impacts of grazing or wildfires, grassland response to climatic changes, and mapped endangered animal habitats, to name a few. Results and information from the presented case studies have helped understand the impacts of current management decisions and can further facilitate future sustainable management policies and plans.

6.3.2 CHALLENGES AND OPPORTUNITIES

6.3.2.1 Image Acquisition and Processing

Remote sensing provides an economical and capable means for detecting, monitoring, and assessing ground features, but challenges still remain. For spaceborne sensors that operate on a predefined orbit and trajectory, the revisit time is fixed. If inclement weather (e.g., heavy cloud cover) is present over the area when the satellite sensor acquires an image, the image data may be unusable. Additionally, it is challenging to obtain spaceborne imagery with both high spatial and temporal resolution. For airborne mounted sensors, the time of image acquisition has to be considered, as well as the high costs of long-term data acquisition. UAV systems have increasingly become an alternative platform for image acquisition owing to their low cost of operation, high flexibility, ability to be deployed quickly and repeatedly, and capability to acquire imagery at very high spatial resolutions (sub-decimeter) (Laliberte and Rango 2011). Since UAV systems are normally operated at a range of low altitudes (50–300 m), it is possible to acquire images of different spatial resolutions, and weather conditions have minimal interference on image acquisition. With the quick response time of grasslands, particularly the mixed-grass prairie ecosystems at the GNP, UAVs offer a high temporal capacity to monitor and quantify the change in ecosystem response to disturbances that otherwise could not be detected with traditional spaceborne and airborne platforms.

Techniques used to analyze image data require specialized training and human error may be introduced. Imagery requires calibration and validated by field data, which are often limited to a few sites around accessible areas, and is time-consuming and therefore expensive to collect. This is not a weakness per se, as local level measurements can be extrapolated and scaled across the image. We point this out here to emphasize the need for collaboration and sharing of resources between agencies, land managers, and other stakeholders to reduce costs (Marsett et al. 2006). In all, careful planning and training in applying remote sensing can help alleviate these limitations.

6.3.2.2 Species-Level Monitoring

Understanding changes in species composition is crucial for conserving biodiversity, managing invasive species, and conducting sustainable grazing practices. For most applications, at least for homogeneous canopies, medium spatial resolution Landsat imagery has the capability to map vegetation at a range of scales (Xie et al. 2008). Conversely, heterogeneous grassland canopies, such as those in the GNP, feature several dominant species in addition to fractional dead material or soil components that compound the ability to accurately map and assess vegetation from the species level. Yet, even with the advent of very high spatial resolution imagery, the mixed pixel problem (i.e., a pixel that contains more than one spectral signature belonging to several ground features) remains a challenge for retrieving ground information from heterogeneous canopies. Radiative transfer modeling may be an avenue to accurately estimate vegetation parameters, but its inherent complexity makes them impractical for land managers.

6.3.2.3   Vegetation Structure Mapping

Optical remote sensing has been applied extensively for investigating grassland bio-physical and biochemical properties. However, extracting canopy structure information (vegetation density profiles) from optical remote sensing images remains a challenge since incoming radiation is primarily intercepted and reflected at the top of the canopy surface. On the other hand, airborne Light Detection and Ranging (LiDAR) systems are active sensors that emit and receive light pulses that can penetrate and reach beneath a canopy and therefore estimate canopy height, LAI, or aboveground biomass effectively. LiDAR systems provide an all-day capability and may be operated in slightly inclement weather conditions for vegetation studies. LiDAR has been widely applied for investigating vegetation structure in forest and agricultural areas (Drake et al. 2002; Hawbaker et al. 2009; Houldcroft et al. 2005) yet has rarely been deployed to study grasslands, especially for investigating vegetation structure. Given the relatively low to sparse vegetation density and canopy heights in grassland environments, the challenge of capturing the subtle structure of grassland canopies remains to be addressed (Bork and Su 2007; Hellesen and Matikainen 2013; Su 2004). To this end, with decreasing acquisition costs, the opportunity to explore the potential of LiDAR for grassland management is fast becoming an attractive option to researchers and land managers. Nevertheless, UAV multispectral images of the same area but from different directions may also be used to produce vegetation structure information in grasslands, using *Structure from Motion* techniques (Smith and Vericat 2015). Specifically, the optical point clouds of vegetation can be produced from multiple fly-overs of the same sites from many different directions.

6.3.2.4   Snow, Topography, and Soil Moisture Mapping

Spaceborne Interferometric Synthetic Aperture Radar (InSAR) sensors are increasingly being explored to resolve ground features. They are active sensors that emit and receive long wavelengths in the microwave portion of the spectrum that are not susceptible to atmospheric scattering, thereby allowing an all-day and all-weather imaging capability. Canada has been a pioneer in the field of InSAR, with the deployment of RADARSAT-1 and RADARSAT-2. For grassland research, RADARSAT-2 shows potential for snow mapping, ground surface topography mapping, and soil moisture detection. To date, RADARSAT-2 has been validated in grassland studies for soil moisture retrieval (e.g., Bertoldi et al. 2014; Xing et al. 2014) and image classification (e.g., Buckley and Smith 2011) with success. However, its application and implementation at an operational level for land managers have yet to be explored.

6.3.2.5   Logistics

Remote sensing may not be well understood by stakeholders involved with management policies and plans, and is thus imperative that researchers are able to engage and demonstrate the utility and application of remote sensing. The extent and focus of research for grassland management using remote sensing has been meager relative to studies that have been conducted for areas with greater perceived economic value (e.g., precision agriculture and forests). To fully appreciate the current trend of remote sensing research, we used Scopus, a comprehensive bibliographic database, and our survey revealed only 488 results inclusive from 1979 to 2015 that were relevant to grassland remote sensing and management. In comparison, forest remote sensing and management returned 3399 results and, similarly, agriculture remote sensing and management returned 3126 results inclusive from 1976 to 2015. Only by actively promoting awareness of the importance of grasslands for research can we better understand the ecological status of the grassland environment. Given the lesser focus of grasslands research within the remote sensing community, it is likely that less emphasis has been placed on educational initiatives and opportunities for stakeholders involved with grassland management policies and plans and is thus imperative that researchers are able to actively engage and demonstrate the utility and application of remote sensing. Through collaborative efforts between both parties, sharing of knowledge, needs, and resources along the way of the research process should be able to lead to the production of spatial products for use at an operational level. Future work may consider integrating remote sensing data with information from other disciplines such as ecology, geography, sociology, or economics to better understand the complexities of human–environmental interactions.

ACKNOWLEDGMENT

The support of the Natural Sciences and Engineering Research Council of Canada Discovery Grants to Dr. Yuhong He and Dr. Xulin Guo is gratefully acknowledged.

REFERENCES

Anderson, R. C. 2006. Evolution and origin of the central grassland of North America: Climate, fire, and mammalian grazers. *Journal of the Torrey Botanical Society* 133(4): 626–647. doi:10.3159/1095-5674(2006)133[626:EAOOTC]2.0.CO;2.

Banerjee, S., Y. He, X. Guo, and B. C. Si. 2011. Spatial relationships between leaf area index and topographic factors in a semiarid grassland: Joint multifractal analysis. *Australian Journal of Crop Science* 5(6): 756–763.

Bertoldi, G., S. Della Chiesa, C. Notarnicola, L. Pasolli, G. Niedrist, and U. Tappeiner. 2014. Estimation of soil moisture patterns in mountain grasslands by means of SAR RADARSAT2 images and hydrological modeling. *Journal of Hydrology* 516: 245–257. doi:10.1016/j.jhydrol.2014.02.018.

Bizikova, L. 2009. Challenges and lessons learned from integrated landscape management (ILM) projects. Winnipeg: International Institute for Sustainable Development.

Bork, E. W. and J. G. Su. 2007. Integrating LIDAR data and multispectral imagery for enhanced classification of rangeland vegetation: A meta analysis. *Remote Sensing of Environment* 111(1): 11–24. doi:10.1016/j.rse.2007.03.011.

Boval, M. and R. M. Dixon. 2012. The importance of grasslands for animal production and other functions: A review on management and methodological progress in the tropics. *Animal* 6(5): 748–762. doi:10.1017/S1751731112000304.

Bradley, A. V. and A. C. Millington. 2006. Spatial and temporal scale issues in determining biomass burning regimes in Bolivia and Peru. *International Journal of Remote Sensing* 27(11): 2221–2253. doi:10.1080/01431160500396550.

Buckley, J. R. and A. M. Smith. 2011. Comparing RADARSAT 2 and TerraSAR-X Quad-Pol SAR imagery of grasslands. doi:10.1109/IGARSS.2011.6050022.

Cui, X., Z. G. Guo, T. G. Liang, Y. Y. Shen, X. Y. Liu, and Y. Liu. 2012. Classification management for grassland using MODIS data: A case study in the Gannan Region, China. *International Journal of Remote Sensing* 33(10): 3156–3175. doi:10.1080/01431161.2011.634861.

Di Bella, C. M., M. A. Fischer, and E. G. Jobbágy. 2011. Fire patterns in north-eastern Argentina: Influences of climate and land use/cover. *International Journal of Remote Sensing* 32(17): 4961–4971. doi:10.1080/01431161.2010.494167.

Drake, J. B., R. O. Dubayah, D. B. Clark, R. G. Knox, J. B. Blair, M. A. Hofton, R. L. Chazdon, J. F. Weishampel, and S. Prince. 2002. Estimation of tropical forest structural characteristics, using large-footprint Lidar. *Remote Sensing of Environment* 79(2–3): 305–319. doi:10.1016/S0034-4257(01)00281-4.

Estrada-Carmona, N., A. K. Hart, F. A. J. DeClerck, C. A. Harvey, and J. C. Milder. 2014. Integrated landscape management for agriculture, rural livelihoods, and ecosystem conservation: An assessment of experience from Latin America and the Caribbean. *Landscape and Urban Planning* 129: 1–11. doi:10.1016/j.landurbplan.2014.05.001.

Flaherty, S., P. W. W. Lurz, and G. Patenaude. 2014. Use of LiDAR in the conservation management of the endangered red squirrel (*Sciurusi vulgaris* L.). *Journal of Applied Remote Sensing* 8(1). doi:10.1117/1.JRS.8.083592.

Girolimetto, D. and V. Venturini. 2013. Water stress estimation from NDVI-Ts plot and the wet environment evapotranspiration. *Advances in Remote Sensing* 2: 283–291. http://dx.doi.org/10.4236/ars.2013.24031.

Guerschman, J. P., M. J. Hill, L. J. Renzullo, D. J. Barrett, A. S. Marks, and E. J. Botha. 2009. Estimating fractional cover of photosynthetic vegetation, non-photosynthetic vegetation and bare soil in the Australian tropical savanna region upscaling the EO-1 Hyperion and MODIS sensors. *Remote Sensing of Environment* 113(5): 928–945. doi:10.1016/j.rse.2009.01.006.

Hawbaker, T. J., N. S. Keuler, A. A. Lesak, T. Gobakken, K. Contrucci, and V. C. Radeloff. 2009. Improved estimates of forest vegetation structure and biomass with a LiDAR-optimized sampling design. *Journal of Geophysical Research: Biogeosciences* 114(3). doi:10.1029/2008JG000870.

He, Y. 2014. The effect of precipitation on vegetation cover over three landscape units in a protected semi-arid grassland: Temporal dynamics and suitable climatic index. *Journal of Arid Environments* 109: 74–82. doi:10.1016/j.jaridenv.2014.05.022.

He, Y., X. Guo, and B. C. Si. 2007a. Detecting grassland spatial variation by a wavelet approach. *International Journal of Remote Sensing* 28(7): 1527–1545. doi:10.1080/01431160600794621.

He, Y., X. Guo, and J. Wilmshurst. 2006a. Studying mixed grassland ecosystems I: Suitable hyperspectral vegetation indices. *Canadian Journal of Remote Sensing* 32(2): 98–107.

He, Y., X. Guo, J. Wilmshurst, and B. C. Si. 2006b. Studying mixed grassland ecosystems II: Optimum pixel size. *Canadian Journal of Remote Sensing* 32(2): 108–115.

He, Y., X. Guo, and J. F. Wilmshurst. 2007b. Comparison of different methods for measuring leaf area index in a mixed grassland. *Canadian Journal of Plant Science* 87(4): 803–813.

He, Y., X. Guo, and J. F. Wilmshurst. 2009. Reflectance measures of grassland biophysical structure. *International Journal of Remote Sensing* 30(10): 2509–2521. doi:10.1080/01431160802552751.

Hellesen, T. and L. Matikainen. 2013. An object-based approach for mapping shrub and tree cover on grassland habitats by use of liDAR and CIR orthoimages. *Remote Sensing* 5(2): 558–583. doi:10.3390/rs5020558.

Houldcroft, C. J., C. L. Campbell, I. J. Davenport, R. J. Gurney, and N. Holden. 2005. Measurement of canopy geometry characteristics using LiDAR laser altimetry: A feasibility study. *IEEE Transactions on Geoscience and Remote Sensing* 43(10): 2270–2282. doi:10.1109/TGRS.2005.856639.

Jun, H. 2006. Effects of integrated ecosystem management on land degradation control and poverty reduction. In *Environment, Water Resources and Agricultural Policies*: Chapter 3, 63–72. Danvers: Organization for Economic Co-operation and Development.

Klimek, S., A. Richter gen. Kemmermann, M. Hofmann, and J. Isselstein. 2007. Plant species richness and composition in managed grasslands: The relative importance of field management and environmental factors. *Biological Conservation* 134(4): 559–570. doi:10.1016/j.biocon.2006.09.007.

Laliberte, A. and A. Rango. 2011. Image processing and classification procedures for analysis of sub-decimeter imagery acquired with an unmanned aircraft over arid rangelands. *GIScience and Remote Sensing* 48(1): 4–23. doi:10.2747/1548-1603.48.1.4.

Laliberte, A. S., M. A. Goforth, C. M. Steele, and A. Rango. 2011. Multispectral remote sensing from unmanned aircraft: Image processing workflows and applications for rangeland environments. *Remote Sensing* 3(11): 2529–2551. doi:10.3390/rs3112529.

Lawes, R. A. and J. F. Wallace. 2008. Monitoring an invasive perennial at the landscape scale with remote sensing. *Ecological Management and Restoration* 9(1): 53–59. doi:10.1111/j.1442-8903.2008.00387.x.

Leisher, C., S. Hess, T. M. Boucher, P. van Beukering, and M. Sanjayan. 2012. Measuring the impacts of community-based grasslands management in Mongolia's Gobi." *PLoS ONE* 7(2). doi:10.1371/journal.pone.0030991.

Listopad, C. M. C. S., R. E. Masters, J. Drake, J. Weishampel, and C. Branquinho. 2015. Structural diversity indices based on airborne LiDAR as ecological indicators for managing highly dynamic landscapes. *Ecological Indicators* 57: 268–279. doi:10.1016/j.ecolind.2015.04.017.

Lu, B., Y. He, and A. Tong. 2015. Evaluation of spectral indices for estimating burn severity in semi-arid grasslands. *International Journal of Wildland Fire* 25(2): 147–157.

Marsett, R. C., J. Qi, P. Heilman, S. H. Biedenbender, M. C. Watson, S. Amer, M. Weltz, D. Goodrich, and R. Marsett. 2006. Remote sensing for grassland management in the arid southwest. *Rangeland Ecology and Management* 59(5): 530–540. doi:10.2111/05-201R.1.

Mirik, M. and R. J. Ansley. 2012. Comparison of ground-measured and image-classified mesquite (*Prosopis glandulosa*) canopy cover. *Rangeland Ecology and Management* 65(1): 85–95. doi:10.2111/REM-D-11-00073.1.

Neumann, W., S. Martinuzzi, A. B. Estes, A. M. Pidgeon, D. Dettki, G. Ericsson, and V. C. Radeloff. 2015. Opportunities for the application of advanced remotely-sensed data in ecological studies of terrestrial animal movement. *Movement Ecology* 3(8). doi:10.1186/s40462-015-0036-7.

Nosetto, M. D., E. G. Jobbágy, and J. M. Paruelo. 2005. Land-use change and water losses: The case of grassland afforestation across a soil textural gradient in Central Argentina. *Global Change Biology* 11(7): 1101–1117. doi:10.1111/j.1365-2486.2005.00975.x.

Olsen, J. L., S. Miehe, P. Ceccato, and R. Fensholt. 2015. Does EO NDVI seasonal metrics capture variations in species composition and biomass due to grazing in semi-arid grassland savannas? *Biogeosciences* 12(14): 4407–4419. doi:10.5194/bg-12-4407-2015.

Olsen, L. M., V. H. Dale, and T. Foster. 2007. Landscape patterns as indicators of ecological change at Fort Benning, Georgia. *Landscape and Urban Planning* 79(2): 137–149. doi:10.1016/j.landurbplan.2006.02.007.

Parks Canada. 2010. Grasslands National Park of Canada Management Plan. http://www.pc.gc.ca/eng/pn-np/sk/grasslands/plan/~/media/pn-np/sk/grasslands/pdf/prairies-grasslands-plan-062010_e.ashx.

Psomas, A., M. Kneubühler, S. Huber, K. Itten, and N. E. Zimmermann. 2011. Hyperspectral remote sensing for estimating aboveground biomass and for exploring species richness patterns of grassland habitats. *International Journal of Remote Sensing* 32(24): 9007–9031. doi:10.1080/01431161.2010.532172.

Rango, A., A. Laliberte, C. Steele, J. E. Herrick, B. Bestelmeyer, T. Schmugge, A. Roanhorse, and V. Jenkins. 2006. Using unmanned aerial vehicles for rangelands: Current applications and future potentials. *Environmental Practice* 8(3): 159–168. doi:10.1017 /S1466046606060224.

Redhead, J., M. Cuevas-Gonzales, G. Smith, F. Gerard, and R. Pywell. 2012. Assessing the effectiveness of scrub management at the landscape scale using rapid field assessment and remote sensing. *Journal of Environmental Management* 97(1): 102–108. doi:10.1016/j.jenvman.2011.12.005.

Ritchie, J. T. 1998. Soil water balance and plant water stress. In *Systems Approaches for Sustainable Agricultural Development: Understanding Options for Agricultural Production*, ed. G. Y. Tsuji, G. Hoogenboom, and P. K. Thornton, 41–54. Dordrecht: Kluwer Academic Publishers.

Samson, F. and E. L. Knopf. 1994. Prairie conservation in North America. *Bioscience* 44(6): 418–421. doi: 10.2307/1312365.

Shen, L., Y. He, and X. Guo. 2013. Exploration of loggerhead shrike habitats in Grassland National Park of Canada based on in situ measurements and satellite-derived Adjusted Transformed Soil-Adjusted Vegetation Index (ATSAVI). *Remote Sensing* 5(1): 432–453. doi:10.3390/rs5010432.

Shorthouse, J. D. and D. J. Larson. 2010. Grasslands and grasslands arthropods of Canada. In *Arthropods of Canadian Grasslands (Volume 1): Ecology and Interactions in Grassland Habitats*, ed. J. D. Shorthouse and K. D. Float, 1–24. Ottawa: Biological Survey of Canada.

Smith, M. W. and D. Vericat. 2015. From experimental plots to experimental landscapes: Topography, erosion and deposition in sub-humid badlands from structure-from-motion photogrammetry. *Earth Surface Processes and Landforms* 40(12): 1656–1671. doi:10.1002/esp.3747.

Su, Q. 2004. DEM modelling, vegetation characterization and mapping of Aspen Parkland Rangeland using LiDAR data. PhD dissertation, University of Alberta.

Tong, A. 2014. Estimating grassland chlorophyll content for a mixed grassland: Exploring the performance of the empirical–statistical and the physical modelling approach. MSc dissertation, University of Toronto.

Tong, A. and Y. He. 2013. Comparative analysis of SPOT, Landsat, MODIS, and AVHRR Normalized Difference Vegetation Index data on the estimation of Leaf Area Index in a mixed grassland ecosystem. *Journal of Applied Remote Sensing* 7(1). doi:10.1117/1 .JRS.7.073599.

Wong, K. and Y. He. 2013. Estimating grassland chlorophyll content using remote sensing data at leaf, canopy, and landscape scales. *Canadian Journal of Remote Sensing* 39(2): 155–166. doi:10.5589/m13-021.

Von Bueren, S. K., A. Burkart, A. Hueni, U. Rascher, M. P. Tuohy, and I. J. Yule. 2015. Deploying four optical UAV-based sensors over grassland: Challenges and limitations. *Biogeosciences* 12(1): 163–175. doi:10.5194/bg-12-163-2015.

Xie, Y., Z. Sha, and M. Yu. 2008. Remote sensing imagery in vegetation mapping: A review. *Journal of Plant Ecology* 1(1): 9–23. doi:10.1093/jpe/rtm005.

Xing, M., B. He, X. Li, and X. Quan. 2014. Soil moisture retrieval using RADARSAT-2 and HJ-1 CCD data in grassland. doi:10.1109/IGARSS.2014.6947164.

Yang, Y., S. Shang, and L. Jiang. 2012. Remote sensing temporal and spatial patterns of evapotranspiration and the responses to water management in a large irrigation district of North China. *Agricultural and Forest Meteorology* 164: 112–122. doi:10.1016/j .agrformet.2012.05.011.

Zhang, Y., Z. Chen, B. Zhu, X. Luo, Y. Guan, S. Guo, and Y. Nie. 2008. Land desertification monitoring and assessment in Yulin of Northwest China using remote sensing and geographic information systems (GIS). *Environmental Monitoring and Assessment* 147(1–3): 327–337. doi:10.1007/s10661-007-0124-2.

7 Classifying Tree Species Using Fine Spatial Resolution Imagery to Support the Conservation of an Endangered Bird Species in Hawaii

Qi Chen

CONTENTS

7.1   INTRODUCTION

Habitat loss and species extinction are two main challenges of conserving biodiversity in the 21st century (Gottschalk et al. 2005) and serious threats to sustainability (Vucetich et al. 2015). Hawaii is the "Endangered Species Capital of the United States," with just 0.2% of the US land area but approximately 25% of the US endangered species (Eldredge and Evenhuis 2003). Because of the introduction of thousands of new species and increased human activities, approximately 90% of the dry land ecosystem in the state is now completely gone and at least 120 Hawaiian plant species under conservation have less than 50 individual plants left. As a result, the avifauna has been decimated and approximately 100 bird species has become extinct (USGS 2006).

Palila (*Loxioides bailleui*) is probably the most-known endangered bird species in Hawaii—a status derived primarily from its ties to a 1979 landmark case redefining the scope and intent of the federal Endangered Species Act (Riddle 2010). Palila feeds on immature seeds of mamane trees (*Sophora chrysophylla*), an endemic dry-forest tree species that occurs widely throughout the main Hawaiian Islands and ranges from near shoreline to tree line (>3000 m elevation) (Banko 2002). Nevertheless, palila and its habitat have experienced a multitude of threats including feral sheep and goats that browse mamane trees (Scowcroft and Giffin 1983), feral cats that depredate nests and adult birds (Hess et al. 2004), fungus such as *Armillaria* that kills mamane trees (Gardner and Trujillo 2001), and alien parasitoid wasps competing with palila for caterpillars that are fed to nestlings (Leonard et al. 2008; Oboyski et al. 2004). Although palila used to be widely distributed in lowlands of the island of Hawaii, these threats reduce its current habitat to an extent that is less than 5% of its historical range. Currently, palila's habitat is limited to a relatively small area (less than 30 km²) on the western slope of Mauna Kea, a high and dormant volcano (Hess et al. 1999; Johnson et al. 2006) (see Figure 7.1).

In recent years, significant efforts have been made to restore palila's habitat and population, for example, by reducing the number of ungulates and trapping rats (Banko 2002; Banko and Farmer 2006). Despite these efforts, a steady decline in palila population has happened, especially during the last 5 years (Leonard et al. 2008). The latest survey showed that there are only approximately 1200 birds left, down from approximately 4400 in 2003 (DLNR 2010). What is particularly perplexing is that no clear causes can be identified for the population decline, making it difficult to take effective measures for palila conservation. Mamane and naio (*Myoporum sandwicense*) are the two tree species in this habitat. However, mamane is more critical for palila's survival since palila almost exclusively eats immature mamane seeds. Therefore, it is important to separate mamane and naio trees so that the abundance of mamane trees and their spatial distribution can be derived to understand the palila population dynamics. Since the crowns of mamane and naio are relatively small (with an average crown diameter of 4–5 m) and they can grow next to each other in the habitat, it is very difficult, if not impossible, to classify these two species using conventional medium spatial resolution satellite imagery such as Landsat.

During the last decade, commercial fine spatial resolution satellite imagery has emerged as a powerful and cost-effective tool for detailed vegetation mapping (e.g., Adelabu and Dube 2015; Boggs 2010; Ji and Wang 2015; Lin et al. 2015; Morales et al. 2012; Murray et al. 2010; Pouteau et al. 2011; Puissant et al. 2014; van Lier et

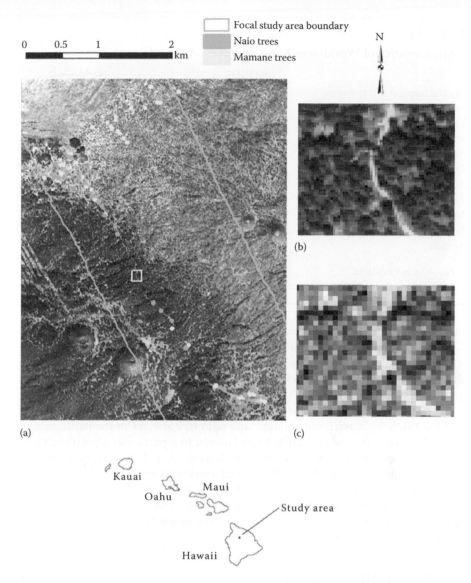

FIGURE 7.1 The study site and sampled trees. (a and b) From Worldview-2 and (c) from IKONOS. (b and c) Correspond to an area outlined by the yellow box in (a).

al. 2009). In October 2009, a new satellite called Worldview-2 from DigitalGlobe Inc. was launched, which has up to 1.84 m spatial resolution in multispectral (MS) bands and 0.46 m for the panchromatic band. One of the main differences between Worldview-2 and other fine spatial resolution satellites such as IKONOS, QuickBird, and Geoeye-1 is that it carries four new MS bands (called coastal, yellow, red-edge, and near-infrared [NIR]2) in addition to the conventional four bands (blue, green, red, and NIR) (Table 7.1). Worldview imagery has recently been used for mapping

TABLE 7.1

Characteristics of Worldview-2 and IKONOS MS Bands

	IKONOS	Worldview-2
Spectral ranges at half-maximum		
Blue	445–516 nm	450–510 nm
Green	506–595 nm	510–580 nm
Red	632–698 nm	630–690 nm
NIR	757–853 nm	770–895 nm
Coastal	N/A	400–450 nm
Yellow	N/A	585–625 nm
Red-edge	N/A	705–745 nm
NIR-2	N/A	860–1040 nm
Spatial resolution[a]		
At-nadir	3.2 m	1.8 m

Note: N/A, not available.

[a] Images were provided with a spatial resolution of 4 m for IKONOS and 2 m for Worldview-2.

detailed vegetation information such as species (e.g., Cho et al. 2015; Heumann 2011; Li et al. 2015; Robinson et al. 2016). However, relatively few studies have compared it with the first-generation commercial satellites such as IKONOS and investigate the potential of its fine spatial resolution and additional spectral bands for mapping vegetation at the species level (Pu and Landry 2012).

The main objective of this study is to compare the capability of Worldview-2 and IKONOS for classifying mamane and naio tree species in the palila habitat of Hawaii. To make the comparison less limited to a particular algorithm, three types of representative classification algorithms are tested: discriminant analysis (DA), support vector machine (SVM), and random forest (RF). The article is organized as follows: the study area and data will be introduced first, followed by an explanation of the classification algorithms and analytical strategy, and then results will be reported and discussed. The major research findings will be summarized in the end.

7.2   STUDY AREA AND DATA

7.2.1   STUDY AREA

The study site is on the western slope of Mauna Kea (elevation, approximately 1900 to 2700 m), which lies above the trade-wind inversion with relatively light cloud cover and rainfall (Juvik and Juvik 1998) (Figure 7.1). Rainfall averaged 35–75 cm and temperature averaged 11.1°C ± 1.5°C annually (Giambelluca et al. 1986). Mamane and naio are the two dominant tree species in the habitat. The majority of mamane occurs at the upper elevation (above 2300 m). At lower elevations, mixed stands of mamane and naio can be found, dominated by the latter (Hess et al. 1999; Scott et al. 1986). The trees are relatively short with a mean height of 4.7 ± 2.0 m. The average canopy cover is around 30%, resembling open-canopy woodland with an understory grass layer. The focal study area is a strip of 2-km swath where airborne lidar data were also acquired to characterize vegetation structure (Figure 7.1).

7.2.2 REMOTELY SENSED IMAGES AND FIELD DATA

Worldview-2 and IKONOS images were acquired at a local time of 10:54 a.m. on December 26, 2009, and 11:03 a.m. on January 8, 2008, respectively. The MS images of Worldview-2 and IKONOS were resampled to 2 and 4 m, respectively, and both types of images were orthorectified and georeferenced with the UTM Zone 5N (WGS84) coordinate system by the data providers.

To collect ground truth for classification, individual mamane and naio trees were located in the field between March 14 and March 21, 2010. A Garmin Venture HC GPS and a 4 ft–by–3 ft hard copy IKONOS false-color map overlain with grid lines of 500-m intervals were used to locate trees (Figure 7.2). A Trimble GeoXT GPS with higher positioning accuracy was carried in the field as well, but it was found that using the Garmin Venture HC GPS gave sufficient information to accurately locate trees in the map owing to the open canopy of the landscape (Riddle 2010). GIS polygon files of individual tree crowns were manually created later in the computer laboratory with the Worldview-2 image as the base layer. As a result, a total of 44 mamane polygons and 40 naio polygons were created within the focal study area, which correspond to 438 mamane pixels and 743 naio pixels in the Worldview-2 image and 112 mamane pixels and 183 naio pixels in the IKONOS image.

7.3 METHODS

A wide range of classification algorithms exist in the literature, which are either parametric or nonparametric. The parametric algorithms typically assume that each class follows a Gaussian distribution while the nonparametric algorithms do not necessarily make such assumption and thus are more flexible to handle classes of different distributions. Among the algorithms tested in this study, DA is parametric and SVM and RF are nonparametric. For completeness, the main ideas of these algorithms are briefly summarized next while RF is introduced in slightly more details since it is relatively new.

7.3.1 DISCRIMINANT ANALYSIS

DA is a classical classification algorithm that derives its discriminant function from Bayes' theorem and assumes that each class's conditional probability is Gaussian and their prior probabilities are equal. The discriminant function has many different variations depending on how to estimate the variance–covariance matrix Σ of each class. The most generic discriminant function is to estimate Σ for individual classes, which is called quadratic discriminant analysis (QDA). When the number of features is large and the sample size is small, estimating Σ could be unstable. One way to handle this problem is to assume that all classes have the same Gaussian distribution

FIGURE 7.2 Examples of mamane and naio trees in the field. (a) An isolated mamane tree, (b) an isolated naio tree, and (c) naio (left) and mamane (right) trees are growing next to each other. (Courtesy of Ryan Riddle.)

so that sample data can be pooled to estimate just one Gaussian distribution, which is linear discriminant analysis (LDA). For DA, the decision boundary of classification is constructed *implicitly* by comparing the discriminant functions of different classes. For QDA, the decision boundary is usually curvilinear, while for LDA, the decision boundary is linear.

7.3.2 SVM

Different from DA, SVM *explicitly* constructs a linear decision boundary (called hyperplane when the number of features is greater than three) to separate two classes (Cortes and Vapnik 1995). It is possible that there exist multiple hyperplanes to separate two classes for the training sample (Duda et al. 2000). SVM finds one unique optimal hyperplane by maximizing the distance from a hyperplane to the nearest point of either class (Hastie et al. 2009). Although SVM is a linear classifier, features can be expanded to a higher feature space to fit a hyperplane, which could correspond to a nonlinear decision boundary in the original lower feature space through a technique called *kernel trick* (Abe 2005). The kernels tested in this study are quadratic (or second-order polynomial) and Gaussian radial basis function.

7.3.3 Random Forest

RF is a recent addition to the nonparametric classifiers (Breiman 2001). It was developed from several techniques including classification and regression tree (CART) and bootstrap aggregation (or bagging). The main idea of RF is to build a large number of de-correlated trees and let them vote for the most popular class (Hastie et al. 2009). When an individual tree is built, RF differs from regular CART in that (1) the tree is based on the bootstrap (random sampling with replacement) instead of the original sample (Breiman 1996) and (2) the splitting of a node is based on a random subset instead of all features (Ho 1998). Such randomization leads to de-correlated trees, which are particularly useful for reducing the variance of prediction. If large individual trees are grown, they can also have low bias. The low bias and variance of estimates make RF popular in many fields (Hastie et al. 2009).

Many benefits are associated with the bootstrapping used in RF. A bootstrap sample usually has repetitive values for some observations and leaves approximately one-third of the observations not included. These observations, called out-of-bag, are thus not used for building the tree from the bootstrap sample. Approximately, an observation is out-of-bag for one-third of the trees. By comparing each observation's true class label with the majority vote of class predictions from these out-of-bag trees, an overall out-of-bag prediction error can be calculated, without further using skills such as cross-validation. The out-of-bag prediction error is useful for tuning parameters such as the number of trees, the number of features randomly selected for node splitting, and the leaf node size.

Out-of-bag observations can also be used to rank the importance of individual features. After an individual tree is grown, it can be used to predict the classes of out-of-bag observations and calculate the prediction accuracy. If the values for a particular feature are randomly permuted among the out-of-bag observations, the prediction accuracy will decrease. The amount of accuracy decrease averaged over all trees can be used as a measure of the importance of that feature.

7.3.4 Classification and Accuracy Assessment Strategy

A total of five specific algorithms, including two DA algorithms (QDA and LDA), SVMs with two different kernels (quadratic and Gaussian radial basis function), and RF, were tested. It is hypothesized that compared to IKONOS, Worldview-2 will produce higher classification accuracy with its (1) finer spatial solution and (2) more spectral bands. To test this hypothesis, three types of classification were conducted: the first was to use all four IKONOS bands (blue, green, red, and NIR), the second was to use the four Worldview-2 conventional bands (blue, green, red, and NIR), and the third was to use all eight Worldview-2 bands including the four new bands (coastal, yellow, red-edge, and NIR2). The comparison between IKONOS and Worldview-2 using the same four conventional MS bands focuses the evaluation on the benefits of Wordlview-2's finer spatial resolution. The comparison of using Worldview-2 four versus eight bands evaluates the usefulness of the new MS bands.

Although both images were orthorectified by the data providers, their geolocation accuracies were further verified by overlaying the images with a digital surface model derived from airborne lidar data and visually checking distinct features such as roads and isolated trees. It was found that the Worldview-2 image matches well with airborne lidar data while IKONOS has some shifts (up to ~5 m in some places). To handle this issue, image-to-image registration was conducted by collecting a total of 158 ground control points well distributed over the study area and warping the IKONOS image to match Worldview-2 using a second-order polynomial transformation in ENVI 4.7 (ITT VIS Corp.). The root mean square error of image registration is 0.4 m. The classification was conducted using the digital numbers without being further calibrated to reflectance because each image was classified separately and radiometric correction usually has minimal effects on classification of a single-date image (Song et al. 2001). Mamane and naio trees are the only two tree species in the study area, with other vegetation types being short scrubs and grass. A canopy height model derived from lidar was used to mask out all pixels that are less than 1 m so that the classification was focused on mamane and naio trees only.

Tenfold cross-validation was used to evaluate the overall classification accuracy. This means that the field data were broken into 10 folds, based on which model development and test were performed for 10 iterations. Each time, one fold was used for testing and the rest were used for model development. For RF, a stepwise approach instead of an exhaustive search of all possible combinations of parameters was used to determine the model parameters based on the out-of-bag prediction errors. The stepwise approach was to calibrate one parameter each time while keeping others fixed. To reduce the computation demand, the minimal leaf node was first determined, followed by the number of subset features for node splitting, and finally the number of trees.

The focus of this study is to perform classification at the pixel level. Nevertheless, the airborne discrete-return lidar data acquired in this site provide the possibility of mapping individual tree crown objects (Chen et al. 2006). To evaluate such potential, I first considered individual pixels within each manually delineated reference tree crown polygon as individual groups (i.e., objects) and then produced the average spectral values of each tree pixel group, which was further used for classification. After the object-level classification was done, all pixels within each reference polygon were labeled as the same tree species and the classification accuracy was summarized at the pixel level to compare with results from the pixel-level classification.

7.4 RESULTS AND DISCUSSION

7.4.1 CLASSIFICATION WITH THE FOUR CONVENTIONAL WORLDVIEW-2 BANDS

Table 7.2 lists the accuracy of individual classifiers and their means when the four conventional MS bands (blue, green, red, and NIR) of either IKONOS or Worldiview-2 are used. It was found that Worldview-2 has consistently higher accuracy than IKONOS. For example, at the pixel level, when the IKONOS image was classified, the mean classification accuracy of different classifiers was 68.6%; when the Worldview-2 image was classified, the mean accuracy increased up to 71.9% (Figure 7.3). For individual classifiers, the accuracy increase varied from 1.3% for QDA to 6.1% for RF.

At the object level, the accuracy increase using Worldview-2 over IKONOS was even larger. When the IKONOS image was classified, the average classification accuracy of different classifiers was 77.1%; however, when the Worldview-2 image was classified, the average accuracy increased by 7.9% (reaching 85.0%) (Figure 7.3). For individual classifiers, the accuracy increase varied from 5.2% for LDA to 12.8% for SVM-GRB. The higher overall classification accuracy of Wordview-2 compared to IKONOS is supported by the spectral scatterplots shown in Figure 7.4b and c: it is evident that the two tree species are more separable in the scatterplots of Worldview-2 than IKONOS.

There are a couple of factors that might contribute to the accuracy difference between Worldview-2 and IKONOS images. For example, there are slight differences between the spectral sensitivity curves of corresponding individual bands of the two sensors. The sun–target–sensor geometry is a little different because these two images were taken from two different dates. The atmospheric conditions could have negligible effects since both images were taken in clear days. Fire is the major natural hazard risk at this study site, but no fires have occurred over this area for years. Therefore, little changes of vegetation are expected between the different image acquisition dates to make a significant contribution to the difference in the classification accuracy. Although it is difficult to quantify the contribution of each factor to the accuracy difference, it is expected that the higher classification accuracy of Worldview-2 is mainly attributed to its finer spatial resolution, which can capture more details of vegetation especially along the edges of the individual tree crowns.

7.4.2 CLASSIFICATION WITH THE INCLUSION OF FOUR NEW WORLDVIEW-2 BANDS

Table 7.2 also summarizes the classification accuracy when all eight bands of Worldview-2 were used. It was found that the inclusion of the four new bands increased the classification accuracy for most classifiers. At the pixel level, the additional four new bands increased the mean classification accuracy from 71.9% to 75.0%. For individual classifiers, the accuracy increase varied from 1.3% of

TABLE 7.2

Tenfold Cross-Validation Overall Accuracy of Species Classification Using Different Algorithms

	Pixel Based			Object Based			
Classifier[a]	IKONOS (4 Bands)	Worldview-2 (4 Bands)	Worldview-2 (8 Bands)	IKONOS (4 Bands)	Worldview-2 (4 Bands)	Worldview-2 (8 Bands)	Mean
QDA	69.1%	70.4%	75.5%	77.6%	82.9%	84.4%	76.7%
LDA	68.9%	70.3%	74.5%	80.2%	85.4%	86.5%	77.6%
SVM-Quad	68.6%	71.5%	73.4%	78.0%	84.0%	84.5%	76.7%
SVM-GRB	68.2%	72.8%	74.1%	74.2%	87.0%	86.5%	77.1%
RF	68.2%	74.3%	77.4%	75.3%	85.9%	84.7%	77.6%
Mean	68.6%	71.9%	75.0%	77.1%	85.0%	85.3%	

Note: The highest accuracy of each column is underlined. The higher accuracy between two specific algorithms (QDA vs. LDA; SVM-Quad vs. SVM-GRB) of the same type of classifier is in boldface. The accuracy of RF is also in boldface to facilitate comparison.

[a] LDA, linear discriminant analysis; QDA, quadratic discriminant analysis; RF, random forest; SVM-GRB, support vector machine with Gaussian radial basis kernel; SVM-Quad, support vector machine with quadratic kernel.

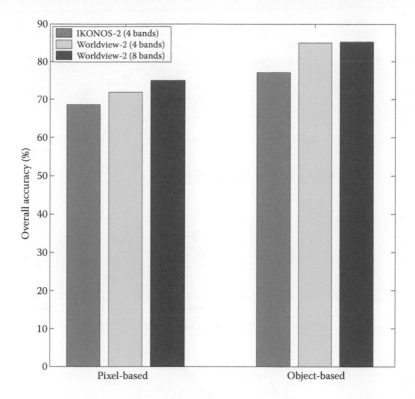

FIGURE 7.3 Comparison of mean classification accuracy (averaged over classifiers) when (1) different sensors and bands are used (four IKONOS bands, four Worldview-2 conventional MS bands, or all eight Worldview-2 bands) and (2) the classification is performed at the pixel level or at the object level.

SVM-GRB to 5.1% of LDA. At the object level, the mean classification accuracy increases by a small margin from 85.0% to 85.3%. However, when the individual classifiers were examined, only three of them (QDA, LDA, and SVM-Quad) had an increase in classification accuracy, and the other two (SVM-GRB and RF) had a minor decrease in classification accuracy. It is not exactly clear why SVM-GRB and RF had slightly worse accuracy. However, these two methods are more complex than the others and thus they may run into the problem of overfitting.

Usually, a classifier with more parameters can model more flexible decision boundaries, but they also need more training data to constrain the classifier. The needed training data also increase with the number of input features (Richard and Jia 2006). SVM-GRB and RF can model more flexible decision boundaries than the other three classifiers (see Figure 7.5c and d), which implies that they essentially need more training data to constrain the construction of decision boundary. However, when the classification is conducted at the object level, the sample size is relatively small and the inclusion of the four additional bands makes the situation even worse, which can explain, to some extent, why SVM-GRB and RF had slightly lower accuracy when the four new bands were added. In other words, the lower

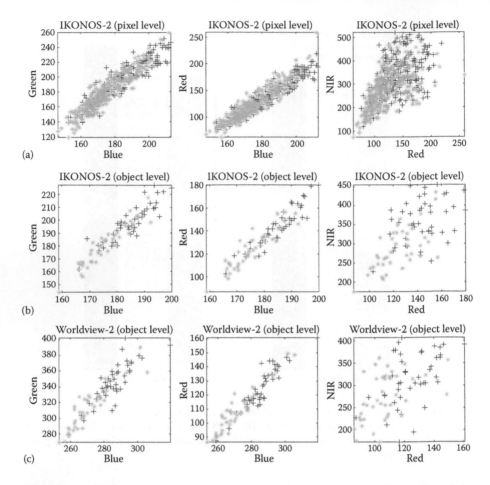

FIGURE 7.4 Scatterplots of (a) object-based mean pixel values for IKONOS bands, (b) object-based mean pixel values for Worldview-2 bands, and (c) individual pixel values for Worldview-2 bands. + represents mamane and * represents naio.

accuracy of these two classifiers is more related to the algorithm itself instead of the new bands.

The out-of-bag observations in RF were also used to rank the importance of individual bands of Worldview-2 for classification. It is interesting to see that the top four bands for tree species classification at the pixel level were coastal, yellow, red-edge, and NIR-2, which are exactly the four new bands (Figure 7.6a). At the object level, the importance of individual bands for RF classification was different. However, the coastal band was still among the top four bands of the highest importance (Figure 7.6b). The analysis above indicates that the four new bands are valuable in improving the classification of tree species at the study site.

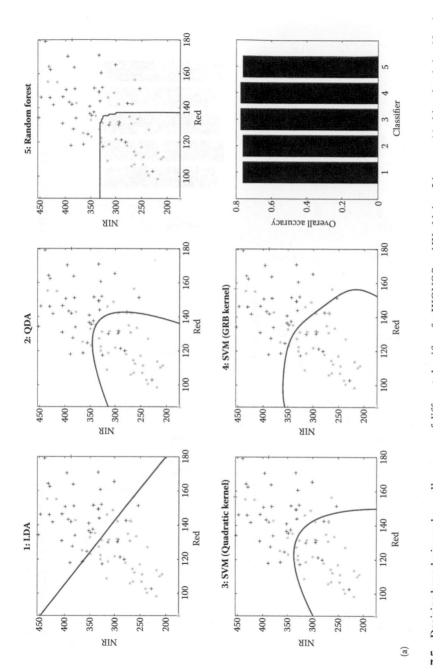

(a)

FIGURE 7.5 Decision boundaries and overall accuracy of different classifiers for IKONOS and Worldview-2 images: (a) object-level classification with IKONOS bands. *(Continued)*

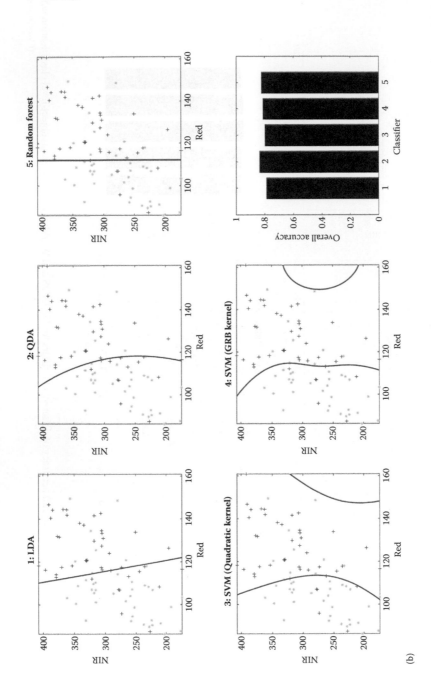

FIGURE 7.5 (CONTINUED) Decision boundaries and overall accuracy of different classifiers for IKONOS and Worldview-2 images: (b) object-level classification with Worldview-2 bands. *(Continued)*

(b)

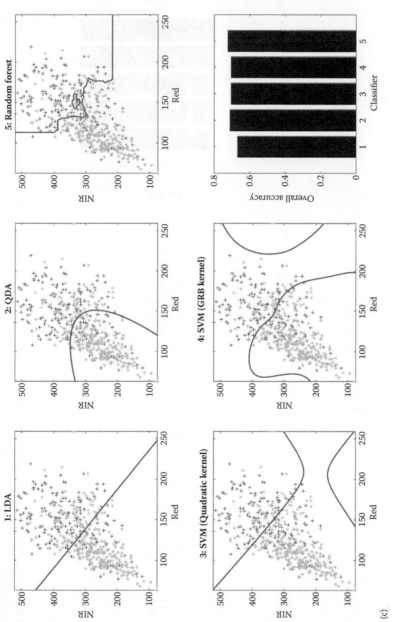

(c)

FIGURE 7.5 (CONTINUED) Decision boundaries and overall accuracy of different classifiers for IKONOS and Worldview-2 images: (c) pixel-level classification with IKONOS bands. *(Continued)*

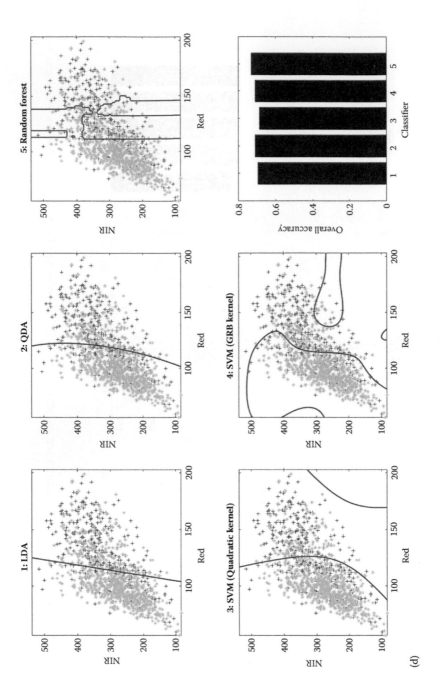

FIGURE 7.5 (CONTINUED) Decision boundaries and overall accuracy of different classifiers for IKONOS and Worldview-2 images: (d) pixel-level classification with Worldview-2 bands. + represents mamane and * represents naio.

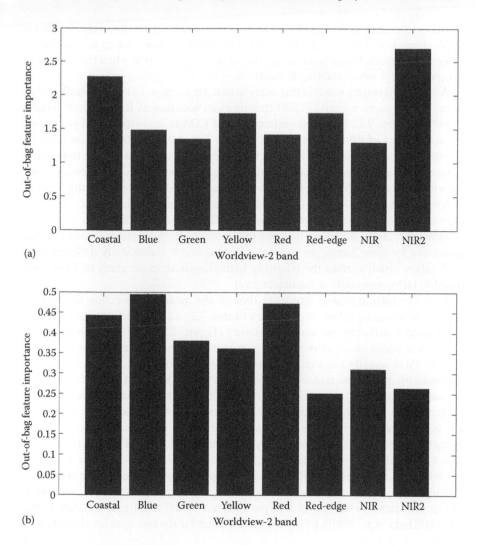

FIGURE 7.6 Feature importance ranked by random forest for (a) pixel-level classification and (b) object-level classification.

7.4.3 COMPARISON OF DIFFERENT CLASSIFIERS

When the two DA methods were compared, it was found that QDA has slightly higher accuracy than LDA at the pixel level (Table 7.2). The accuracy increase varied from 0.1% for four Worldview-2 conventional bands to 1.0% for all eight Worldview-2 bands. However, LDA was clearly better than QDA at the object level. The accuracy increase varied from 2.1% for eight Worldview-2 bands to 2.6% for four IKONOS bands. This again is likely related to the fewer observations for training at the object-level classification. LDA pools the two classes to estimate the variance–covariance matrix while QDA estimates the variance–covariance matrix for each class separately. Therefore, LDA has a higher ratio of training sample size to the number of model parameters, which is advantageous in model estimation when the number of observations for model training is small.

Another interesting result in this study is that, on average, LDA (the simplest classifier) has the same highest (77.6%) classification accuracy as RF (the most flexible classifier) (Table 7.2). The good performance of LDA is not surprising if we inspect the scatterplots of the two tree species (Figure 7.4). Although there is considerable amount of overlay between two classes for some bands, each class seems to have one unique mean and Gaussian-like distribution. Despite the fact that the more complex algorithms (such as SVM and RF) can more easily model classes of multiple-mode or non-Gaussian distributions, LDA and QDA could achieve high classification accuracy as long as the assumption of Gaussian distribution for each class is satisfied (Hastie et al. 2009). Although it seems that mamane trees are more spread out than naio trees for some bands, the degree of spread is not dramatically different from each other, which justifies the relatively high classification accuracy of LDA compared to QDA, especially at the object level.

It is worthwhile to note that even though the mean classification accuracy is very similar among different classifiers (Table 7.2), each classifier could generate dramatically different decision boundaries (Figure 7.5). The decision boundary for LDA is linear owing to the common variance–covariance matrix between two classes. QDA usually has a curvilinear decision boundary because each class has different variance–covariance matrices. Both SVM-Quad and SVM-GRB generate smooth nonlinear boundaries, but it seems that SVM-GRB has more flexibility in generating different shapes of decision boundaries. Among these classifiers, RF is most versatile in terms of generating decision boundaries varying from a straight line (e.g., Figure 7.5b) to very complex and localized curves (see Figure 7.5c and d). If we assume that each class follows a Gaussian-like distribution, it seems that, at the pixel level, RF has overfitting problems caused by the significant overlay between two classes in the spectral space (Figure 7.5c and d). According to *Occam's razor* (which states that the simplest model explaining the data is preferred) (Duda et al. 2000), LDA is a good choice for the tree species classification at this study site.

There are divergent results in the literature regarding the relative performance of parametric and nonparametric classifiers. For example, Camps-Valls and Bruzzone (2005) compared LDA with several kernel-based nonparametric classifiers including SVM, regularized radial basis function neural networks, kernel Fisher discriminant analysis, and regularized AdaBoost for classifying the nine classes in the AVIRIS NW Indiana's Indian Pines 1992 data set. It was found that LDA had significantly lower accuracy than the nonparametric classifiers. However, a recent study by Szuster et al. (2011) showed that Maximum Likelihood Classifier (equivalent to QDA in this study) had similar performance with neural network and SVM when ASTER images were used for classifying seven land cover and land use types in the coastal zone of Koh Tao island, Thailand. These contrasting results agree with the *No Free Lunch Theorem*: there is no context- or problem-independent reason to favor one classifier over another (Duda et al. 2000). The apparent superiority of one algorithm over another in a particular study usually depends on the distribution of data and the *match* between an algorithm and the problem it addresses.

7.4.4 PIXEL-BASED VERSUS OBJECT-BASED CLASSIFICATION

It was found that the object-based classification can significantly increase the classification accuracy compared to pixel-based classification. On average, the increase in the overall classification accuracy was 8.5%, 13.1%, and 10.3% when the four IKONOS bands, four conventional Worldview-2 bands, and all eight Worldview-2 bands were used, respectively. Note that the accuracy increase is larger for Worldview-2 than for IKONOS, probably because the finer spatial resolution of Worldview-2 has a better characterization of pixels at the crown edge.

Although object-based image analysis (OBIA) has become popular over the past decade owing to the increasing use of fine spatial resolution remotely sensed data (Blaschke 2010), only a few studies have compared the classification accuracy of object-based classification with pixel-based classification (e.g., Cleve et al. 2008; Gao 2008). For example, Cleve et al. (2008) found that the object-based classification approach provided a 17.97% higher overall accuracy than the pixel-based approach for mapping the wildland–urban interface with aerial photos. It is encouraging that the object-level classification at this study also produces positive results. One of the main reasons why the object-based approach outperforms the pixel-based approach is the reduction of spectral heterogeneity (e.g., caused by shade) of tree canopy using the object-based approach. More research is needed in the future to conduct object-level classification with individual tree objects automatically delineated using fine spatial resolution imagery (e.g., Chopping 2011) or airborne lidar data (e.g., Chen et al. 2006).

7.5 CONCLUSIONS

This study compared Worldview-2 and IKONOS for identifying mamane and naio trees in the palila habitat of the island of Hawaii using both parametric and nonparametric classifiers. Classification was conducted at the pixel and object levels. The major findings are as follows:

- On average, the four conventional Worldview-2 bands (blue, green, red, and NIR) can achieve 3.3% and 7.9% higher overall accuracy than the same four IKONOS bands at the pixel level and object level, respectively. If individual classifiers are considered, the highest accuracy improvement is 6.1% at the pixel level and 12.8% at the object level.
- The inclusion of four new Worldview-2 bands (coastal, yellow, red-edge, and NIR2) enhanced the average overall accuracy by 3.1% and 0.3% at the pixel level and object level, respectively. The analysis of feature importance using RF indicated that the four new bands were among the top four bands (out of the eight Worldview-2 bands) for classifying tree species at the pixel level. Coastal band was among the top four bands for classification at the object level.

- The five classifiers (QDA, LDA, SVM-Quad, SVM-GRB, and RF) had similar average overall accuracies varying from 76.7% (QDA and SVM-Quad) to 77.6% (LDA and RF). However, each classifier could generate dramatically different decision boundaries. Among these classifiers, LDA has the simplest linear decision boundary while achieving the highest classification accuracy. This indicates that LDA is a good choice for tree species at this study site.
- Classification at the object level, compared to that at the pixel level, can significantly increase the overall accuracy by 8.5%, 13.1%, and 10.3% when the four IKONOS bands, four conventional Worldview-2 bands, and all eight Worldview-2 bands were used, respectively. In a nutshell, it was found that both the finer spatial resolution and the additional spectral bands of Worldview-2 can improve the classification of tree species at such a critical habitat.

Conserving endangered species and maintaining nature's intrinsic values for future generations are an important obligation of human beings and a key aspect of sustainability. However, in recent years, species conservation has experienced a rapid decline in the dialogue of global sustainability, which has shifted its focus to issues such as climate change. One driver of this concerning trend is the insufficient funding and high cost of monitoring species and their habitat in the field. The technology innovation of high-resolution yet low-cost satellite imagery can greatly increase the efficiency in habitat characterization and thus can help avert this alarming trend, especially in remote areas and developing countries where the accessibility of high-quality remotely sensed data is very limited. The encouraging findings of this research will further motivate the use of state-of-the-art remote sensing technology to address pressing sustainability issues in our society.

ACKNOWLEDGMENTS

The field data of this study were collected by Ryan Riddle. The Worldview-2 and IKONOS images were provided by DigitalGlobe Inc. and the GeoEye Foundation, respectively. Thanks are also extended to Greg Asner and Paul Banko for their discussion of the palila project.

REFERENCES

Abe, S., 2005. *Support Vector Machines for Pattern Classification*. London: Springer.
Adelabu, S., Dube, T., 2015. Employing ground and satellite-based QuickBird data and random forest to discriminate five tree species in a Southern African Woodland. *Geocarto International* 30(4), 457–471.
Banko, P., 2002. Restoring palila on the big island. *Environmental Review* 9(6), 1–8.
Banko, P., Farmer, C., 2006. Palila restoration: Lessons from long-term research. USGS Fact Sheet FS-2006-3104.
Blaschke, T., 2010. Object based image analysis for remote sensing. *ISPRS Journal of Photogrammetry and Remote Sensing* 65(1), 2–16.

Boggs, G. S., 2010. Assessment of SPOT 5 and QuickBird remotely sensed imagery for mapping tree cover in savannas. *International Journal of Applied Earth Observation and Geoinformation* 12(4), 217–224.

Breiman, L., 1996. Bagging predictors. *Machine Learning* 26(2), 123–140.

Breiman, L., 2001. Random forests. *Machine Learning* 45(1), 5–32.

Camps-Valls, G., Bruzzone, L., 2005. Kernel-based methods for hyperspectral image classification. *IEEE Transactions on Geoscience and Remote Sensing* 43(6), 1352–1362.

Chen, Q., Baldocchi, D. D., Gong, P., Kelly, M., 2006. Isolating individual trees in a savanna woodland using small footprint LIDAR data. *Photogrammetric Engineering and Remote Sensing* 72(8), 923–932.

Cho, M. A., Malahlela, O., Ramoelo, A., 2015. Assessing the utility WorldView-2 imagery for tree species mapping in South African subtropical humid forest and the conservation implications: Dukuduku forest patch as case study. *International Journal of Applied Earth Observation and Geoinformation* 38, 349–357.

Chopping, M., 2011. CANAPI: Canopy analysis with panchromatic imagery. *Remote Sensing Letters* 2(1), 21–29.

Cleve, C., Kelly, M., Kearns, F. R., Moritz, M., 2008. Classification of the wildland–urban interface: A comparison of pixel- and object-based classifications using high-resolution aerial photography. *Computers Environment and Urban Systems* 32(4), 317–326.

Cortes, C., Vapnik, V., 1995. Support-vector networks. *Machine Learning* 20(3), 273–297.

DLNR, 2010. Hawaii's endangered palila bird numbers dropping. http://www.staradvertiser.com/news/breaking/102911719.html (accessed on October 1, 2010).

Duda, R., Hart, P., Stork, D., 2000. *Pattern Classification* (2nd ed.). New York: John Wiley & Sons.

Eldredge, L. G., Evenhuis, N. L., 2003. Hawaii's biodiversity: A detailed assessment of the numbers of species in the Hawaiian Islands. *Bishop Museum Occasional Papers* 76, 1–28.

Gao, J., 2008. Mapping of land degradation from ASTER data: A comparison of object-based and pixel-based methods. *GIScience & Remote Sensing* 45(2), 149–166.

Gardner, D. E., Trujillo, E. E., 2001. Association of *Armillaria mellea* with mämane decline at Pu'u La'au. *Newsletter of the Hawaiian Botanical Society* 40, 33–34.

Giambelluca, T. W., Nullet, M. A., Schroeder, T. A., 1986. Hawaii Rainfall Atlas, Report R76, Hawai'i Division of Water and Land Development, Department of Land and Natural Resources, Honolulu.

Gottschalk, T., Huettmann, F., Ehlers, M., 2005. Thirty years of analysing and modelling avian habitat relationships using satellite imagery data: A review. *International Journal of Remote Sensing* 26(12), 2631–2656.

Hastie, T., Tibshirani, R., Friedman, J., 2009. *The Elements of Statistical Learning: Data Mining, Inference, and Prediction* (2nd ed.). New York: Springer.

Hess, S. C., Banko, P. C., Brenner, G. J., Jacobi, J. D., 1999. Factors related to the recovery of subalpine woodland on Mauna Kea, Hawai'i. *Biotropica* 31(2), 212–219.

Hess, S. C., Banko, P. C., Goltz, D. M., Danner, R. M., Brinck, K. W., 2004. Strategies for reducing feral cat threats to endangered Hawaiian birds. In Timms, R. M., Gorensel, W. P. (Eds.), *Proceedings of the 21st Vertebrate Pest Conference*, University of California, Davis, CA.

Heumann, B. W., 2011. An object-based classification of mangroves using a hybrid decision tree—Support vector machine approach. *Remote Sensing* 3(11), 2440–2460.

Ho, T. K., 1998. The random subspace method for constructing decision forests. *IEEE Transactions on Pattern Analysis and Machine Intelligence* 20(8), 832–844.

Ji, W., Wang, L., 2015. Discriminating Saltcedar (*Tamarix ramosissima*) from sparsely distributed Cottonwood (*Populus euphratica*) using a summer season satellite image. *Photogrammetric Engineering & Remote Sensing* 81(10), 795–806.

Johnson, L., Camp, R. J., Brinck, K. W., Banko, P. C., 2006. Long-term population monitoring: Lessons learned from an endangered passerine in Hawai'i. *Wildlife Society Bulletin* 34(4), 1055–1063.

Juvik, J., Juvik, S., 1998. *Atlas of Hawaii* (3rd ed.). Honolulu: University of Hawai'i Press.

Leonard Jr., D. L., Banko, P. C., Brinck, K. W., Farmer, C., Camp, R. J., 2008. Recent surveys indicate rapid decline of palila population. *Elepaio* 68(4), 27–30.

Li, D., Ke, Y., Gong, H., Li, X., 2015. Object-based urban tree species classification using bi-temporal WorldView-2 and WorldView-3 images. *Remote Sensing* 7(12), 16917–16937.

Lin, C., Popescu, S.C., Thomson, G., Tsogt, K., Chang, C.I., 2015. Classification of tree species in overstorey canopy of subtropical forest using QuickBird images. *PLoS One* 10(5), e0125554.

Morales, R. M., Idol, T., Chen, Q., 2012. Differentiation of Acacia koa forest stands across an elevation gradient in Hawai'i using fine-resolution remotely sensed imagery. *International Journal of Remote Sensing* 33(11), 3492–3511.

Murray, H., Lucieer, A., William, R., 2010. Texture-based classification of sub-Antarctic vegetation communities on Heard Island. *International Journal of Applied Earth Observation and Geoinformation* 12(3), 138–149.

Oboyski, P. T., Slotterback, J. W., Banko, P. C., 2004. Differential parasitism of seed-eating Cydia (Lepidoptera: Tortricidae) by native and alien wasp species relative to elevation in subalpine Sophora (Fabaceae) forests on Mauna Kea, Hawaii. *Journal of Insect Conservation* 8, 229–240.

Pouteau, R., Meyer, J. Y., Stoll, B., 2011. A SVM-based model for predicting distribution of the invasive tree *Miconia calvescens* in tropical rainforests. *Ecological Modelling* 222(15), 2631–2641.

Pu, R., Landry, S., 2012. A comparative analysis of high spatial resolution IKONOS and WorldView-2 imagery for mapping urban tree species. *Remote Sensing of Environment* 124, 516–533.

Puissant, A., Rougier, S., Stumpf, A., 2014. Object-oriented mapping of urban trees using Random Forest classifiers. *International Journal of Applied Earth Observation and Geoinformation* 26, 235–245.

Richards, J., Jia, X., 2006. *Remote Sensing Digital Image Analysis: An Introduction* (4th ed.). Berlin, Germany: Springer-Verlag.

Riddle, R., 2010. Remote sensing techniques for classifying the habitat of an endangered bird species on Mauna Kea. MA thesis, Department of Geography, University of Hawaii at Manoa.

Robinson, T. P., Wardell-Johnson, G. W., Pracilio, G., Brown, C., Corner, R., van Klinken, R. D., 2016. Testing the discrimination and detection limits of WorldView-2 imagery on a challenging invasive plant target. *International Journal of Applied Earth Observation and Geoinformation* 44, 23–30.

Scott, J. M., Mountainspring, S., Ramsey, F. L., Kepler, C. B., 1986. Forest bird communities of the Hawaiian Islands: Their dynamics, ecology, and conservation. *Studies in Avian Biology* 9, 1–431.

Scowcroft, P. G., Giffin, J. G., 1983. Feral herbivores suppress māmane and other browse species on Mauna Kea, Hawai'i. *Journal of Range Management* 36(5), 638–645.

Song, C., Woodcock, C. E., Seto, K. C., Pax-Lenney, M., Macomber, S. A., 2001. Classification and change detection using Landsat TM data: When and how to correct atmospheric effects? *Remote Sensing of Environment* 75, 230–244.

Szuster, B., Chen, Q., Borger, M., 2011. A comparison of classification techniques to support land cover and land use analysis in tropical coastal zones. *Applied Geography* 31, 525–532.

USGS, 2006. Hawaii forest bird interagency database project: Collecting, understanding, and sharing population data on Hawaiian forest birds. USGS Fact Sheet. http://biology.usgs .gov/pierc/Native_Birds.html (accessed on October 1, 2008).

van Lier, O. R., Fournier, R. A., Bradley, R. L., Thiffault, N., 2009. A multi-resolution satellite imagery approach for large area mapping of ericaceous shrubs in Northern Quebec, Canada. *International Journal of Applied Earth Observation and Geoinformation* 11(5), 334–343.

Vucetich, J. A., Bruskotter, J. T., Nelson, M. P., 2015. Evaluating whether nature's intrinsic value is an axiom of or anathema to conservation. *Conservation Biology* 29(2), 321–332.

8 Remote Sensing of Forest Damage by Diseases and Insects

Gang Chen and Ross K. Meentemeyer

CONTENTS

8.1 INTRODUCTION

Forests are an integral part of natural ecosystems, providing numerous ecological, economic, social, and cultural services (Boyd et al. 2013; Chen et al. 2015a). For example, they store approximately 45% of terrestrial carbon (C) and remain as a large net C sink by capturing one-quarter of the anthropogenic carbon dioxide (CO_2) each year (Bonan 2008; Pan et al. 2011). However, environmental change (e.g., severe drought) and global trade have increased forest vulnerability to a range of natural disturbances, including diseases and insects (Asner 2013; Boyd et al. 2013; Wang et al. 2008; Wingfield et al. 2015). Forest diseases are caused by pathogens that are infectious and transmissible, such as bacteria, fungi, viruses, and helminths. Insects attack different parts of the tree, with defoliators feeding on leaves or needles, and bark/wood borers boring into the bark/wood. While some pathogen and insect species are native to local ecosystems, many of the recent disturbances arise from the nonindigenous species that may pose more pernicious and unpredictable threats to forest health (Boyd et al. 2013). Over the past few decades, the frequency and intensity of disease- and insect-caused forest disturbances have dramatically increased, leading to extensive tree mortality in key forest biomes worldwide. Examples include the sudden oak death epidemic in western United States, outbreaks of mountain pine beetle in Canada's boreal forest, bronze bug damage in plantation forests in South Africa, and the spread of bark beetles in central Europe and Scandinavia (Fassnacht et al. 2014; Meentemeyer et al. 2015; Oumar and Mutanga 2014; Wulder et al. 2009). Figure 8.1 illustrates two typical symptoms of forest damage attributed to the outbreaks of mountain pine beetles and the infectious disease sudden oak death, respectively.

Sustainable forest management is essential to mitigating the destructive impacts of diseases or insects on forest ecosystems. This is especially true when major disturbance events have the potential to reduce the dominant native species, causing a permanent change in forest structure. One prerequisite for effective management is to understand the spatial distribution and severity of forest damage. Consequently, mitigation efforts can be performed to limit the population and the spread of pathogens or insects on infected or susceptible host trees. Although conventional field mensuration remains the most accurate way to quantify stages of infestation, it becomes time-consuming and costly when pathogen or insect populations reach epidemic levels. Remote sensing provides a timely and accurate approach to scale up field measurements and characterize spatially explicit information about the Earth's surface at landscape to regional scales. Recent developments in spaceborne and airborne sensors have further

(a)   (b)

FIGURE 8.1 Landscape-scale forest mortality caused by (a) mountain pine beetle and (b) sudden oak death. The infected trees show distinct symptoms of (a) red needles and (b) brown-to-gray leaf lesions, respectively. (Courtesy of (a) Ministry of Forests, Lands and Natural Resource Operations, http://www2.gov.bc.ca/gov/content/governments/organizational-structure/ministries -organizations/ministries/forest-and-natural-resource-operations, and (b) California Oak Mortality Task Force.)

advanced our ability to collect Earth observation data across multiple spatial, tempo-ral, and spectral scales, making remote sensing feasible to monitor forest disturbances (e.g., variations in forest biophysical and biochemical parameters) in response to the disease and insect outbreaks of varying stages of invasion. Such rapid and accurate delineation of large-area forest damage allows decision makers to take prompt and informed actions, supporting the sustainable management of forests.

The main objective of this chapter is to provide a brief survey of remote sens-ing assessment of forest damage by diseases and insects. Emphasis is directly laid on mapping forest disturbances with satellite and airborne Earth observation data. The following sections are organized to (i) summarize the recent trends of applying remote sensing to detect forest disease and insect outbreaks, (ii) investigate remote sensing characteristics and its qualifications for studying the topic, (iii) provide a brief review of remote sensing algorithms, and (iv) discuss several remaining chal-lenges that face researchers and decision makers in sustainable forest management.

8.2 TRENDS OF REMOTELY DETECTING FOREST DISEASE AND INSECT OUTBREAKS

While the idea of applying remote sensing to detect disease- and insect-induced for-est damage was considered as early as the 1970s and 1980s (e.g., Heller and Bega 1973; Nelson 1983; Rock et al. 1986), only recently (since the late 1990s) did the topic receive considerable attention for managing emerging outbreak (Table 8.1). Two reasons possibly explain slow adoption. First, a growing number of studies showed that the frequency and intensity of forest disease and insect attacks signifi-cantly increased over the past two decades as a result of climate change and glo-balization (see a brief review by Boyd et al. [2013]). There was a growing need to understand the mechanisms (e.g., spatial patterns) of the landscape-scale disease and insect progression informing effective mitigation strategies. Second, the collected Earth observation data have increased immensely during the same time period. The large volumes of data sets with relatively cheap acquisition costs, for example, the opening of more than four decades of Landsat archive (Woodcock et al. 2008), made it easier to systematically analyze the impact of certain diseases or insects in specific areas of interest. Ironically, one of the recent challenges facing many researchers is how to better handle such *big data*.

Geographically, research hotspots were primarily located in North America (e.g., Canada and the United States) and Europe (e.g., Germany, Norway, Spain, Sweden, and the United Kingdom), with several other studies conducted in Australia, China, and South Africa. Please note that the case studies cited in Table 8.1 were collected by searching Elsevier's ScienceDirect database with the following formula: *remote sensing* AND *forest* AND (*disease* OR *insect*). We also respectively substituted *pathogen* for *disease*, and *pest* for *insect* in the search. The results were further refined by removing the studies that did not contain a significant remote sensing component or did not target specific disease or insect types.

Compared to forest diseases, insects appeared to be more intensively studied using remote sensing (Table 8.1). This reflects the fact of high tree mortality induced by insects as well as their globally widespread occurrence. For example, among the

TABLE 8.1
Types of Diseases and Insects, and the Corresponding Regions, Countries, and Case Studies

Region	Type of Disease or Insect	Country	Case Study
Europe	Autumnal moth (*Epirrita autumnata*)	Sweden	Babst et al. 2010
	Bark beetle *Ips grandicollis*	Germany	Fassnacht et al. 2014
	Bark beetle *Ips typographus* L.	Germany	Kautz 2014
	Beech leaf-miner weevil (*Rhynchaenus fagi*)	Spain	Rullán-Silva et al. 2015
	Fungal spore *Ganoderma* sp.	United Kingdom	Sadyś et al. 2014
	Insect *Physokermes inopinatus*	Sweden	Olsson et al. 2012
	Pine processionary moth (*Thaumetopoea pityocampa*)	Spain	Sangüesa-Barreda et al. 2014
	Pine sawfly (*Neodiprion sertifer* (Geoffrey))	Norway	Solberg et al. 2006
North America	Black-headed budworm (*Acleris gloverana* (Walsingham))	Canada	Luther et al. 1997
	Blister rust fungus (*Cronartium ribicola*)	United States	Hatala et al. 2010
	Eastern hemlock looper (*Lambdina fiscellaria*)	Canada	Fraser and Latifovic 2005
	Eastern spruce budworm (*Choristoneura fumiferana*)	United States	Wolter et al. 2009
	Emerald ash borer (*Agrilus planipennis* Fairmaire)	United States	Pontius et al. 2008
	Gypsy moth (*Lymantria dispar* L.)	United States	de Beurs and Townsend 2008; Townsend et al. 2012; Thayn 2013
	Hemlock woolly adelgid (*Adelges tsugae* Annand)	United States	Siderhurst et al. 2010
	Jack pine budworm (*Choristoneura pinus pinus* (Free.))	Canada, United States	Leckie et al. 2005; Radeloff et al. 1999
	Mountain pine beetle (*Dendroctonus ponderosae* Hopkins)	Canada, United States	Assal et al. 2014; Bright et al. 2012; Cheng et al. 2010; Coops et al. 2009; Goodwin et al. 2008; Hatala et al. 2010; Meddens et al. 2011; Meigs et al. 2011, 2015; Raffa et al. 2013; Skakun et al. 2002; Walter and Platt 2013; Wulder et al. 2008, 2009
	Spruce budworm (*Choristoneura fumiferana*)	Canada, United States	Wolter et al. 2008

(Continued)

TABLE 8.1 (CONTINUED)
Types of Diseases and Insects, and the Corresponding Regions, Countries, and Case Studies

Region	Type of Disease or Insect	Country	Case Study
	Sudden oak death (*Phytophthora ramorum*)	United States	Kelly and Meentemeyer 2002; Lamsal et al. 2011; Liu et al. 2006, 2007; Meentemeyer et al. 2008; Pu et al. 2008
	Western spruce budworm (*Choristoneura freemani*)	United States	Meigs et al. 2011, 2015
Others	Aphid (*Essigella californica*)	Australia	Goodwin et al. 2005
	Bark beetle (*Ips grandicollis*)	Australia	Verbesselt et al. 2009
	Fungal pathogen (*Sphaeropsis sapinea*)	Australia	Goodwin et al. 2005
	Insect (*Thaumastocoris peregrinus*)	South Africa	Oumar and Mutanga 2014; Oumar et al. 2013
	Mopane worm (*Gonimbrasia belina*)	South Africa	Adelabu et al. 2014
	Pine caterpillar (*Dendrolimus superans* Butler, Dendrolimus: Lasiocampidae, Lepidoptera)	China	Huang et al. 2010

United States' 20 major diseases and insects that caused 6.4 million acres of tree mortality in 2011, 60% were insects; mountain pine beetle (*Dendroctonus ponderosae* Hopkins) alone killed 3.8 million acres of trees (USDA Forest Service 2012). Several other insects, such as bark beetle *Ips grandicollis*, gypsy moth (*Lymantria dispar* L.), and jack pine budworm (*Choristoneura pinus pinus* [Free.]), have also been well studied across forest biomes (Table 8.1). In contrast, remote detection of the disease impacts on forest ecosystems was less studied. One exception is sudden oak death caused by the invasive plant pathogen *Phytophthora ramorum* (Rizzo et al. 2005), which received considerable attention as a result of rapid transmission and widespread mortality of oak and tanoak trees in coastal forests of California and Oregon (Table 8.1).

8.3 REMOTE SENSING CHARACTERISTICS AND QUALIFICATIONS

The premise of utilizing remote sensing to detect disease- or insect-infested forests is that the damaged trees show distinct symptoms capable of being observed remotely. Depending on the type or stage of damage, the symptoms may indicate the decline in chlorophyll/water quantity in foliage, leaf discoloration, defoliation, or treefall gaps. For effective monitoring, Earth observation data acquired from satellite or airborne sensors are expected to capture the differences in the reflected radiation from damaged versus healthy trees. In this section, we base our discussion on the previous research efforts to demonstrate the qualifications of remote sensing for monitoring forest disturbances attributed to diseases and insects.

8.3.1 SPECTRAL CHARACTERISTICS

The spectral values in a forest image scene are often biased to representing the upper layer traits of tree canopies. While the top-down manner of photographing vegetation lacks the ability to characterize the entire tree, it is possible to link the status of canopy to forest health because diseases or insects substantially affect a tree's ability to photosynthesize and store moisture in foliage. One consequence is the noticeable change in foliage color (i.e., discoloration). For example, needles on pine trees turn red in the red-attack stage by mountain pine beetle (Wulder et al. 2006). Oak trees visually appear brown and freeze-dried as a result of sudden oak death (Kelly and Meentemeyer 2002). Remote sensors with the capacity to record the visible portion of the electromagnetic spectrum (wavelengths from approximately 400 to 700 nm) are able to detect these symptoms, which appear similarly in the human visual system. However, disease- and insect-mediated forest mortality is a gradual process. Some early-stage symptoms cannot be easily observed; for instance, unhealthy trees with reduced chlorophylls may only appear to be slightly brighter than the healthy trees in the visible spectral range owing to reduced absorbance of the blue and red wavelengths by foliage (Knipling 1970). Sensors with the capacity to further detect the near-infrared spectrum (wavelengths from approximately 700 to 1300 nm) are probably more sensitive to such physiological stress. Similarly, the amount of energy reflected in the short-wave infrared range (wavelengths from approximately 1300 to 2500 nm) is correlated with vegetation moisture (Laurent et al. 2005). Today's remote sensing technologies are already capable of recording the radiation reflected in those spectral ranges. To further advance the performance of remote detection, researchers utilized a variety of spectral indices (i.e., combinations of spectral bands) and have repetitively confirmed their effectiveness in monitoring forest damage subject to disease and insect attacks (see case studies in Table 8.1). Examples of the indices include normalized difference vegetation index (NDVI; Tucker 1979), enhanced vegetation index (Liu and Huete 1995), disturbance index (DI; Healey et al. 2005), normalized difference moisture index (NDMI; Jin and Sader 2005), normalized difference infrared index (Jackson et al. 2004), and enhanced wetness difference index (EWDI; Skakun et al. 2003).

While multispectral imagery has proven its potential to assess the status of damaging diseases and insects, previous studies discovered that the subtle spectral discrepancies between healthy and damaged trees (e.g., during the previsual green mortality stage) can be better detected by fine-spectral resolution data, that is, dozens to hundreds of narrow and contiguous spectral bands acquired through hyperspectral imaging (Coops et al. 2003; Hatala et al. 2010). On the basis of this technology, researchers have further developed narrowband vegetation indices, some of which were freshly designed (e.g., transformed chlorophyll absorption reflectance index; Haboudane et al. 2002), while the others were simple modifications of the traditional vegetation indices by means of substituting narrowband for broadband reflectance (e.g., red edge NDVI; Gitelson and Merzlyak 1994). Although not as common as the broadband indices yet, narrowband indices have shown the potential to explain the physiological changes in the forests suffering damage from insects (Fassnacht et al. 2014; Oumar et al. 2013).

8.3.2 Spatial Characteristics

Recent development in remote sensing allows us to perceive spatial details on the Earth's surface at varying scales, for example, 1 km/500 m/250 m MODIS, 30 m/15 m Landsat, 10 m/5 m SPOT-5, 4 m/1 m IKONOS, 1.2 m/0.3 m Worldview-3, and centimeter-level aerial photos. This offers forest practitioners a range of choices for balancing the accuracy of detecting disease or insect occurrence and data acquisition cost. Typically, coarse to moderate-resolution imagery has been traditionally applied to measure forest structural change at the landscape scale. For example, de Beurs and Townsend (2008) applied MODIS data with a 250-m spatial resolution to monitor more than 16,000 km^2 of insect defoliation of hardwood forests by gypsy moth. Fraser and Latifovic (2005) showed that 1-km-resolution SPOT VEGETATION data were sufficient for mapping a 350,000-km^2 area of coniferous forest mortality in Quebec, Canada, caused by the eastern hemlock looper. A higher-severity disturbance event may lead to a more satisfactory detection result, because the infected tree patches tend to be larger on average.

However, challenges arise if the majority of the damaged trees are within small, discrete patches. High–spatial resolution satellite and airborne imagery are more suitable for fine-scale detection and have proven to be feasible in previous studies (e.g., Adelabu et al. 2014; Cheng et al. 2010; Kautz 2014; Meddens et al. 2011; Wulder et al. 2008). It should be noted that a unique consideration of processing such type of data sets is the recent paradigm shift from pixel-based to object-based image analysis, that is, geographic object-based image analysis (GEOBIA; Blaschke et al. 2014). Because a high-resolution pixel often covers a portion of a tree or a small tree cluster, the corresponding pixel value may contain a high spectral variation as a result of the complex forest 3D structure and sun–tree–sensor geometry (Chen et al. 2011). Compared to the traditional pixel-based modeling, GEOBIA extracts image objects (groups of pixels) to represent meaningful geographic objects, for the purpose of reducing spectral noises and increasing mapping accuracies.

8.3.3 Temporal Characteristics

The size of Earth observation data archives is growing at an unprecedented pace. With rich time series data, it becomes feasible to extract the trajectories of disease and insect progression over a long term (e.g., Meigs et al. 2011; Vogelmann et al. 2009; Walter and Platt 2013). Because most of the infected trees do not die instantly, many forest disease or insect studies tend to apply annual or biannual imagery to characterize the spatiotemporal patterns of forest change. To mitigate the impact of seasonal variation, multidate images are preferably collected in the same months or the same seasons. Of the variety of date archives, Landsat time series have been the most widely used (see case studies in Table 8.1). This is possibly attributed to the features of four decades of data storage with minimized temporal gaps, free data access, and global coverage (Woodcock et al. 2008). However, as we are entering the remote sensing big data era, we expect to see an increasing application of diverse data archives for long-term forest health monitoring in the near future.

8.4   A REVIEW OF REMOTE SENSING ALGORITHMS

To date, a variety of remote sensing algorithms have been developed to measure forest damage caused by diseases and insects. The main principle is to extract the differences in spectral reflectance between healthy and infected trees, as well as among the infected trees during varying stages of decline. Here, we provide a brief review of those algorithms and categorize them into five groups: thresholding, classification, change detection, statistical regression, and the others, with details described below.

8.4.1   THRESHOLDING

Compared to healthy trees, damaged trees have distinct symptoms, such as reduced moisture, discolored foliage, and defoliated canopy. A thresholding method defines one or multiple thresholds to extract the pixels representing damaged trees from the entire forest image scene. While the operation appears simple, the success of applying thresholding largely depends on the effective description of forest symptoms and the accurate definition of threshold(s).

Describing the symptoms of forest damage has been primarily relying on image spectral indices. Some of those indices were specifically designed to assess forest disturbances. For example, Coops et al. (2006) created a red–green index, the ratio of QuickBird red to green wavelengths, to extract the red-attack damage (i.e., foliage color turning red from green) in the mountain pine beetle–infested coniferous forests. Their results confirmed the potential of using a simple threshold to red–green index values for separating the infected from the healthy trees. For many other studies, however, thresholding methods often directly employed or modified the existing indices that had not been intentionally developed for monitoring infestation. For example, multiple thresholds were applied to Landsat NDMI for extracting beetle-infested trees and forest regrowth after disturbance events (Coops et al. 2010; Goodwin et al. 2008). Similarly, Coops et al. (2009) calculated DI using 1-km-resolution MODIS images covering a part of the terrestrial land base of Canada. They found that those DI pixel values larger than ±1 standard deviation of the long-term mean were consistent with the areas flagged as infested using aerial survey. To further improve the thresholding performance, Skakun et al. (2003) created an EWDI through combing three different dates of wetness bands (derived from the Landsat TM tasseled cap transformation). Likewise, Olsson et al. (2012) modified the classic NDVI index by substituting the green band for the red band in the equation. The new index GNDVI was found to outperform NDVI, and negative GNDVI values indicated damage. Overall, the thresholding methods are simple to implement, with thresholds typically defined with assistance of field survey and manual photo interpretation. One major limitation for thresholding is that it is only suitable to identify major stages of forest disturbances, for example, extracting heavily damaged trees from healthy ones.

8.4.2 Classification

Land-cover classification using imagery to differentiate between land-cover types was developed almost immediately after the advent of remote sensing. The suitability of using image classification to measure forest damage is based on the fact that the distinct symptoms of infected forests make them appear as *new* land-cover types. It also seems to be consistent and convenient to apply one classification framework to extract not only the damaged/healthy forests but also the other land-cover types coexisting with forests, for example, grasses, shrubs, built-ups, and water.

Of the variety of classification algorithms, the classic supervised maximum likelihood classifier (MLC) demonstrated continued success in forest disease and insect monitoring. For example, MLC was effectively applied to Landsat imagery for differentiating mountain pine beetle–induced red attacks from non-red attacks (Walter and Platt 2013). MLC and Landsat imagery were also used to extract gypsy moth–caused defoliation from the nondefoliated trees (Thayn 2013). In addition, previous studies suggested that the application of MLC to classify high–spatial resolution imagery has the potential to detect forest damage of multiple stages. For example, Leckie et al. (2005) was able to estimate jack pine budworm–induced four classes of discoloration (nil–trace, light, moderate, and severe) through the application of MLC and 2.5-m-resolution aerial imagery acquired from the multispectral electro-optical imaging sensor. Meddens et al. (2011) and Bright et al. (2012) independently used aerial photography and MLC to classify beetle-caused tree mortality into green, red (dead trees with red needles), and gray (dead trees without needles) tree classes with the same overall accuracy of 87%. When integrated with hyperspectral imagery, MLC was found to be a viable solution to estimate forest stress during the early previsual stage of a sudden oak death outbreak (Pu et al. 2008).

Novel machine learning methods, as a complement to classic classifiers, have been introduced to the domain of remote sensing classification since the 1990s. Support vector machines (SVMs) are a successful example, which have proven to be feasible to detect three levels of insect defoliation ranging from nonimpacted undefoliated plants to partly defoliated plants and finally refoliating plants after severe defoliation in an African savanna (Adelabu et al. 2014). When applied to classify hyperspectral imagery acquired from HyMap, SVMs were found to have notable high overall accuracies mapping bark beetle–caused tree mortality, with the best result reaching as high as 97% accuracy (Fassnacht et al. 2014). Random forests (RFs) act as another popular machine learning method in classification. In a case study of mapping insect defoliation levels with RapidEye 5-m-resolution imagery, Adelabu et al. (2014) compared RFs and SVMs, and found comparable results. It should be noted that one outstanding feature of RF is that it can rank all the input variables based on their importance (Breiman 2001), which facilitates result analysis by identifying the most crucial spectral bands or indices in disease and insect mapping.

A subpixel classification scheme is needed if the spatial resolution of image pixels is too coarse to detect small, fragmented disturbances in a patchy distribution. To do so, spectral mixture analysis (SMA) provides a viable means, which is typically based on the assumption that the spectral value of each pixel is a linear combination of the reflectance from surface materials (endmembers) weighted by their factions. For example, Radeloff et al. (1999) performed SMA on Landsat TM imagery to classify jack pine budworm defoliation levels in a mixed forest stand and found a strong negative correlation between SMA-derived green needle fraction and field-measured budworm population ($r = -0.94$). With SMA and 0.5-m-resolution multispectral imagery, Goodwin et al. (2005) quantified the fractional abundance of three endmembers: sunlit canopy, shadow, and soil. Their results suggested a possibility of using the sunlit canopy image fraction to describe crown/leader color in the forests affected by damaging agents. When it comes to classifying hyperspectral imagery, the high spectral noises in data often challenge the performance of classifiers. To address the issue when using HyMap imagery, Hatala et al. (2010) employed the mixture-tuned matched-filter algorithm, an improved SMA through maximizing the target response and minimizing background spectral signatures, to classify whitebark pine stress and mortality.

8.4.3 STATISTICAL REGRESSION

Statistical regression analysis allows practitioners to estimate not only the discrete stages of forest disturbances (e.g., damaged vs. healthy) but also continuous defoliation or tree mortality levels from none to 100%. Compared to most classification methods, regression has the capacity to demonstrate the significance of the selected explanatory variables derived from remote sensing imagery. Such information can inform sustainable forest management, for example, predicting forest vulnerability in response to disease or insect attacks.

Logistic regression has been shown as a simple solution for identifying forest status of being damaged or not. For example, this model was applied to estimate an outbreak of black-headed budworm in Western Newfoundland, Canada, with a proven success to distinguish susceptible trees from those that were not (Luther et al. 1997). However, such analysis may not be sufficient for developing effective mitigation strategies. Researchers have expressed higher interests in understanding the detailed (i.e., continuous) tree damage levels. To do so, classic multiple linear regression was widely used to link remote sensing–derived metrics (e.g., spectral bands, spectral indices, and topographic variables) with field-measured damage indicators, such as defoliation intensity (de Beurs and Townsend 2008; Pontius et al. 2008), basal area (Siderhurst et al. 2010), leaf area index (Solberg et al. 2006), foliar nitrogen and plant growth vigor (McNeil et al. 2007), concentration of total chlorophyll (Cheng et al. 2010), and leaf water content (Cheng et al. 2010). Their studies also indicated the suitability of applying multiple regression to analyze a wide range of remote sensing data types (e.g., MODIS, Landsat, lidar, and the hyperspectral).

Recent sensor development has increased image spectral resolution and extended the coverage of data spectral range. However, this poses a challenge to regression modeling, that is, high dimensionality and collinearity of remotely sensed explanatory variables. To address this issue, Verbesselt et al. (2009) applied the least absolute shrinkage and selection operator (LASSO) to model bark beetle–induced tree mortality in *Pinus radiata* plantations. Compared to the standard data fitting method of least squares, LASSO is an alternative regularized version to minimizing the residual sum of squares "under a constraint on the sum of the absolute values of regression coefficient estimates" (Verbesselt et al. 2009). Another solution is partial least squares regression (also known as projection to latent structures), which finds new hyperplanes for minimizing the variance between impendent and dependent variables (Geladi and Kowalski 1986). Researchers have confirmed its effectiveness of mitigating the variable multicollinearity effects in studying insect-caused forest damage (Oumar and Mutanga 2014; Oumar et al. 2013; Wolter et al. 2008).

The aforementioned regression models are considered as fixed effects, that is, treating all the variables as nonrandom. However, Rullán-Silva et al. (2015) argued that a mixed-effects model, containing both fixed and random effects, is more appropriate for estimating the percentage of defoliation caused by beech leaf-miner weevil. The addition of random effects to a fixed-effects model was found to better account for the variability possibly introduced by environmental uncertainties (Rullán-Silva et al. 2015). While the mixed-effects models are relatively new to the field of remote sensing, we note that their merits have been increasingly recognized in forest ecology (Bolker et al. 2009).

8.4.4 Change Detection

Change detection employs multitemporal imagery (i.e., time series data) to measure the spatial patterns of forest disturbances through time. In contrast with using single-date imagery to identify damaged trees, this approach analyzes shifts in spectral reflectance across multiple dates. Accordingly, extra considerations are required to deal with spectral variation through time that arises from both forest disturbances and differences in atmospheric conditions and the sun–view–tree geometries (Chen et al. 2011; Song et al. 2001).

Previous efforts showed two ways of conducting change detection. First, the spectral discrepancies between multidate images are calculated through differencing the same spectral bands or indices from the base year (before disturbance) and the disturbance year(s). This is followed by applying thresholding, statistical regression, or classification to extract the pixels containing higher spectral variation (indicating damaged trees) than the others (e.g., de Beurs and Townsend 2008; Townsend et al. 2012; Wulder et al. 2008). Second, change detection focuses on measuring forest damage directly through all the spectral bands or indices. For example, Babst et al. (2010) applied principal component analysis to transfer multidate NDVI images (derived from Landsat time series) into new principal components. They discovered that the second principal component contained crucial information representing the change of NDVI, which was correlated with the level of defoliation caused by autumnal moth. Additionally, because the spectral discrepancies among the Landsat time series include both real and noisy false changes, Kennedy et al. (2010) developed a LandTrendr temporal segmentation algorithm to capture only the salient features of the trajectory (representing real changes) using a multilevel model fitting strategy. This algorithm was employed by Meigs et al. (2011) to successfully characterize the impacts of bark beetle on tree mortality.

8.4.5 Additional Approaches

In addition to the aforementioned mainstream methods, several other algorithms have been developed for unique considerations in disease and insect monitoring. For example, the occurrence of tree dieback is associated with specific forest environmental factors (e.g., distance from hosts to target trees; Kelly and Meentemeyer 2002). Liu et al. (2006) modeled such ecological compatibility with Markov random field, which was used to refine the results from a noncontextual SVM classification.

To deal with nontraditional data types, such as lidar for characterizing forest 3D structure (Chen and Hay 2011; Lim et al. 2003), Zhang (2008) applied mathematical morphology to process lidar point clouds for identifying small gaps in mangrove forests owing to natural disturbances, including the outbreaks of insects. Bright et al. (2012) employed lidar to estimate forest aboveground carbon. When integrated with the beetle-caused tree mortality map from an MLC classification, the carbon storage map was able to clearly reveal the impact of insect severity on forest carbon loss.

8.5 CHALLENGES AND OPPORTUNITIES

Advancements in remote sensing data acquisition and analysis have remarkably improved the feasibility of assessing landscape-scale forest disturbances induced by diseases or insects. However, challenges remain. In this section, we identify some of those challenges and suggest potential solutions.

8.5.1 Early Warning of Forest Damage

Forests that are infected by diseases or insects do not die instantly. The detection of early stage forest damage offers forest managers an opportunity to perform efficient disease and insect control. During this stage, the infected trees may only show a slight decline in chlorophyll levels and leaf water content. Previous efforts have confirmed the potential of applying hyperspectral remote sensing to assist with early detection of tree stress (e.g., Fassnacht et al. 2014; Pu et al. 2008). However, most of the sensors were mounted on airborne platforms (e.g., CASI, HyMap, and AVIRIS), making data acquisition an expensive process. To date, only a few satellite sensors (e.g., EO-1 Hyperion) are operational, although their application has been restrained because of limited spatial coverage and high spectral noises. To address the challenge, developing Landsat-like hyperspectral sensors is a promising solution. For example, NASA's hyperspectral infrared imager (HyspIRI) mission will mount two instruments on a satellite in low Earth orbit. Once launched, HyspIRI will deliver global coverage hyperspectral imagery at the 10-nm spectral resolution from the visible, short-wave infrared range to the thermal infrared range (NASA 2015). Another potential solution is to assemble a small, inexpensive hyperspectral unmanned aircraft system (UAS; see a recent review by Pajares [2015]). While such a system still has small spatial coverage, its highly operational flexibility combined with a proper sampling strategy makes early warning feasible. One limitation, however, is the obligation to meet UAS regulations and policies that may vary considerably from region to region.

8.5.2 Consistent Monitoring of Long-Term, Historical Forest Damage

While several remote sensing programs (e.g., AVHRR, Landsat, or SPOT) have been operational for three to four decades, many new types of sensors appeared only recently, such as those featuring high spatial resolution, hyperspectral resolution, and the ability to characterize forest 3D structure. These new sensors do have a higher capacity to detect forest stress and mortality; however, their data archives often have limited temporal and spatial coverage. This poses a challenge for consistently monitoring the long-term, historical impacts of diseases and insects on forests. One dilemma facing many researchers is that the study area was only partially covered by the data acquired from high-performance sensors for limited periods. Choosing the data that have full coverage (e.g., Landsat) can be one solution, while combining data from multiple sensors can be another solution (e.g., using Landsat data to fill in the gaps that lack hyperspectral imagery). In the latter case, the developed algorithms should have the capacity to accommodate varying types of remote sensing data across spatial, spectral, and temporal scales, so that all the results can be compared using consistent criteria.

8.5.3 Differentiating among Compound Disturbances

Forests are a natural ecosystem. The disturbances affecting the same forested regions may come from a range of sources. Besides insect and disease, other natural disasters (e.g., wildfire and wind) or anthropogenic activities (e.g., logging) can lead to compound disturbances. It is also possible that one disturbance regime (e.g., wildfire) may influence forest responses to another disturbance (e.g., disease), resulting in interacting disturbances (Turner 2010). Recent remote sensing studies have been limited on the topic of differentiating between disease-/insect-caused forest damage and other types of damage. One major challenge is that single sensors are typically not suitable to complete this task. For example, in a study of estimating burn severity in a forest that had experienced pre-fire disease outbreaks, Chen et al. (2015b) found similar spectral reflectance in burned and diseased trees using Landsat imagery. Therefore, a likely solution is the development of a multisensor approach, taking advantage of the strengths from individual sensors, for example, Landsat time series for temporal analysis of disease and insect progression, hyperspectral imaging for tracking the early signs of forest damage, and lidar for assessing the change in forest vertical profiles. Data integration maximizes practitioners' ability to estimate changes in forest biophysical and biochemical parameters, augmenting accurate assessments of forest damage.

8.6 CONCLUSION

Global forest ecosystems face high frequencies of landscape-level disturbances resulting from disease and insect epidemics. Over the past decades, remote sensing tools have improved detection of forest disturbances in a timely and cost-effective manner. As sensor technologies advance, richer Earth observation data with higher spatial, spectral, and temporal resolutions are expected to offer better choices to assess varying stages of disease/insect invasion in a range of forest biomes.

Accordingly, algorithms for modeling spectra–disturbance relationships will need to be continually refined or redeveloped to take advantage of new data and novel landscape changes caused by nonnative, invasive pathogens and insects.

ACKNOWLEDGMENT

This research was supported by the University of North Carolina at Charlotte through a Junior Faculty Development Award.

REFERENCES

Adelabu, S., Mutanga, O., & Adam, E. (2014). Evaluating the impact of red-edge band from Rapideye image for classifying insect defoliation levels. *ISPRS Journal of Photogrammetry and Remote Sensing*, *95*, 34–41.

Asner, G. P. (2013). Geography of forest disturbance. *Proceedings of the National Academy of Sciences of the United States of America*, *110*, 3711–2.

Assal, T. J., Sibold, J., & Reich, R. (2014). Remote sensing of environment modeling a historical mountain pine beetle outbreak using Landsat MSS and multiple lines of evidence. *Remote Sensing of Environment*, *155*, 275–288.

Babst, F., Esper, J., & Parlow, E. (2010). Landsat TM/ETM+ and tree-ring based assessment of spatiotemporal patterns of the autumnal moth (*Epirrita autumnata*) in northernmost Fennoscandia. *Remote Sensing of Environment*, *114*, 637–646.

Blaschke, T., Hay, G. J., Kelly, M. et al. (2014). Geographic object-based image analysis—Towards a new paradigm. *ISPRS Journal of Photogrammetry and Remote Sensing*, *87*, 180–191.

Bolker, B. M., Brooks, M. E., Clark, C. J., Geange, S. W., Poulsen, J. R., Stevens, M. H. H., & White, J. S. S. (2009). Generalized linear mixed models: A practical guide for ecology and evolution. *Trends in Ecology and Evolution*, *24*, 127–135.

Bonan, G. B. (2008). Forests and climate change: Forcings, feedbacks, and the climate benefits of forests. *Science*, *320*, 1444–1449.

Boyd, I. L., Freer-Smith, P. H., Gilligan, C. A, & Godfray, H. C. J. (2013). The consequence of tree pests and diseases for ecosystem services. *Science*, *342*(6160), 1235773. doi:10.1126/science.1235773.

Breiman, L. (2001). Random forests. *Machine Learning*, *45*, 5–32.

Bright, B. C., Hicke, J. A., & Hudak, A. T. (2012). Estimating aboveground carbon stocks of a forest affected by mountain pine beetle in Idaho using lidar and multispectral imagery. *Remote Sensing of Environment*, *124*, 270–281.

Chen, G., & Hay, G. J. (2011). A support vector regression approach to estimate forest biophysical parameters at the object level using airborne lidar transects and QuickBird data. *Photogrammetric Engineering and Remote Sensing*, *77*, 733–741.

Chen, G., Hay, G. J., Castilla, G., & St-Onge, B. (2011). A multiscale geographic object-based image analysis to estimate lidar-measured forest canopy height using Quickbird imagery. *International Journal of Geographical Information Science*, *25*, 877–893.

Chen, G., Metz, M. R., Rizzo, D. M., Dillon, W. W., & Meentemeyer, R. K. (2015a). Object-based assessment of burn severity in diseased forests using high-spatial and high-spectral resolution MASTER airborne imagery. *ISPRS Journal of Photogrammetry and Remote Sensing*, *102*, 38–47.

Chen, G., Metz, M. R., Rizzo, D. M., & Meentemeyer, R. K. (2015b). Mapping burn severity in a disease-impacted forest landscape using Landsat and MASTER imagery. *International Journal of Applied Earth Observation and Geoinformation*, *40*, 91–99.

Cheng, T., Rivard, B., Sánchez-Azofeifa, G. A., Feng, J., & Calvo-Polanco, M. (2010). Continuous wavelet analysis for the detection of green attack damage due to mountain pine beetle infestation. *Remote Sensing of Environment, 114,* 899–910.

Coops, N., Stanford, M., Old, K., Dudzinski, M., Culvenor, D., & Stone, C. (2003). Assessment of dothistroma needle blight of *Pinus radiata* using airborne hyperspectral imagery. *Ecology and Epidemiology, 93,* 1524–1532.

Coops, N. C., Johnson, M., Wulder, M. A., & White, J. C. (2006). Assessment of QuickBird high spatial resolution imagery to detect red attack damage due to mountain pine beetle infestation. *Remote Sensing of Environment, 103,* 67–80.

Coops, N. C., Wulder, M. A., & Iwanicka, D. (2009). Large area monitoring with a MODIS-based Disturbance Index (DI) sensitive to annual and seasonal variations. *Remote Sensing of Environment, 113*(6), 1250–1261.

Coops, N. C., Gillanders, S. N., Wulder, M. A., Gergel, S. E., Nelson, T., & Goodwin, N. R. (2010). Assessing changes in forest fragmentation following infestation using time series Landsat imagery. *Forest Ecology and Management, 259,* 2355–2365.

de Beurs, K. M., & Townsend, P. A. (2008). Estimating the effect of gypsy moth defoliation using MODIS. *Remote Sensing of Environment, 112*(10), 3983–3990.

Fassnacht, F. E., Latifi, H., Ghosh, A., Joshi, P. K., & Koch, B. (2014). Assessing the potential of hyperspectral imagery to map bark beetle-induced tree mortality. *Remote Sensing of Environment, 140,* 533–548.

Fraser, R. H., & Latifovic, R. (2005). Mapping insect-induced tree defoliation and mortality using coarse spatial resolution satellite imagery. *International Journal of Remote Sensing, 26,* 193–200.

Geladi, P., & Kowalski, B. R. (1986). Partial least-squares regression: A tutorial. *Analytica Chimica Acta, 185,* 1–17.

Gitelson, A., & Merzlyak, M. (1994). Spectral reflectance changes associated with autumn senescence of *Aesculus hippocastanum* L. and *Acer platanoides* L. leaves. *Journal of Plant Physiology, 143,* 286–292.

Goodwin, N., Coops, N. C., & Stone, C. (2005). Assessing plantation canopy condition from airborne imagery using spectral mixture analysis and fractional abundances. *International Journal of Applied Earth Observation and Geoinformation, 7,* 11–28.

Goodwin, N. R., Coops, N. C., Wulder, M. A., Gillanders, S., Schroeder, T. A., & Nelson, T. (2008). Estimation of insect infestation dynamics using a temporal sequence of Landsat data. *Remote Sensing of Environment, 112,* 3680–3689.

Haboudane, D., Miller, J. R., Tremblay, N., Zarco-Tejada, P. J., & Dextraze, L. (2002). Integrated narrow-band vegetation indices for prediction of crop chlorophyll content for application to precision agriculture. *Remote Sensing of Environment, 81,* 416–426.

Hatala, J. A., Crabtree, R. L., Halligan, K. Q., & Moorcroft, P. R. (2010). Landscape-scale patterns of forest pest and pathogen damage in the Greater Yellowstone Ecosystem. *Remote Sensing of Environment, 114,* 375–384.

Healey, S. P., Cohen, W. B., Yang, Z., & Krankina, O. N. (2005). Comparison of Tasseled Cap-based Landsat data structures for use in forest disturbance detection. *Remote Sensing of Environment, 97,* 301–310.

Heller, R.C., & Bega, R. V. (1973). Detection of forest diseases by remote sensing. *Journal of Forestry, 71,* 18–21.

Huang, L., Ning, Z., & Zhang, X. (2010). Impacts of caterpillar disturbance on forest net primary production estimation in China. *Ecological Indicators, 10,* 1144–1151.

Jackson, T. L., Chen, D., Cosh, M., Li, F., Anderson, M., Walthall, C., Doriaswamy, P., & Hunt, E. R. (2004). Vegetation water content mapping using Landsat data derived normalized difference water index for corn and soybeans. *Remote Sensing of Environment, 92,* 475–482.

Jin, S., & Sader, S. A. (2005). MODIS time-series imagery for forest disturbance detection and quantification of patch size effects. *Remote Sensing of Environment*, *99*, 462–470.

Kautz, M. (2014). On correcting the time-lag bias in aerial-surveyed bark beetle infestation data. *Forest Ecology and Management*, *326*, 157–162.

Kelly, M., & Meentemeyer, R. K. (2002). Landscape dynamics of the spread of sudden oak death. *Photogrammetric Engineering & Remote Sensing*, *68*, 1001–1009.

Kennedy, R. E., Yang, Z., & Cohen, W. B. (2010). Detecting trends in forest disturbance and recovery using yearly Landsat time series: 1. LandTrendr—Temporal segmentation algorithms. *Remote Sensing of Environment*, *114*, 2897–2910.

Knipling, E. B. (1970). Physical and physiological basis for the reflectance of visible and near-infrared radiation from vegetation. *Remote Sensing of Environment*, *1*, 155–159.

Lamsal, S., Cobb, R. C., Hall Cushman, J., Meng, Q., Rizzo, D. M., & Meentemeyer, R. K. (2011). Spatial estimation of the density and carbon content of host populations for *Phytophthora ramorum* in California and Oregon. *Forest Ecology and Management*, *262*, 989–998.

Laurent, E. J., Shi, H., Gatziolis, D., LeBouton, J. P., Walters, M. B., & Liu, J. (2005). Using the spatial and spectral precision of satellite imagery to predict wildlife occurrence patterns. *Remote Sensing of Environment*, *97*, 249–262.

Leckie, D. G., Cloney, E., & Joyce, S. P. (2005). Automated detection and mapping of crown discolouration caused by jack pine budworm with 2.5 m resolution multispectral imagery. *International Journal of Applied Earth Observation and Geoinformation*, *7*, 61–77.

Lim, K., Treitz, P., Wulder, M., & Flood, M. (2003). LiDAR remote sensing of forest structure. *Progress in Physical Geography*, *1*, 88–106.

Liu, H., & Huete, A. (1995). A feedback based modification of the NDVI to minimize canopy background and atmospheric noise. *IEEE Transactions on Geoscience and Remote Sensing*, *33*, 457–465.

Liu, D., Kelly, M., & Gong, P. (2006). A spatial–temporal approach to monitoring forest disease spread using multi-temporal high spatial resolution imagery. *Remote Sensing of Environment*, *101*(2), 167–180.

Liu, D., Kelly, M., Gong, P., & Guo, Q. (2007). Characterizing spatial–temporal tree mortality patterns associated with a new forest disease. *Forest Ecology and Management*, *253*, 220–231.

Luther, J. E., Franklin, S. E., Hudak, J., & Meades, J. P. (1997). Forecasting the susceptibility and vulnerability of balsam fir stands to insect defoliation with Landsat thematic mapper data. *Remote Sensing of Environment*, *59*, 77–91.

McNeil, B. E., de Beurs, K. M., Eshleman, K. N., Foster, J. R., & Townsend, P. A. (2007). Maintenance of ecosystem nitrogen limitation by ephemeral forest disturbance: An assessment using MODIS, Hyperion, and Landsat ETM+. *Geophysical Research Letters*, *34*, L19406.

Meddens, A. J. H., Hicke, J. A., & Vierling, L. A. (2011). Evaluating the potential of multispectral imagery to map multiple stages of tree mortality. *Remote Sensing of Environment*, *115*, 1632–1642.

Meentemeyer, R. K., Dorning, M. A., Vogler, J. B., Schmidt, D., & Garbelotto, M. (2015). Citizen science helps predict risk of emerging infectious disease. *Frontiers in Ecology and the Environment*, *13*, 189–194.

Meentemeyer, R. K., Rank, N. E., Shoemaker, D. A., Oneal, C. B., Wickland, A. C., Frangioso, K. M., & Rizzo, D. M. (2008). Impact of sudden oak death on tree mortality in the Big Sur ecoregion of California. *Biological Invasions*, *10*, 1243–1255.

Meigs, G. W., Kennedy, R. E., & Cohen, W. B. (2011). A Landsat time series approach to characterize bark beetle and defoliator impacts on tree mortality and surface fuels in conifer forests. *Remote Sensing of Environment*, *115*, 3707–3718.

Meigs, G. W., Kennedy, R. E., Gray, A. N., & Gregory, M. J. (2015). Spatiotemporal dynamics of recent mountain pine beetle and western spruce budworm outbreaks across the Pacific Northwest Region, USA. *Forest Ecology and Management*, *339*, 71–86.

NASA (2015). *HyspIRI Mission Study*. Available on: https://hyspiri.jpl.nasa.gov.

Nelson, R. F. (1983). Detecting forest canopy change due to insect activity using Landsat MSS. *Photogrammetric Engineering and Remote Sensing, 49*, 1303–1314.

Olsson, P.-O., Jönsson, A. M., & Eklundh, L. (2012). A new invasive insect in Sweden— *Physokermes inopinatus*: Tracing forest damage with satellite based remote sensing. *Forest Ecology and Management, 285*, 29–37.

Oumar, Z., & Mutanga, O. (2014). Integrating environmental variables and WorldView-2 image data to improve the prediction and mapping of *Thaumastocoris peregrinus* (bronze bug) damage in plantation forests. *ISPRS Journal of Photogrammetry and Remote Sensing, 87*, 39–46.

Oumar, Z., Mutanga, O., & Ismail, R. (2013). Predicting *Thaumastocoris peregrinus* damage using narrow band normalized indices and hyperspectral indices using field spectra resampled to the hyperion sensor. *International Journal of Applied Earth Observation and Geoinformation, 21*, 113–121.

Pajares, G. (2015). Overview and current status of remote sensing applications based on unmanned aerial vehicles (UAVs). *Photogrammetric Engineering & Remote Sensing, 81*, 281–329.

Pan, Y., Birdsey, R. A, Fang, J. et al. (2011). A large and persistent carbon sink in the world's forests. *Science, 333*, 988–93.

Pontius, J., Martin, M., Plourde, L., & Hallett, R. (2008). Ash decline assessment in emerald ash borer-infested regions: A test of tree-level, hyperspectral technologies. *Remote Sensing of Environment, 112*, 2665–2676.

Pu, R., Kelly, M., Anderson, G. L., & Gong, P. (2008). Using CASI hyperspectral imagery to detect mortality and vegetation stress associated with a new hardwood forest disease. *Photogrammetric Engineering & Remote Sensing, 74*, 65–75.

Radeloff, V. C., Mladenoff, D. J., & Boyce, M. S. (1999). Detecting Jack Pine budworm defoliation using spectral mixture analysis: Separating effects from determinants. *Remote Sensing of Environment, 169*, 156–169.

Raffa, K. F., Powell, E. N., & Townsend, P. A. (2013). Temperature-driven range expansion of an irruptive insect heightened by weakly coevolved plant defenses. *Proceedings of the National Academy of Sciences of the United States of America, 110*, 2193–2198.

Rizzo, D. M., Garbelotto, M. & Hansen, E. M. (2005). *Phytophthora ramorum*: Integrative research and management of an emerging pathogen in California and Oregon forests. *Annual Review of Phytopathology, 43*, 309–335.

Rock, B. N., Vogelmann, J. E., Williams, D. L., Vogelmann, A. F., & Hoshizaki, T. (1986). Remote detection of forest damage. *BioScience, 36*, 439–445.

Rullán-Silva, C., Olthoff, A. E., Pando, V., Pajares, J. A., & Delgado, J. A. (2015). Remote monitoring of defoliation by the beech leaf-mining weevil *Rhynchaenus fagi* in northern Spain. *Forest Ecology and Management, 347*, 200–208.

Sadyś, M., Skjøth, C. A., & Kennedy, R. (2014). Back-trajectories show export of airborne fungal spores (*Ganoderma* sp.) from forests to agricultural and urban areas in England. *Atmospheric Environment, 84*, 88–99.

Sangüesa-Barreda, G., Camarero, J. J., García-Martín, A., Hernández, R., & de la Riva, J. (2014). Remote-sensing and tree-ring based characterization of forest defoliation and growth loss due to the Mediterranean pine processionary moth. *Forest Ecology and Management, 320*, 171–181.

Siderhurst, L. A., Griscom, H. P., Hudy, M., & Bortolot, Z. J. (2010). Changes in light levels and stream temperatures with loss of eastern hemlock (*Tsuga canadensis*) at a southern Appalachian stream: Implications for brook trout. *Forest Ecology and Management, 260*, 1677–1688.

Skakun, R. S., Wulder, M. A., & Franklin, S. E. (2003). Sensitivity of the thematic mapper enhanced wetness difference index to detect mountain pine beetle red-attack damage. *Remote Sensing of Environment, 86*, 433–443.

Solberg, S., Næsset, E., Holt, K., & Christiansen, E. (2006). Mapping defoliation during a severe insect attack on Scots pine using airborne laser scanning, *102*, 364–376.

Song, C., Woodcock, C. E., Seto, K. C., Lenney, M. P., & Macomber, S. A. (2001). Classification and change detection using Landsat TM data: When and how to correct atmospheric effects? *Remote Sensing of Environment*, *75*, 230–244.

Thayn, J. B. (2013). Using a remotely sensed optimized Disturbance Index to detect insect defoliation in the Apostle Islands, Wisconsin, USA. *Remote Sensing of Environment*, *136*, 210–217.

Townsend, P. A., Singh, A., Foster, J. R., Rehberg, N. J., Kingdon, C. C., Eshleman, K. N., & Seagle, S. W. (2012). A general Landsat model to predict canopy defoliation in broadleaf deciduous forests. *Remote Sensing of Environment*, *119*, 255–265.

Tucker, C. (1979). Red and photographic infrared linear combinations for monitoring vegetation. *Remote Sensing of Environment*, *8*, 127–150.

Turner, M. G. (2010). Disturbance and landscape dynamics in a changing world. *Ecology*, *91*, 2833–2849.

USDA Forest Service (2012). *Major Forest Insect and Disease Conditions in the United States: 2011*. Available online: http://www.fs.fed.us/foresthealth/publications/Conditions Report_2011.pdf.

Verbesselt, J., Robinson, A., Stone, C., & Culvenor, D. (2009). Forecasting tree mortality using change metrics derived from MODIS satellite data. *Forest Ecology and Management*, *258*, 1166–1173.

Vogelmann, J. E., Tolk, B., & Zhu, Z. (2009). Monitoring forest changes in the southwestern United States using multitemporal Landsat data. *Remote Sensing of Environment*, *113*, 1739–1748.

Walter, J. A., & Platt, R. V. (2013). Multi-temporal analysis reveals that predictors of mountain pine beetle infestation change during outbreak cycles. *Forest Ecology and Management*, *302*, 308–318.

Wang, C., He, H. S., & Kabrick, J. M. (2008). A risk rating study to predict oak decline and recovery in the Missouri Ozark Highlands, USA. *GIScience and Remote Sensing*, *45*, 406–425.

Wingfield, M. J., Brockerhoff, E. G., Wingfield, B. D., & Slippers, B. (2015). Planted forest health: The need for a global strategy. *Science*, *349*, 832–836.

Wolter, P. T., Townsend, P. A., & Sturtevant, B. R. (2009). Estimation of forest structural parameters using 5 and 10 meter SPOT-5 satellite data. *Remote Sensing of Environment*, *113*, 2019–2036.

Wolter, P. T., Townsend, P. A., Sturtevant, B. R., & Kingdon, C. C. (2008). Remote sensing of the distribution and abundance of host species for spruce budworm in Northern Minnesota and Ontario. *Remote Sensing of Environment*, *112*(10), 3971–3982.

Woodcock, C. E., Allen, A. A., Belward, A. S. et al. (2008). Free access to Landsat imagery. *Science*, *320*, 1011.

Wulder, M. A., Dymond, C. C., White, J. C., Leckie, D. G., & Carroll, A. L. (2006). Surveying mountain pine beetle damage of forests: A review of remote sensing opportunities. *Forest Ecology and Management*, *221*, 27–41.

Wulder, M. A., Ortlepp, S. M., White, J. C., Coops, N. C., & Coggins, S. B. (2009). Monitoring the impacts of mountain pine beetle mitigation. *Forest Ecology and Management*, *258*, 1181–1187.

Wulder, M. A., White, J. C., Coops, N. C., & Butson, C. R. (2008). Multi-temporal analysis of high spatial resolution imagery for disturbance monitoring. *Remote Sensing of Environment*, *112*, 2729–2740.

Zhang, K. (2008). Identification of gaps in mangrove forests with airborne LIDAR. *Remote Sensing of Environment*, *112*, 2309–2325.

9 Monitoring Water Quality with Remote Sensing Image Data

Bunkei Matsushita, Wei Yang,
Lalu Muhamad Jaelani, Fajar Setiawan,
and Takehiko Fukushima

CONTENTS

9.1 INTRODUCTION

Water covers approximately 74% of the Earth's surface, and it plays many important roles in the lives of all human being. Inland and coastal waters in particular have a direct interface with human society by providing value for food supplies, industrial uses, transportation, commerce, and human health (UNEP 2006). However, many water bodies have encountered severe environmental problems (e.g., eutrophication) in recent decades as a result of human interventions and climate change (Ayres et al. 1996; Haddeland et al. 2014). It is thus crucial to monitor and understand the amount and quality of these water bodies as well as their biogeochemical processes in order to achieve the effective management and sustainable use of the water resources (United Nations Open Working Group 2014).

Since conventional water-monitoring methods (e.g., water sampling from a boat) are very time-, labor-, and cost-consuming, the maintenance of steady monitoring is difficult for local and national governments with meager financial resources, especially in developing countries. For example, even when a lake is monitored by an in situ sampling method, the number of sampling locations and the frequency of water sampling may not accurately represent the actual status of the entire lake, because the water quality can dramatically change both spatially and temporally (Kiefer et al. 2015). However, with the progress of satellite sensors and water quality estimation algorithms, remote sensing is regarded as a quite useful technique for studying bodies of water, as it can provide synoptic observations at very frequent intervals for water bodies, at relatively low cost (Dekker and Hestir 2012).

In this chapter, we focus on how to obtain water quality data from remote sensing techniques that can be used toward achieving the Sustainable Development Goal 6.3 (Monitoring Ambient Water Quality; United Nations Open Working Group 2014). As noted above, we will describe major satellite sensors and the advances needed to enable further progress in optical water-quality retrieval, several representative algorithms for water area delineation, atmospheric correction, and water quality estimations, plus a hybrid approach for the universal application of these algorithms.

9.2 MAJOR SATELLITE SENSORS FOR MONITORING WATER QUALITY

Since the late 1970s, many attempts have been made to observe water quality from space. The first ocean-color sensor, Coastal Zone Color Scanner (CZCS), was launched in October 1978 aboard the Nimbus-7 satellite (Hovis et al. 1980). With four spectral bands in visible range (443, 520, 550, and 670 nm), and one spectral band in the thermal infrared spectrum (10.5–12.5 μm), the CZCS was the only spaceborne sensor that observed global ocean colors during the period 1978–1986 (Table 9.1 and Figure 9.1). However, since the CZCS was just a proof-of-concept sensor for the first generation of ocean-color observation at that time, there were many limitations in its instrument design such as a lack of a sufficient number of spectral bands in the visible and near-infrared (NIR) range as well as relatively low radiometric sensitivity (Gordon and Wang 1994; Hu et al. 2012a). In light of these limitations, the data provided by the CZCS should be used with caution. For example, Tassan (1988) reported that when CZCS data were used, high concentrations of gelbstoff and suspended minerals reduced the accuracy of both chlorophyll-a estimations and atmospheric corrections.

The second generation of ocean-color sensor, Sea-viewing Wide Field-of-View Sensor (SeaWiFS), was launched in 1997 and ended its mission in 2010 (Table 9.1 and Figure 9.1). The SeaWiFS has four additional bands for ocean observations: two in the visible range (412 and 490 nm) and two in the NIR region (765 and 865 nm). The band around 412 nm is useful for separating the detrital and viable phytoplankton signals (Ciotti and Bricaud 2006), and the band centered on 490 nm can assist in the detection of accessory pigments (Pan et al. 2010). The two NIR bands centered on 765 and 865 nm contribute to atmospheric correction (Gordon and Wang 1994). Compared to the CZCS, the radiometric sensitivity of the SeaWiFS was also largely improved (the mean signal-to-noise ratio [SNR] is between 424 and 790 for six visible bands; Hu et al. 2012a). However, the spatial and temporal resolutions of the SeaWiFS are still similar to those of the CZCS.

TABLE 9.1

Summary of Indicative Satellite Sensors Used for Studying Waters

Satellite Sensors	Orbit	Active Period	No. of Bands Used for Water Quality Monitoring	Resolution (m)	Revisit Time	Mean SNR
CZCS	Sun-synchronous	1978–1986	VIS 4	800	1–2 days	193–220
SeaWiFS	Sun-synchronous	1997–2010	VIS 6, NIR 2	1100	1–2 days	424–790 for VIS and 183–219 for NIR
MODISA	Sun-synchronous	2002 to present	VIS 7, NIR 2	1000	1–2 days	1366–2401 for VIS and 806–995 for NIR
MERIS	Sun-synchronous	2002–2012	VIS 8, NIR 7	300	3 days	513–1088 for VIS and 218–667 for NIR (excluding band 11 of 54)
GOCI	Geostationary	2010 to present	VIS 6, NIR 2	500	1 hour	579–833 for VIS and 596–587 for NIR
TM	Sun-synchronous	1984–2011	VIS 3, NIR 1, SWIR 2	30	16 days	29–72 for VIS, 17 for NIR, and 6–10 for SWIR
ETM+	Sun-synchronous	1999 to present	VIS 3, NIR 1, SWIR 2	30	16 days	41–78 for VIS, 13 for NIR, and 6–10 for SWIR
OLI	Sun-synchronous	2013 to present	VIS 4, NIR 1, SWIR 2	30	16 days	144–478 for VIS, 67 for NIR, and 14–30 for SWIR

Note: Mean SNR values are obtained from Hu et al. (2012a) except for SNR of OLI (obtained from Franz et al. 2015).

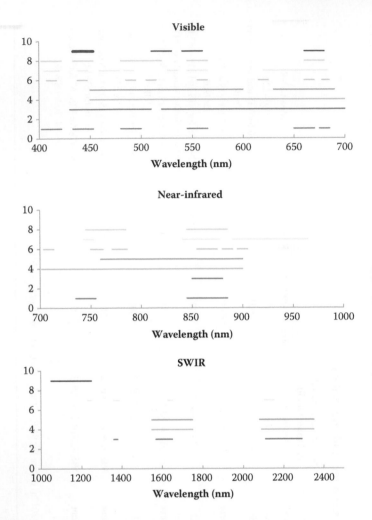

FIGURE 9.1 Location of the spectral bands of several satellite sensors that have been used for studying waters. 1: GOCI, 3: OLI, 4: ETM+, 5: TM, 6: MERIS, 7: MODIS, 8: SeaWiFS, 9: CZCS.

The Moderate Resolution Imaging Spectroradiometer (MODIS), launched in December 1999 (aboard the Terra satellite MODIST) and May 2002 (aboard the Aqua satellite MODISA), is a sensor for both land and water observations. However, the water-leaving radiance obtained from MODIST is almost unusable for ocean-color applications owing to the uncertainties and instabilities of the sensor observed since the start of the mission (Franz et al. 2008). The MODISA has 36 spectral bands in total, of which 9 bands are used for ocean-color remote sensing (Table 9.1 and Figure 9.1). Generally, the MODISA ocean-color bands are similar to those of the SeaWiFS, with three bands in blue and two bands in green, but two bands in red wavelengths are different from those of the SeaWiFS (which has only one red band). The additional red band of the MODISA (around 678 nm) contributes to the detection of chlorophyll fluorescence. The radiometric sensitivity of the MODISA ocean-color bands were further improved with the mean SNR values of 1366–2401 in the visible range and 806–995 in the NIR region (Hu et al. 2012a). The seven MODISA land bands are often applied to high-turbidity inland and coastal waters because of their higher spatial resolutions (250–500 m) and longer-wavelength bands in the short-wave infrared region (SWIR; Hu 2009; Hu et al. 2010; Wang and Shi 2007).

During a similar active period of the MODIS, the European Space Agency launched another ocean-color sensor (the Medium Resolution Imaging Spectroradiometer [MERIS]) in March 2002. This mission was ended in 2012. Compared to the three ocean-color sensors described above, the MERIS provided higher spectral and spatial resolutions, but slightly lower temporal resolution and radiometric sensitivity (Table 9.1 and Figure 9.1). The new visible band around 620 nm made the MERIS a unique sensor that is able to distinguish cyanobacterial blooms from other algal blooms (Kutser et al. 2006). Another unique band of the MERIS is around 709 nm. This band has been demonstrated to be very useful for estimating chlorophyll-a concentrations and detecting algal blooms in turbid inland and coastal waters (Dall'Olmo et al. 2003; Gilerson et al. 2010; Gitelson et al. 2008; Gower et al. 2005).

Although the spectral characteristics, temporal resolutions, and radiometric sensitivities of the current ocean-color sensors are adequate to monitor water quality, the relatively coarse spatial resolutions of these data sets (300–1000 m) often limit many applications in inland waters with small size. Therefore, several land sensors with relatively fine spatial resolution (30 m) such as the TM (Thematic Mapper), the ETM+ (Enhanced Thematic Mapper Plus), and the OLI (Operational Land Imager) aboard Landsat satellites (Table 9.1 and Figure 9.1) have also been used for assessing water quality parameters (e.g., Dekker et al. 2002; Oyama et al. 2009; Tyler et al. 2006; Wu et al. 2015). In addition to the relatively fine spatial resolution of the sensors aboard Landsat, another advantage of the use of Landsat data is the longevity of its successive missions (1972 to present; Lobo et al. 2015). However, the wide bandwidth, infrequent revisit time, and lower radiometric sensitivity of Landsat are substantial limitations for applying Landsat data to water quality monitoring.

Unlike all of the abovementioned sensors, the Geostationary Ocean Color Imager (GOCI), launched in June 2010 by the Korea Aerospace Research Institute (KARI), is the first and so far the only sensor in geostationary orbit for monitoring ocean color from space (Ryu et al. 2012). As a benefit owing to its geostationary orbit, GOCI can provide eight images during daylight with 500 m spatial resolution and higher radiometric sensitivity (>579, Table 9.1). The bands of GOCI are similar to those of MODIS ocean bands, but it uses less one green band than MODIS, centered at 531 nm (Figure 9.1). However, its lack of bands around SWIR wavelengths is a significant limitation for atmospheric correction in turbid inland and coastal waters (Wang et al. 2012). In addition, the lack of bands around 620 nm makes the GOCI unable to distinguish cyanobacterial blooms from other algal blooms (Kutser et al. 2006), and its lack of bands around 709 nm makes it difficult to accurately retrieve the chlorophyll-a concentrations in waters with higher concentrations of colored dissolved organic matter (CDOM) and tripton (Gitelson et al. 2008). Another limitation of GOCI is its spatial coverage. GOCI covers only the regions covering Korea,

China, and Japan, and it cannot provide a global data set because it is designed to monitor the ocean color around the Korean Peninsula (130°E, 36°N at the center) on a geostationary orbit (Ryu et al. 2012).

Thus, all past and present ocean-color or land sensors have limitations for monitoring water quality. The following four aspects of future satellite sensors for remotely retrieving water quality parameters should be improved: temporal resolution, spatial resolution, spectral resolution, and radiometric sensitivity (Mouw et al. 2015). First, higher temporal resolution such as acquiring one or more satellite image per hour during daylight would allow not only observations of fast processes of water bodies (e.g., the tidal and diurnal variability of water quality) but also the accumulation of more available satellite data without cloud contaminations (Ruddick et al. 2014). This high temporal resolution could be achieved by a geostationary platform.

Second, finer spatial resolution is required for monitoring water quality in inland and coastal waters and for detecting algal blooms in open oceans. Bissett et al. (2004) showed that the optimal spatial resolution for monitoring near-shore waters (within 200 m of shore) is 100 m. This spatial resolution may also meet the requirement of some inland waters larger than 0.09 km^2—corresponding to 9 pixels (i.e., 100 m × 100 m × 9 pixels)—as Verpoorter et al. (2012) reported that it is difficult to verify water bodies smaller than 9 pixels relative to image noise. However, there are numerous water bodies smaller than 0.09 km^2 (Verpoorter et al. 2014) and water bodies with spatially heterogeneous water quality parameters (Kutser 2004). Further improvements in spatial resolution are thus necessary for these water bodies.

Third, good spectral resolution is highly desirable for many water-quality remote sensing studies such as those concerning the retrieval of water quality parameters, the detection of cyanobacterial blooms or *Trichodesmium* distribution, and the removal of bottom-reflectance effects (Mouw et al. 2015). However, satellite-based measurements with too small spectral intervals often face the problems of how to accurately remove gas absorption effects, how to select appropriate band for efficient data processing, and how to store a large volume of data (Lee et al. 2014). Lee et al. (2014) found that the remote sensing reflectance of the neighboring bands is highly correlated when the spectral distance between two neighboring spectral bands is less than 10 nm, and even when this distance is increased to 20 nm, the spectral information loss is still very limited. Accordingly, they suggested that future ocean-color satellite sensors should have 15 bands space almost evenly between 400 and 700 nm at 10-nm bandwidth intervals. For the wavelength beyond 700 nm, at least two NIR bands and two SWIR bands are necessary for atmospheric correction in clear and turbid waters, respectively (Gordon and Wang 1994; Wang and Shi 2007). In addition, a band around 709 nm is also necessary for estimating chlorophyll-a concentrations and detecting algal blooms in turbid inland and coastal waters (Dall'Olmo et al. 2003; Gilerson et al. 2010; Gitelson et al. 2008; Gower et al. 2005).

Finally, previous studies have stated that the radiometric sensitivity of future ocean-color sensors should have SNRs larger than 1000 for wavelengths between 350 and 720 nm, larger than 600 for wavelengths between 720 and 900 nm, and between 100 and 200 nm for SWIR bands (Hu et al. 2012a; Wang 2007). Since the SNR is in proportion to the pixel size, the bandwidth for a specific band, and the integration time, a longer integration time is necessary if we want to ensure high SNRs as well as fine spatial and spectral resolutions. A sensor on a geostationary satellite would probably achieve this goal because "the sensor can stare at the desired pixels" (Lee et al. 2014). A detailed review of the opportunities and challenges presented by geostationary sensors was provided by Ruddick et al. (2014). Summaries of past, present, and planned ocean-color sensors such as the Pre-Aerosol, Clouds, and ocean Ecosystem (PACE) and the Ocean and Land Color Imager (OLCI) can be found in an appendix of a report published by the International Ocean Colour Coordinating Group (IOCCG 2012).

9.3 WATER DELINEATION FROM SATELLITE IMAGERY

To monitor water quality from satellite data, the first step is the extraction of the water body from the data. This step can also serve to detect cloud-contaminated pixels in water bodies. Many algorithms and techniques have been developed for delineating water bodies from optical remote sensing data. These methods can be summarized as belonging to three categories: (1) digitizing water bodies via a visual assessment of satellite images; (2) using image classification techniques such as supervised or unsupervised classification procedures, spectral transformation, and texture analysis; and (3) using a single or multiple algebraically operated bands such as the band ratio and spectral water index, combined with appropriate threshold values (Verpoorter et al. 2012). Methods based on the spectral water index have been widely used to delineate water bodies from satellite imagery, because not only can this approach efficiently enhance water features while suppressing or even eliminating non-water features, it is also easy to use and less computationally time-consuming than other methods.

The first spectral water index devised for delineating water from land is the normalized difference water index (NDWI), which was proposed by McFeeters (1996) and can be expressed as follows:

$$NDWI = (d_{green} - d_{NIR})/(d_{green} + d_{NIR}), \qquad (9.1)$$

where d_{green} and d_{NIR} are the digital number (DN) at green and NIR bands, respectively. The NDWI values can range from -1 to 1. The selection of the green band is done to maximize the typical reflectance of water features. The use of NIR band in the NDWI can not only minimize the water reflectance (strong absorption by water), but also take advantage of the high reflectance at the NIR band by terrestrial vegetation and soil features (even higher than the reflectance of the green band). Accordingly, water features will have positive NDWI values, and land features will have zero or negative values (McFeeters 1996). Zero was thus set as the threshold to distinguish water bodies from land by McFeeters (1996) (i.e., if NDWI > 0, then = water; otherwise, non-water).

Xu (2006) found that the NDWI cannot efficiently suppress the signal from built-up land, and thus extracted water bodies were still mixed with built-up land noise. For example, Figure 9.2a through c show a Landsat5/TM sub-scene for a part of Wuhan City (China), an NDWI image of the same place, and extracted water bodies from the NDWI image, respectively, showing that many built-up lands near Lake Dong (surrounded by a circle in Figure 9.2a) show positive NDWI values in Figure 9.2b and thus were misclassified as water bodies (white color in Figure 9.2c).

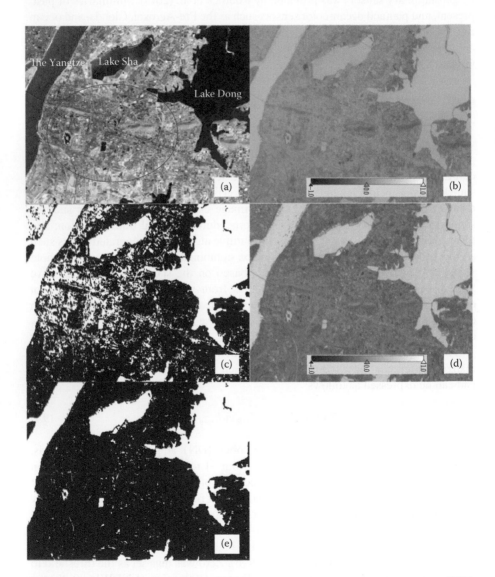

FIGURE 9.2   NDWI and MNDWI images for part of Wuhan City, China. (a) Landsat5/TM image of part of Wuhan City (RGB: 541), acquired on October 24, 2003. (b) NDWI image of the same area. (c) Extracted water bodies from the NDWI image. (d) MNDWI image of the same area. (e) Extracted water bodies from the MNDWI image.

To further suppress these built-up land noises, Xu (2006) proposed a modified NDWI (MNDWI), which uses a SWIR band instead of the NIR band in the NDWI. The MNDWI can be expressed as follows:

$$MNDWI = (d_{green} - d_{SWIR})/(d_{green} + d_{SWIR}), \qquad (9.2)$$

where d_{SWIR} is the DN value at a SWIR band such as Landsat5/TM band 5. Figure 9.2d and e show the MNDWI image for the same Landsat5/TM sub-scene and extracted water bodies from the MNDWI image, respectively. Compared to the NDWI image (Figure 9.2b), most of the MNDWI values of the built-up lands became zero or negative values and thus were removed from the water bodies (Figure 9.2e). This is because both built-up land and water reflect green light more than they reflect NIR light, but built-up land also reflects much greater light at the SWIR band than that at the green band, which is different from water features (Xu 2006). In addition, since the soil and vegetation still reflect more SWIR light than green light, the MNDWI can keep negative values for them like the NDWI does.

Although the MNDWI has taken the normalization form, which can reduce shadow noise (Xu 2006), the MNDWI still suffers from mountain and cloud shadows when a satellite image includes these types of surfaces. Figure 9.3a through c show a Landsat5/TM sub-scene of Lake Maninjau (Indonesia), an MNDWI image of the lake and its surrounding area, and the extracted water bodies from the MNDWI image, respectively, showing that many cloud and mountain shadows are misclassified as water bodies. This is because spectral features of cloud and mountain shadows at green and SWIR bands are similar to those of water bodies and thus show positive MNDWI values (Figure 9.4).

Verpoorter et al. (2012) suggested the use of digital elevation model data with the hillshade algorithm and ArcGIS hydrology tools in ArcMap program to reduce the noise attributed to cloud and mountain shadows. It was also found that these cloud and mountain shadowed pixels still show negative NDWI values, indicating that the NDWI also has the potential to suppress these shadow noises (Figure 9.3d and e). This is because the background of these shadowed pixels is vegetation (Figure 9.3a), which reflects more NIR light than SWIR light (Figure 9.4). Table 9.2 shows value ranges of NDWI and MNDWI for several different land cover types; only water features have positive values for both NDWI and MNDWI.

These findings suggest that the combined use of the MNDWI and NDWI can efficiently suppress not only built-up noise but also vegetation shadow noise attributed to clouds and mountains. Figure 9.3f shows the extracted water bodies by combining the MNDWI image with the NDWI image (i.e., if the MNDWI > 0 and the NDWI > 0, then = waters; otherwise, non-waters). The results demonstrate that most of the cloud and mountain noises were markedly suppressed. However, a misclassification owing to a different type of cloud remained (white pixels surrounded by a circle in Figure 9.3f).

An example spectrum of this type of cloud is also shown in Figure 9.4 (i.e., cloud 2 in the figure). Similar to the water features, this cloud type showed NDWI and MNDWI values that were both larger than zero. However, the spectral brightness of both cloud types is greater than those of other land cover types, and thus it is easy to differentiate the cloud from other land cover types using a brightness index such as the Modified Brightness Index proposed by Verpoorter et al. (2012).

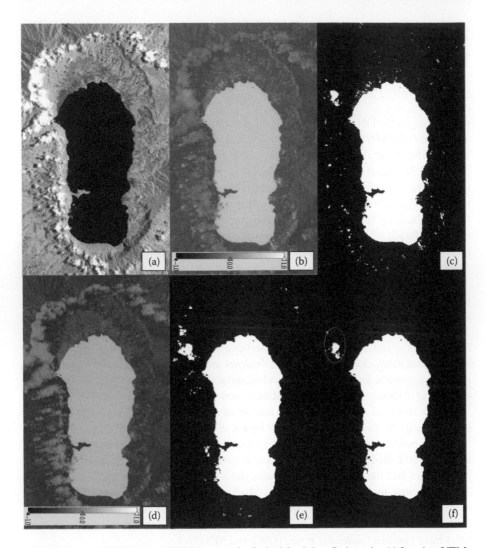

FIGURE 9.3 MNDWI and NDWI images for Lake Maninjau, Indonesia. (a) Landsat5/TM image of Lake Maninjau (RGB 541), acquired on September 11, 1989. (b) MNDWI image of the area. (c) Extracted water bodies from the MNDWI image. (d) NDWI image of the same area. (e) Extracted water bodies from the NDWI image. (f) Extracted water bodies from the combination of MNDWI and NDWI images.

Setting an appropriate threshold for the spectral water index is a challenge for accurately delineating water bodies from other land cover features. McFeeters (1996) and Xu (2006) set zero as the default threshold value for water delineation. However, Xu (2006) also found that a more accurate water body extraction could be achieved if the zero threshold value was adjusted slightly. The variation of threshold value is mainly attributed to the mixed pixel problem, in which a pixel is considered to contain both water and land features.

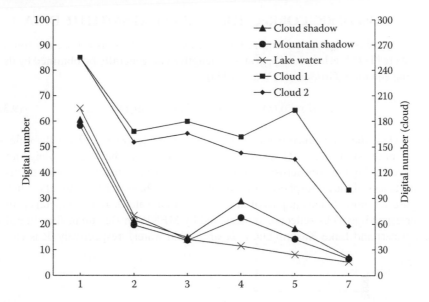

FIGURE 9.4   Spectral reflectance patterns of cloud shadow, mountain shadow, lake water, and clouds (the right *y* axis) obtained from Landsat5/TM image of Lake Maninjau, Indonesia, acquired on September 11, 1989.

TABLE 9.2
NDWI and MNDWI Value Ranges for Different Land Cover Types

	Water	Vegetation	Soil	Built-Up	Cloud Shadow	Mountain Shadow
NDWI	+	0 or −	0 or −	+	0 or −	0 or −
MNDWI	+	0 or −	0 or −	0 or −	+	+

Using simulation data, Ji et al. (2009) extensively investigated how the threshold values of several spectral water indices (including McFeeters' NDWI and Xu's MNDWI) were influenced by the fractions of soil and vegetation at a given water fraction. They found that the MNDWI shows the most stable threshold at a given water fraction, and a zero threshold value was suitable only for extracting pixels with a water fraction close to 100% for both the NDWI and MNDWI. It can thus be considered that the zero threshold value of the MNDWI or NDWI probably provides an underestimated water area compared to the true water body, because the shoreline of a water body is probably a mix of water and land features owing to the image's spatial resolution, especially for the coarse spatial resolution satellite images (e.g., those provided by SeaWiFS, MODIS, and MERIS). It should also be noted that optical shallow water areas, water areas covered by aquatic vegetation, and water areas with high turbidity caused by strong algal blooms and mineral particles can also provide negative values of spectral water indices and thus influence accurate water body delineations (Verpoorter et al. 2012).

9.4 ATMOSPHERIC CORRECTION FOR USING SATELLITE DATA

For a water pixel, the optical signal received by a satellite sensor (i.e., the top of atmosphere [TOA] reflectance, ρ_{toa}) at wavelength λ can generally be obtained by the following equation (Gordon and Wang 1994):

$$\rho_{toa}(\lambda) = \rho_r(\lambda) + [\rho_a(\lambda) + \rho_{ra}(\lambda)] + t(\lambda)\rho_w(\lambda), \qquad (9.3)$$

where $\rho_r(\lambda)$ is the reflectance from Rayleigh scattering, $[\rho_a(\lambda) + \rho_{ra}(\lambda)]$ is the reflectance from the sum of aerosol scattering and the interaction between Rayleigh and aerosol scattering (i.e., the aerosol multiple-scattering reflectance), $t(\lambda)$ is the diffuse transmittances of the atmospheric column, and $\rho_w(\lambda)$ is the water-leaving reflectance.

Figure 9.5 shows two examples of in situ–measured water-leaving reflectance and the corresponding TOA reflectance measured by MERIS in Deadman Bay, Florida (clear water) and Lake Kasumigaura, Japan (turbid water), respectively. It is clear

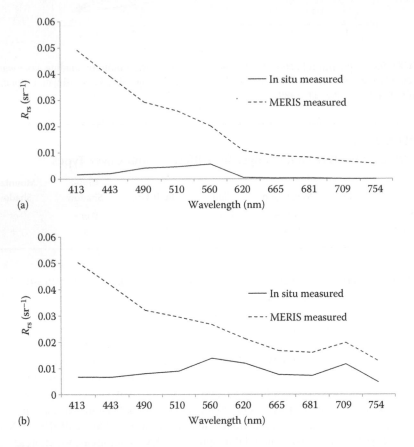

FIGURE 9.5 Examples of in situ–measured water-leaving reflectance and corresponding TOA reflectance measured by MERIS in Deadman Bay, Florida (a) and Lake Kasumigaura, Japan (b).

in these examples that the atmosphere contributes much more signal than the water body itself. According to the IOCCG (2000, 2010), in most water areas, only 10% of the signals recorded by the satellite sensors originate from water bodies in the visible wavelengths (e.g., Figure 9.5a). It is therefore essential to carry out atmospheric correction before applying the satellite data to quantitative estimations of water quality parameters.

The key point for atmospheric correction is to estimate the aerosol type and amount for a given pixel because the Rayleigh-scattering component (ρ_r) can be predicted a priori. The most general approach for estimating the aerosol contribution is to use two specific bands (or reference bands) for which the water-leaving reflectance (ρ_w) can be assumed to be zero. The aerosol contribution at these two specific bands can be estimated as follows:

$$[\rho_a(\lambda_1) + \rho_{ra}(\lambda_1)] = \rho_{toa}(\lambda_1) - \rho_r(\lambda_1) \tag{9.4}$$

$$[\rho_a(\lambda_2) + \rho_{ra}(\lambda_2)] = \rho_{toa}(\lambda_2) - \rho_r(\lambda_2). \tag{9.5}$$

Two specific bands are required because we need to estimate not only the magnitude of the aerosol's contribution but also its dependence on wavelength for extrapolating (or interpolating) the aerosol contribution to the other shorter bands. For this, the estimated aerosol multiple scatterings must first be converted to aerosol single scatterings (ρ_{as}) at the same bands through a look-up table (LUT) as follows (Gordon and Wang 1994):

$$[\rho_a(\lambda_1) + \rho_{ra}(\lambda_1)] \rightarrow \rho_{as}(\lambda_1) \tag{9.6}$$

$$[\rho_a(\lambda_2) + \rho_{ra}(\lambda_2)] \rightarrow \rho_{as}(\lambda_2). \tag{9.7}$$

The atmospheric correction parameter epsilon (i.e., $\varepsilon(\lambda_1, \lambda_2)$) is then calculated as follows:

$$\varepsilon(\lambda_1, \lambda_2) = \rho_{as}(\lambda_1)/\rho_{as}(\lambda_2). \tag{9.8}$$

The values of $\varepsilon(\lambda_1, \lambda_2)$ correspond to different aerosol models. For a given aerosol model, epsilon values for other bands can be extrapolated using the following equation:

$$\varepsilon(\lambda, \lambda_2) = \exp[\ln(\varepsilon(\lambda_1, \lambda_2))/(\lambda_2 - \lambda_1)^*(\lambda_2 - \lambda)]. \tag{9.9}$$

Aerosol single scatterings for all wavelengths ($\rho_{as}(\lambda)$) can then be estimated using the following equation:

$$\rho_{as}(\lambda) = \rho_{as}(\lambda_2)^*\varepsilon(\lambda, \lambda_2). \tag{9.10}$$

The obtained aerosol single scatterings can be reconverted to the corresponding aerosol multiple scatterings for all wavelengths through the LUT prepared in advance:

$$\rho_{as}(\lambda) \rightarrow [\rho_a(\lambda) + \rho_{ra}(\lambda)]. \tag{9.11}$$

Finally, the water-leaving reflectance ($\rho_w(\lambda)$) at all wavelengths can be derived as follows:

$$\rho_w(\lambda) = (\rho_{toa}(\lambda) - \{\rho_r(\lambda) + [\rho_a(\lambda) + \rho_{ra}(\lambda)]\})/t(\lambda). \tag{9.12}$$

For clear waters (e.g., open oceans, or Figure 9.5a), the water-leaving reflectance at NIR bands can be assumed to be zero owing to the very high absorption of pure water and little or no contribution from suspended particles in these bands. Generally, SeaWiFS bands 7 and 8, MODIS bands 15 and 16, MERIS bands 12 and 13, and GOCI bands 7 and 8 can be used as two specific bands (i.e., λ_1 and λ_2) for the purpose of atmospheric correction. However, for turbid waters (e.g., some inland and coastal water, or Figure 9.5b), these two specific NIR bands should be shifted to longer wavelengths such as SWIR bands because high concentrations of particulate matters make the water-leaving reflectance in the NIR bands nonnegligible (Wang and Shi 2005). At the SWIR bands, even for turbid water, water-leaving reflectance can still be assumed to be zero because the absorption of pure water at these wavelengths is much higher than that at NIR bands (Shi and Wang 2009).

Unfortunately, there is no existing satellite sensor with SWIR bands for ocean-color remote sensing. Although the MODIS land bands at SWIR wavelengths (e.g., bands 5 and 7) have shown potential for atmospheric correction over highly turbid water (Wang and Shi 2007), SNR values of the two SWIR bands that are too low make atmospherically corrected water-leaving reflectance at visible bands quite noisy (Hu et al. 2012a; Wang and Shi 2012).

Another approach for solving the atmospheric correction problem over turbid waters is to estimate the water-leaving reflectance at two NIR reference bands by a bio-optical model instead of assuming them to be zero (e.g., Bailey et al. 2010; Hu et al. 2000; Jaelani et al. 2015; Stumpf et al. 2003; Wang et al. 2012). The estimated water-leaving reflectance values at the two NIR reference bands are then removed from the corresponding TOA reflectance to meet Equations 9.4 and 9.5. This approach is available for the satellite sensors without SWIR bands (e.g., SeaWiFS, MERIS, and GOCI).

A summary of existing atmospheric correction algorithms for turbid waters can be found in papers by the IOCCG (2010) and Jaelani et al. (2013). Most of these algorithms have been integrated into widely used software packages such as SeaDAS (SeaWiFS Data Analysis System; http://seadas.gsfc.nasa.gov/) and BEAM (Basic ERS & ENVISAT AATSR and MERIS; http://www.brockmann-consult.de/cms /web/beam), which are easily available and well-supported platforms.

9.5 ESTIMATION OF WATER QUALITY PARAMETERS FROM REMOTE SENSING DATA

9.5.1 BIO-OPTICAL MODEL

The fundamental relationship for retrieving water quality parameters is the bio-optical model, which can be expressed by the following equations (Gordon et al. 1988):

$$r_{rs}(\lambda) = g_0\mu(\lambda) + g_1\mu(\lambda)^2 \tag{9.13}$$

$$\mu(\lambda) = b_b(\lambda)/[a(\lambda) + b_b(\lambda)], \tag{9.14}$$

where $g_0 = 0.089$, $g_1 = 0.125$, and $a(\lambda)$ and $b_b(\lambda)$ are the total absorption and backscattering coefficients, respectively. $r_{rs}(\lambda)$ is the subsurface remote sensing reflectance, which can be calculated from the above-surface remote sensing reflectance ($R_{rs}(\lambda)$) as follows (Lee et al. 2002):

$$r_{rs}(\lambda) = R_{rs}(\lambda)/[0.52 + 1.7\,R_{rs}(\lambda)]. \tag{9.15}$$

$a(\lambda)$ and $b_b(\lambda)$ can be further expressed as the sum of the water constituents' absorption and the backscattering coefficients, respectively, as follows (Mobley 1994):

$$a(\lambda) = a_w(\lambda) + a_{ph}(\lambda) + a_{dg}(\lambda) \tag{9.16}$$

$$b_b(\lambda) = b_{bw}(\lambda) + b_{bp}(\lambda), \tag{9.17}$$

where $a_w(\lambda)$, $a_{ph}(\lambda)$, and $a_{dg}(\lambda)$ are the absorption coefficients of pure water, phytoplankton, and the combination of detritus (often referred to as a_{NAP}) and gelbstoff (often referred to as a_{CDOM}), respectively, and $b_{bw}(\lambda)$ and $b_{bp}(\lambda)$ are the backscattering coefficients of pure water and particulate matter, respectively. The parameter $b_{bp}(\lambda)$ is often further expressed as the sum of backscattering coefficients of phytoplankton ($b_{bph}(\lambda)$) and nonalgal particles ($b_{bNAP}(\lambda)$).

9.5.2 ESTIMATIONS OF THE BIOGEOCHEMICAL PARAMETERS OF WATER BODIES

Mouw et al. (2015) recommended a minimum list for desired biogeochemical parameters from remote sensing: the concentrations of chlorophyll-a (Chl-a), total suspended matter, particulate organic matter, particulate inorganic matter, dissolved organic matter, and dissolved inorganic matter. Algorithms for estimating Chl-a have been well developed (Matthews 2011; Odermatt et al. 2012).

Two types of algorithms have been widely used for estimating Chl-a from remote sensing data: band ratio–based and baseline-based algorithms (Table 9.3). The band ratio–based algorithm can be generally expressed as

$$\text{Chl-a} = f(\text{band ratio}), \tag{9.18}$$

TABLE 9.3

Summary for Several Representative Algorithms for Estimating Chlorophyll-a (Chl-a) Concentration

Category	Algorithm	Band Combination (Band Center, nm)	Advantage	Disadvantage	References
Band ratio based	Generally		Can reduce effects from non-phytoplankton matters	Intolerance to spectrally related errors of remote sensing reflectance (e.g., instrument noise, imperfect atmospheric correction)	
	OCx	443, 490, 510, 555	Suitable for low Chl-a between 0.3 and 10 mg m^{-3}	Intolerance to optically complex waters	O'Reilly et al. 1998
	Two-band model	665, 709	Suitable for moderately turbid waters (e.g., remote sensing reflectance at 865 nm between 0.0001 and 0.001)	Other turbid waters	Gilerson et al. 2010
	Three-band model	665, 709, 754	Suitable for very turbid waters (e.g., remote sensing reflectance at 865 nm between 0.001 and 0.01)	Other turbid waters	Gitelson et al. 2008
	SAMO-LUT	550, 665, 709, 754	Suitable for very turbid and even for extremely turbid waters	Other turbid waters	Yang et al. 2011
Baseline based	Generally		Tolerance to spectrally related errors of remote sensing reflectance	Intolerance to optically complex waters	
	CIA	443, 555, 670	Suitable for very low Chl-a ($\leq$0.25 mg m^{-3})	Other Chl	Hu et al. 2012b
	FLH	665, 681, 709	Suitable for moderate Chl-a (10–30 mg m^{-3})	Other Chl	Gower 1980
	MCI	681, 709, 754	Suitable for high Chl-a (>25 mg m^{-3})	Other Chl	Gower et al. 2005
	FAI	667, 859, 1240	Suitable for extremely high Chl-a (e.g., surface scums)	Other Chl	Hu 2009

where f represents a function that can be linear or nonlinear. For clear or optically simple waters (e.g., open oceans), the band ratio of R_{rs} at blue and green bands is widely used for estimating Chl-a (e.g., OCx form-based algorithms; O'Reilly et al. 1998). This is because the R_{rs} at the blue band is strongly related to chlorophyll absorption, whereas at the green band, it is relatively insensitive to changes in Chl-a values owing to the weak absorption of chlorophyll at this band, and it can thus be used as a reference.

However, for optically complex waters with moderate turbidities, the R_{rs} values in the red and NIR bands are preferred for the retrieval of Chl-a (Dekker 1993; Gitelson and Kondratyev 1991; Han and Rundquist 1997). This is because the R_{rs} values at the blue and green bands are affected not only by phytoplankton but also by other constituents (i.e., NAP and CDOM) (Gitelson et al. 2009). The use of the band ratio of $R_{rs}(\text{NIR})/R_{rs}(\text{red})$ can minimize the effects from NAP and CDOM based on four assumptions: (i) at the red band, $a_{ph} + a_w \gg a_{NAP} + a_{CDOM}$; (ii) at the NIR band, $a_w \gg a_{ph} + a_{NAP} + a_{CDOM}$; (iii) at both the red and NIR bands, $a \gg b_b$; and (iv) for backscattering coefficients, $b_b(\text{red}) \approx b_b(\text{NIR})$ (Gitelson et al. 2008). According to the above assumptions, the $R_{rs}(\text{NIR})/R_{rs}(\text{red})$ can be approximated as

$$R_{rs}(\text{NIR})/R_{rs}(\text{red}) \approx [a_{ph}(\text{red}) + a_w(\text{red})]/a_w(\text{NIR}). \quad (9.19)$$

Since $a_w(\lambda)$ can be considered constant, the $R_{rs}(\text{NIR})/R_{rs}(\text{red})$ value varies only with $a_{ph}(\text{red})$, which is strongly related to Chl-a. For example, Gilerson et al. (2010) proposed a two-band model, which uses the band ratio of R_{rs} at 709 and 665 nm (i.e., $R_{rs}(709)/R_{rs}(665)$) to estimate Chl-a in moderately turbid waters.

For optically complex waters with higher turbidities, the above assumptions from (i) to (iii) will become invalid (Dall'Olmo et al. 2005; Gitelson et al. 2008). Therefore, a three-band index that uses one R_{rs} at the red band (around 665 nm) and two R_{rs} at NIR bands (around 709 and 754 nm) was designed to address the effects of NAP and CDOM in these waters (Dall'Olmo et al. 2003; Gitelson et al. 2008). First, the R_{rs} around the wavelengths 665 and 709 nm were inverted, and their difference was calculated to minimize the absorption effects of NAP and CDOM (i.e., $R_{rs}^{-1}(665) - R_{rs}^{-1}(709)$). This step is based on two assumptions: (i) $a_{NAP}(665) \approx a_{NAP}(709)$ and $a_{CDOM}(665) \approx a_{CDOM}(709)$, and (ii) $b_b(665) \approx b_b(709)$. Therefore, $R_{rs}^{-1}(665) - R_{rs}^{-1}(709)$ can be approximated as $[a_{ph}(665) + a_w(665) - a_{ph}(709) - a_w(709)]/b_b$. Then, the R_{rs} at the third wavelength around 754 nm is used to further minimize the effects by b_b. This step is based on another two assumptions: (i) $b_b(665) \approx b_b(709) \approx b_b(754)$, and (ii) $a_w(754) \gg a_{ph}(754) + a_{NAP}(754) + a_{CDOM}(754) + b_b(754)$. The three-band index is thus designed as $\left[R_{rs}^{-1}(665) - R_{rs}^{-1}(709) \right] \times R_{rs}(754)$, and can be approximated according to the above assumptions as

$$\left[R_{rs}^{-1}(665) - R_{rs}^{-1}(709) \right] \times R_{rs}(754) \approx [a_{ph}(665) + a_w(665)$$
$$- a_{ph}(709) - a_w(709)]/a_w(754). \quad (9.20)$$

Similar to the two-band index (Equation 9.19), the three-band index varies only with a_{ph} and thus can provide accurate Chl-a estimations in more turbid waters.

However, the assumption of $a_w(754) \gg a_{ph}(754) + a_{NAP}(754) + a_{CDOM}(754) + b_b(754)$ will be invalid in extremely turbid waters (Le et al. 2009; Yang et al. 2010). A four-band index (Le et al. 2009), an enhanced three-band index (Yang et al. 2010), and the Semi-Analytical Model Optimizing and Look-Up Tables (SAMO-LUT) algorithm (Yang et al. 2011) were then proposed for this type of waters. Since the independent variable in the SAMO-LUT algorithm is still a three-band index, the SAMO-LUT algorithm is considered a type of band ratio–based algorithm.

The second type of Chl-a estimation algorithm, that is, the baseline-based algorithms, can be generally expressed as

$$\text{Chl-a} = f(\text{RH}), \tag{9.21}$$

where RH is the relative height of a reflectance peak (or trough, $R_{rs}(\lambda_2)$) from a baseline formed linearly between R_{rs} values at two adjacent bands (i.e., $R_{rs}(\lambda_1)$ and $R_{rs}(\lambda_3)$ with $\lambda_1 < \lambda_2 < \lambda_3$):

$$\text{RH} = R_{rs}(\lambda_2) - [R_{rs}(\lambda_1) + (\lambda_2 - \lambda_1)/(\lambda_3 - \lambda_1)*(R_{rs}(\lambda_3) - R_{rs}(\lambda_1))]. \tag{9.22}$$

A number of baseline-based algorithms have been proposed for estimating Chl-a in waters with various trophic states (Table 9.3). For example, Hu et al. (2012b) proposed a color index–based algorithm (CIA) for ultra-oligotrophic oceans; Gower (1980) developed a fluorescence line height algorithm for mesotrophic and eutrophic waters; Gower et al. (2005) and Hu (2009) suggested a maximum chlorophyll index (MCI) and a floating algae index, respectively, for hypertrophic waters.

Generally, the tolerance to spectrally related errors of R_{rs} (e.g., instrument noise and imperfect atmospheric correction) is a great advantage of these baseline-based algorithms (Hu 2009; Hu et al. 2012b). However, the baseline-based algorithms often suffer from effects of non-phytoplankton matter (e.g., NAP and CDOM). In contrast, some band ratio–based algorithms can reduce the effects from NAP and CDOM, but they are intolerant to spectrally related errors of R_{rs}. This is probably because of their form designs (Equations 9.19, 9.20, and 9.22; Hu et al. 2012b). Table 9.3 gives a summary of several representative algorithms for Chl-a estimation in various waters.

Other algorithms for estimating water quality parameters are described in two comprehensive reviews (Matthews 2011; Odermatt et al. 2012).

9.5.3 ESTIMATIONS OF THE INHERENT OPTICAL PROPERTIES OF WATER BODIES

The recommended standard products of inherent optical properties (IOPs) from remote sensing are the total absorption coefficient ($a(\lambda)$), the absorption coefficient of each constituent (i.e., $a_{ph}(\lambda)$, $a_{NAP}(\lambda)$, and $a_{CDOM}(\lambda)$), and the backscattering coefficient of particulate matters ($b_{bp}(\lambda)$) (Mouw et al. 2015). The $b_{bp}(\lambda)$ is usually expressed as follows (Lee et al. 2002):

$$b_{bp}(\lambda) = b_{bp}(\lambda_0)(\lambda_0/\lambda)^Y, \tag{9.23}$$

where λ_0 is the reference wavelength and Y is the spectral slope. For clear waters, $\lambda_0 = 555$ nm was used (Lee et al. 2002). However, for turbid waters, $\lambda_0 = 750$ nm was suggested (Yang et al. 2013). From Equations 9.14 and 9.17, $b_{bp}(\lambda_0)$ can be expressed as

$$b_{bp}(\lambda_0) = \mu(\lambda_0)a(\lambda_0)/(1 - \mu(\lambda_0)) - b_{bw}(\lambda_0). \tag{9.24}$$

For clear waters, $a(\lambda_0)$ can be obtained from an empirical relationship, which is the function of r_{rs} at 443, 490, 555, and 667 nm (Lee et al. 2002). For turbid waters, since $\lambda_0 = 750$ nm, $a(\lambda_0) \approx a_w(\lambda_0)$ can be assumed because the water absorption at this wavelength is much larger than that from other constituents (Yang et al. 2013).

The spectral slope Y can be estimated by an empirical equation for clear waters (Equation 9.25; Lee et al. 2002) and a semi-analytical equation for turbid waters (Equation 9.26; Yang et al. 2013):

$$Y = 2.2(1 - 1.2\exp(-0.9r_{rs}(443)/r_{rs}(555))) \tag{9.25}$$

$$Y = -372.99(\log[\mu(750)/\mu(780)])^2 + 37.286(\log[\mu(750)/\mu(780)]) + 0.84. \tag{9.26}$$

Then, the total absorption coefficient can be obtained as

$$a(\lambda) = (1 - \mu(\lambda))*(b_{bw}(\lambda) + b_{bp}(\lambda))/\mu(\lambda). \tag{9.27}$$

It is still much more challenging to separate the absorption coefficient of each constituent from the total absorption coefficient because more assumptions are required. A detailed review of IOP retrievals was provided by the IOCCG (2006).

9.5.4 ESTIMATIONS OF THE APPARENT OPTICAL PROPERTIES OF WATER BODIES

Except for $R_{rs}(\lambda)$, the recommended standard products of apparent optical properties from remote sensing are the diffuse attenuation coefficient ($K_d(\lambda)$) and the euphotic zone depth (Z_{eu}) (Mouw et al. 2015). Both of these parameters are of great importance in studying and modeling the physical, chemical, and biological processes in water bodies (Kirk 1994; Mobley 1994). Semi-analytical algorithms have been proposed to remotely estimate $K_d(\lambda)$ and Z_{eu} based on the retrievals of the total absorption and backscattering coefficients (i.e., a and b_b; Lee et al. 2005a, 2007).

$K_d(\lambda)$ is commonly defined as the vertically averaged value of the spectral diffuse attenuation coefficient at depth z (i.e., $K_d(\lambda, z)$) over the water layer (Lee et al. 2005a). $K_d(\lambda, z)$ is defined as follows (Gordon et al. 1980):

$$K_d(\lambda, z) = -(1/E_d(\lambda, z))*(dE_d(\lambda, z)/dz), \tag{9.28}$$

where $E_d(\lambda, z)$ is the downwelling irradiance at depth z. The maximum depth for vertically averaging $K_d(\lambda, z)$ values is generally defined as the depth where the downwelling irradiance is reduced to 10% of the surface irradiance (Lee et al. 2005a).

A general model of $K_d(\lambda)$ was suggested by Lee et al. (2005a) through numerical simulations of radiative transfer in water bodies:

$$K_d(\lambda) = (1 + 0.005\theta_a)\, a(\lambda) + 4.18(1 - 0.52e^{-10.8\,a(\lambda)})\, b_b(\lambda),\qquad (9.29)$$

where θ_a is the solar zenith angle in air, which can be derived from the location and measuring time.

Z_{eu} is practically defined as the depth at which photosynthetic available radiation (PAR) is 1% of its surface value (Kirk 1994). In other words, if the vertical transmittance of the PAR at a depth z (i.e., $T_{PAR}(z)$) reaches 1%, this depth is the Z_{eu}. $T_{PAR}(z)$ can be derived as follows (Lee et al. 2007):

$$T_{PAR}(z) = \exp[-K_d(PAR, z)^* z],\qquad (9.30)$$

where $K_d(PAR, z)$ is the attenuation coefficient of PAR between the water surface and the depth z, which can be estimated from a semi-analytical model (Lee et al. 2005b):

$$K_d(PAR, z) = K_1 + K_2/(1 + z)^{0.5},\qquad (9.31)$$

where K_1 and K_2 are functions of the total absorption and backscattering coefficients at 490 nm (i.e., $a(490)$ and $b_b(490)$) as well as the solar zenith angle (θ_a):

$$K_1 = [-0.057 + 0.482\sqrt{a(490)} + 4.221b_b(490)][1 + 0.09\sin(\theta_a)]\qquad (9.32)$$

$$K_2 = [0.183 + 0.702a(490) - 2.567b_b(490)][1.465 - 0.667\cos(\theta_a)]\,.\qquad (9.33)$$

Since $a(\lambda)$ and $b_b(\lambda)$ can be derived from R_{rs} (Lee et al. 2002; Yang et al. 2013), the $K_d(\lambda)$ and Z_{eu} values can also be remotely derived. The above models (Equations 9.29 through 9.33) have been successfully applied to both clear oceanic waters (Lee et al. 2005b, 2007) and turbid inland waters (Yang et al. 2014, 2015).

9.6 A HYBRID APPROACH FOR DEVELOPING A UNIVERSAL ALGORITHM

A universal algorithm should have the ability to address various water types ranging from clear ultra-oligotrophic to turbid hypertrophic systems. However, previous studies have shown that each algorithm has its strengths and limitations for different water types. For example, Wang and Shi (2007) reported that the standard NIR atmospheric correction algorithm can provide good-quality ocean-color products for the global open oceans and offshore waters, but the SWIR atmospheric correction algorithm should be used for turbid coastal waters to produce similar quality. Hu et al. (2012b) reported that the CIA can only be applied to estimate Chl-a for ultra-oligotrophic oceans with Chl-a concentrations lower than 0.25 mg m^{-3}, whereas for oceanic waters with Chl-a concentrations higher than 0.3 mg m^{-3}, the OC4 algorithm should be used. Matsushita et al. (2015) also reported that a blue-green algorithm such as OC4 can be used in optically simple lakes, but a red-NIR algorithm such as the two-band model or a three-band model must be used for optically complex turbid lakes to minimize effects from non-phytoplankton matter. These findings indicate that a hybrid approach is desirable to address waters with various optical properties.

Figure 9.6 illustrates a conceptual diagram for a hybrid approach. In addition to candidate estimation algorithms for atmospheric correction and water quality parameters (e.g., Chl-a), another type of algorithm is required for selecting the most appropriate algorithm for different water types. Several studies have attempted to address this issue. For example, Wang and Shi (2007) proposed the use of a turbid water index to identify productive or turbid waters for which the SWIR atmospheric correction should be used instead of the standard NIR atmospheric correction algorithm. Gomez et al. (2011) developed two normalized difference indices, that is, $[R_{rs}(560) - R_{rs}(443)]/[R_{rs}(560) + R_{rs}(443)] < 0$ and $[R_{rs}(709) - R_{rs}(665)]/[R_{rs}(709) + R_{rs}(665)] > 0$ for first separating Mediterranean lakes of the European Union into two water types, and then different Chl-a retrieval algorithms are selected for each

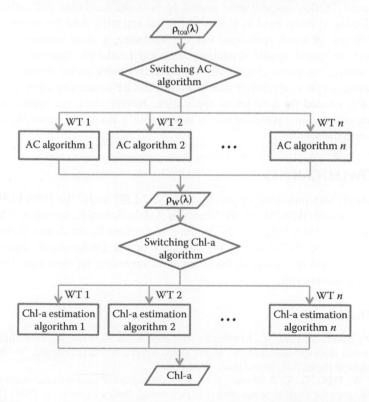

FIGURE 9.6 Conceptual diagram for a hybrid approach. AC, atmospheric correction; Chl-a, chlorophyll-a concentration; WT, water type.

water type to obtain the most accurate Chl-a concentrations for the whole study area. Moore et al. (2014) proposed an optical water type (OWT) framework for selecting and blending two Chl-a retrieval algorithms. Matsushita et al. (2015) suggested a hybrid algorithm that uses an MCI to switch three different types of Chl-a estimation algorithms for clear, moderate-turbid, and high-turbid waters, respectively. These attempts indicate that the hybrid approach has significant potential toward the development of a universal algorithm for monitoring water quality with remote sensing image data.

9.7   CONCLUSIONS

The remote sensing technique has been widely used for detecting, mapping, and quantifying water bodies. In this chapter, we introduced algorithms that can be used to obtain water quality data from remote sensing, and we described major suitable satellite sensors, water body delineation, atmospheric correction, and water quality estimation. These remotely obtained water quality data may be useful for meeting Sustainable Development Goal 6.3 proposed by the United Nations Open Working Group. Although the usefulness of remote sensing data is widely recognized, a few cases remain in which these data have been operationally used in water resource management. Difficulties still exist regarding both the hardware and software of remote sensing systems used to obtain additional scientific data for studies of the sustainable use of water resources. For the hardware, a more powerful satellite sensor with sufficient spatial resolution, temporal resolution, spectral resolution, and radiometric sensitivity is necessary for collecting more useful information. For the software, a hybrid algorithm that combines the advantages of each individual algorithm is needed for a universal application. Nevertheless, the remote sensing technique will remain a powerful tool in sustainability studies by providing images of Earth on a daily basis.

ACKNOWLEDGMENTS

This research was supported in part by JSPS and LIPI under the JSPS-LIPI Joint Research Program (BDB28034), the Grants-in-Aid for Scientific Research of MEXT from Japan (No. 25420555), and also by the Environment Research and Technology Development Fund (S-9-4-(1)) of the Ministry of the Environment, Japan. The authors also would like to thank the anonymous reviewers for their valuable comments on the manuscript.

REFERENCES

Ayres, W., Busia, A., Hirji, R., Lintner, S., McCalla, A., & Robelus, R., 1996. Integrated lake and reservoir management: World Bank Approach and Experience, World Bank Technical Paper, 358, World Bank, Washington.
Bailey, S. W., Franz, B. A., & Werdell, P. J., 2010. Estimation of near-infrared water-leaving reflectance for satellite ocean color data processing. *Optics Express*, 18, 7521–7527.
Bissett, W. P., Amone, R. A., Davis, C. O., Dickey, T. D., Dye, D., Kohler, D. D. et al., 2004. From meters to kilometers. *Oceanography*, 17, 32–42.

Ciotti, A. M., & Bricaud, A., 2006. Retrievals of a size parameter for phytoplankton and spectral light absorption by colored detrital matter from water-leaving radiances at SeaWiFS channels in a continental shelf region off Brazil. *Limnology and Oceanography: Methods*, 4, 237–253.

Dall'Olmo, G., Gitelson, A., & Rundquist, D. C., 2003. Towards a unified approach for remote estimation of chlorophyll-a in both terrestrial vegetation and turbid productive waters. *Geophys. Res. Lett.*, 30.

Dall'Olmo, G., Gitelson, A., Rundquist, D. C., Leavitt, B., Barrow, T., & Holz, J. C., 2005. Assessing the potential of SeaWiFS and MODIS for estimating chlorophyll concentration in turbid productive waters using red and near-infrared bands. *Remote Sensing of Environment*, 96, 176–187.

Dekker, A. G., 1993. Detection of water quality parameters for eutrophic waters by high resolution remote sensing. Ph.D. dissertation, Vrije Universiteit, Amsterdam, The Netherlands.

Dekker, A. G. & Hestir, E. L., 2012. Evaluating the feasibility of systematic inland water quality monitoring with satellite remote sensing, CSIRO: Water for a Healthy Country National Research Flagship.

Dekker, A. G., Vos, R. J., & Peters, S. W. M., 2002. Analytical algorithms for lake water TSM estimation for retrospective analyses of TM and SPOT sensor data. *International Journal of Remote Sensing*, 23, 15–35.

Franz, B., Bailey, S. W., Kuring, N., & Werdell, P. J., 2015. Ocean color measurements with the Operational Land Imager on Landsat-8: Implementation and evaluation in SeaDAS. *Journal of Applied Remote Sensing*, 9, 096070, 1–16.

Franz, B., Kwiatowska, E. J., Meister, G., & McClain, C. R., 2008. Moderate resolution imaging spectroradiometer on Terra: Limitations for ocean color applications. *Journal of Applied Remote Sensing*, 2, 023525, 1–17.

Gilerson, A., Gitelson, A., Zhou, J., Gulrin, D., Moses, W., Ioannou, I., & Ahmed, S., 2010. Algorithms for remote sensing of chlorophyll-a in coastal and inland waters using red and near infrared bands. *Optics Express*, 18, 24109–24125.

Gitelson, A., Dall'Olmo, G., Moses, W., Rundquist, D. C., Barrow, T., Fisher, T. R. et al., 2008. A simple semi-analytical model for remote estimation of chlorophyll-a in turbid waters: Validation. *Remote Sensing of Environment*, 112, 3582–3593.

Gitelson, A., Gurlin, D., Moses, W., & Barrow, T., 2009. A bio-optical algorithm for the remote estimation of the chlorophyll-a concentration in case-2 waters. *Environment Research Letters*, 4 (045003), 5.

Gitelson, A., & Kondratyev, K. Y., 1991. On the mechanism of formation of maximum in the reflectance spectra near 700 nm and its application for remote monitoring of water quality. *Trans. Doklady USSR Acad. Sci.: Earth Sci. Sections* 306, 1–4.

Gomez, J. A. D., Alonso, C. A., & Garcia, A. A., 2011. Remote sensing as a tool for monitoring water quality parameters for Mediterranean Lakes of European Union water framework directive (WFD) and as a system of surveillance of cyanobacterial harmful algae blooms (SCyanoHABs). *Environ. Monit. Assess.*, 181, 317–334.

Gordon, H., Brown, O., Evans, R., Brown, R., Smith, R., Baker, K. et al., 1988. A semi-analytic radiance model of ocean color. *Journal of Geophysical Research*, 93, 10909–10924.

Gordon, H. R., Smith, R. C., & Zaneveld, J. R. V., 1980. Introduction to ocean optics. *Proc. SPIE Soc. Pot. Eng.*, 6, 1–43.

Gordon, H. R., & Wang, M., 1994. Retrieval of water-leaving radiance and aerosol optical thickness over the oceans with SeaWiFS: A preliminary algorithm. *Applied Optics*, 33, 443–452.

Gower, J., 1980. Observations of in-situ fluorescence of chlorophyll a in Saanich Inlet. *Boundary-Layer Meteorology*, 18, 235–245.

Gower, J., King, S., Borstad, G., & Brown, L., 2005. Detection of intense plankton blooms using the 709 nm band of the MERIS imaging spectrometer. *International Journal of Remote Sensing*, 26, 2005–2012.

Haddeland, I., Heinke, J., Biemans, H., Eisner, S., Flörke, M., Hanasaki, N., Konzmann, M., Ludwig, F., Masaki, Y., Schewe, J., Stacke, T., Tessler, Z. D., Wada, Y., & Wisser, D., 2014. Global water resources affected by human interventions and climate change. *PNAS*, 111, 3251–3256.

Han, L. H., & Rundquist, D. C., 1997. Comparison of NIR/RED ratio and first derivative of reflectance in estimating algal-chlorophyll concentration: A case study in a turbid reservoir. *Remote Sensing of Environment*, 62, 253–261.

Hovis, W. A., Clark, D. K., Anderson, F., Austin, R. W., Wilson, W. H., Baker, E. T., Ball, D., Gordon, H. R., Mueller, J. L., Al-Sayed, S. Z., Sturm, B., Wrigley, R. C., & Yentsch, C. S., 1980. Nimbus-7 Coastal Zone Color Scanner: System description and initial imagery. *Science*, 210, 60–63.

Hu, C., 2009. A novel ocean color index to detect floating algae in the global oceans. *Remote Sensing of Environment*, 113, 2118–2129.

Hu, C., Carder, K. L., & Muller-Karger, F. E., 2000. Atmospheric correction of SeaWiFS imagery over turbid coastal waters: A practical method. *Remote Sensing of Environment*, 74, 195–206.

Hu, C., Feng, L., Lee, Z., Davis, C. O., Mannino, A., McClain, C. R. et al., 2012a. Dynamic range and sensitivity requirements of satellite ocean color sensors: Learning from the past. *Applied Optics*, 51, 6045–6062.

Hu, C., Lee, Z., & Franz, B., 2012b. Chlorophyll a algorithms for oligotrophic oceans: A novel approach based on three-band reflectance difference. *Journal of Geophysical Research*, 117, C01011 (doi:01010.01029/02011JC007395).

Hu, C. et al., 2010. Moderate Resolution Imaging Spectroradiometer (MODIS) observations of cyanobacteria blooms in Taihu Lake, China. *Journal of Geophysical Research—Oceans*, 115, C04002.

IOCCG, 2000. Remote sensing of ocean colour in coastal and other optically-complex waters. In S. Sathyendranath (Ed.), *Report of the International Ocean-Colour Coordinating Group, No. 3*. Dartmouth, Canada: IOCCG.

IOCCG, 2006. Remote sensing of inherent optical properties: Fundamentals, tests of algorithms, and applications. In Z. P. Lee (Ed.), No. 5. Dartmouth, Canada: IOCCG.

IOCCG, 2010. Atmospheric correction for remotely-sensed ocean-colour products. In M. Wang (Ed.), *Reports of the International Ocean-Colour Coordinating Group, No. 10*. Dartmouth, Canada: IOCCG.

IOCCG, 2012. Mission Requirements for Future Ocean-Colour Sensors. In McClain, C. R. and Meister, G. (Eds.), *Reports of the International Ocean-Colour Coordinating Group, No. 13*. Dartmouth, Canada: IOCCG.

Jaelani, L. M., Matsushita, B., Yang, W., & Fukushima, T., 2013. Evaluation of four MERIS atmospheric correction algorithms in Lake Kasumigaura, Japan. *International Journal of Remote Sensing*, 34, 8967–8985.

Jaelani, L. M., Matsushita, B., Yang, W., & Fukushima, T., 2015. An improved atmospheric correction algorithm for applying MERIS data to very turbid inland waters. *International Journal of Applied Earth Observation and Geoinformation*, 39, 128–141.

Ji, L., Zhang, L., & Wylie, B., 2009. Analysis of dynamic thresholds for the normalized difference water index. *Photogrammetric Engineering & Remote Sensing*, 75, 1307–1317.

Kiefer, I., Odermatt, D., Anneville, O., Wuest, A., & Bouffard, D., 2015. Application of remote sensing for the optimization of in-situ sampling for monitoring of phytoplankton abundance in a large lake. *Science of the Total Environment*, 527–528, 493–506.

Kirk, J. T. O., 1994. *Light and Photosynthesis in Aquatic Ecosystems*. Cambridge: University Press.

Kutser, T., 2004. Quantitative detection of chlorophyll in cyanobacterial blooms by satellite remote sensing. *Limnology and Oceanography*, 49, 2179–2189.

Kutser, T., Metsamaa, L., Strombeck, N., & Vatmae, E., 2006. Monitoring cyanobacterial blooms by satellite remote sensing. *Estuarine, Coastal and Shelf Science*, 67, 303–312.

Le, C., Li, Y., Zha, Y., Sun, D., Huang, C., & Lu, H., 2009. A four-band semi-analytical model for estimating chlorophyll a in highly turbid lakes: The case of Taihu Lake, China. *Remote Sensing of Environment*, 113, 1175–1182.

Lee, Z. P., Carder, K. L., & Arnone, R., 2002. Deriving inherent optical properties from water color: A multi-band quasi-analytical algorithm for optically deep waters. *Applied Optics*, 41, 5755–5772.

Lee, Z. P., Darecki, M., Carder, K. L., Davis, C. O., Stramski, D., & Rhea, W. J., 2005b. Diffuse attenuation coefficient of downwelling irradiance: An evaluation of remote sensing methods. *Journal of Geophysical Research*, 110, C2, C02017.

Lee, Z. P., Du, K., & Arnone, R., 2005a. A model for the diffuse attenuation coefficient of downwelling irradiance. *Journal of Geophysical Research*, 110, C2, C02016.

Lee, Z. P., Shang, S., Hu, C., & Zibordi, G., 2014. Spectral interdependence of remote-sensing reflectance and its implications on the design of ocean color satellite sensors. *Applied Optics*, 53, 3301–3310.

Lee, Z. P., Weidemann, A., Kindle, J., Arnone, R., Carder, K. L., & Davis, C., 2007. Euphotic zone depth: Its derivation and implication to ocean-color remote sensing. *Journal of Geophysical Research*, 112, C03009.

Lobo, F. L., Costa, M. P. F., & Novo, E. M. L. M., 2015. Time-series analysis of Landsat-MSS/TM/OLI images over Amazonian waters impacted by gold mining activities. *Remote Sensing of Environment*, 157, 170–184.

Matsushita, B., Yang, W., Yu, G., Oyama, Y., Yoshimura, K., & Fukushima, T., 2015. A hybrid algorithm for estimating the chlorophyll-a concentration across different trophic states in Asian inland waters. *ISPRS Journal of Photogrammetry and Remote Sensing*, 102, 28–37.

Matthews, M., 2011. A current review of empirical procedures of remote sensing in inland and near-coastal transitional waters. *International Journal of Remote Sensing*, 32, 6855–6899.

McFeeters, S. K., 1996. The use of normalized difference water index (NDWI) in the delineation of open water features. *International Journal of Remote Sensing*, 17, 1425–1432.

Mobley, C., 1994. Light and water: *Radiative Transfer in Natural Waters*. New York: Academic Press.

Moore, T. S., Dowell, M. D., Bradt, S., & Verdu, A. R., 2014. An optical water type framework for selecting and blending retrievals from bio-optical algorithms in lakes and coastal waters. *Remote Sensing of Environment*, 143, 97–111.

Mouw, C. B. et al., 2015. Aquatic color radiometry remote sensing of coastal and inland waters: Challenges and recommendations for future satellite missions. *Remote Sensing of Environment*, 160, 15–30.

Odermatt, D., Gitelson, A., Brando, V. E., & Schaepman, M., 2012. Review of constituent retrieval in optically deep and complex waters from satellite imagery. *Remote Sensing of Environment*, 118, 116–126.

O'Reilly, J., Maritorena, S., Mitchell, B., Siegel, D., Carder, K., Garver, S. et al., 1998. Ocean color chlorophyll algorithms for SeaWiFS. *Journal of Geophysical Research*, 103, 24937–24953.

Oyama, Y., Matsushita, B., Fukushima, T., Matsushige, K., & Imai, A., 2009. Application of spectral decomposition algorithm for mapping water quality in a turbid lake (Lake Kasumigaura, Japan) from Landsat TM data. *ISPRS Journal of Photogrammetry and Remote Sensing*, 64, 73–85.

Pan, X., Mannino, A., Russ, M. E., Hooker, S. B., & Harding, L. W., 2010. Remote sensing of phytoplankton pigment distribution in the United States northeast coast. *Remote Sensing of Environment*, 114, 2403–2416.

Ruddick, K., Neukermans, G., Vanhellemont, Q., & Jolivet, D., 2014. Challenges and opportunities for geostationary ocean colour remote sensing of regional seas: A review of recent results. *Remote Sensing of Environment*, 146, 63–76.

Ryu, J. H., Han, H. J., Cho, S., Park, Y. J., & Ahn, Y. H., 2012. Overview of Geostationary Ocean Color Imager (GOCI) and GOCI Data Processing System (GDPS). *Ocean Science Journal*, 47, 223–233.

Shi, W., & Wang, M., 2009. An assessment of the black ocean pixel assumption for MODIS SWIR bands. *Remote Sensing of Environment*, 113, 1587–1597.

Stumpf, R. P., Arnone, R. A., Gould, R. W., Martinolich, P. M., & Ransibrahmanakul, V., 2003. A partly coupled ocean-atmosphere model for retrieval of water-leaving radiance from SeaWiFS in coastal waters. In SeaWiFS Postlaunch Technical Report Series, NASA Technical Memorandum 2003-206892, Vol. 22, S. B. Hooker and E. R. Firestone (Eds.), 51–59. Greenbelt, MD: NASA Goddard Space Flight Center.

Tassan, S., 1988. The effect of dissolved yellow substance on the quantitative retrieval of chlorophyll and total suspended sediment concentrations from remote measurements of water colour. *International Journal of Remote Sensing*, 9, 787–797.

Tyler, A. N., Svab, E., Preston, T., Presing, M., & Kovacs, W. A., 2006. Remote sensing of the water quality of shallow lakes: A mixture modeling approach to quantifying phytoplankton in water characterized by high-suspended sediment. *International Journal of Remote Sensing*, 27, 1521–1537.

UNEP, 2006. Marine and coastal ecosystems and human wellbeing: A synthesis report based on the findings of the Millennium Ecosystem Assessment. UNEP, 76 pp.

United Nations Open Working Group, 2014. Outcome document—Open Working Group on Sustainable Development Goals (19th July 2014). United Nations, New York, New York. URL: http://sustainabledevelopment.un.org/content/documents/1579SDGs%20Proposal.pdf.

Verpoorter, C., Kutser, T., Seekell, D. A., & Tranvik, L. J., 2014. A global inventory of lakes based on high-resolution satellite imagery. *Geophysical Research Letters*, 41, 6396–6402.

Verpoorter, C., Kutser, T., & Tranvik, L. J., 2012. Automated mapping of water bodies using Landsat multispectral data. *Limnology and Oceanography: Methods*, 10, 1037–1050.

Wang, M., 2007. Remote sensing of the ocean contributions from ultraviolet to near-infrared using the shortwave infrared bands: Simulations. *Applied Optics*, 46, 1535–1547.

Wang, M., & Shi, W., 2005. Estimation of ocean contribution at the MODIS near-infrared wavelengths along the east coast of the US: Two case studies. *Geophysical Research Letters*, 32, L13606.

Wang, M., & Shi, W., 2007. The NIR–SWIR combined atmospheric correction approach for MODIS ocean color data processing. *Optics Express*, 15, 15722–15733.

Wang, M., & Shi, W., 2012. Sensor noise effects of the SWIR bands on MODIS-derived ocean color products. *IEEE Transactions on Geoscience and Remote Sensing*, 50, 3280–3292.

Wang, M., Shi, W., & Jiang, L., 2012. Atmospheric correction using near-infrared bands for satellite ocean color data processing in the turbid Western Pacific Region. *Optics Express*, 20, 741–753.

Wu, G., Cui, L., Liu L., Chen, F., Fei, T., & Liu, Y., 2015. Statistical model development and estimation of suspended particulate concentrations with Landsat 8 OLI images of Dongting Lake, China. *International Journal of Remote Sensing*, 36, 343–360.

Xu, H., 2006. Modification of normalized difference water index (NDWI) to enhance open water features in remotely sensed imagery. *International Journal of Remote Sensing*, 27, 3025–3033.

Yang, W., Matsushita, B., Chen, J., & Fukushima, T., 2011. Estimating constituent concentrations in case II waters from MERIS satellite data by semi-analytical model optimizing and look-up tables. *Remote Sensing of Environment*, 115, 1247–1259.

Yang, W., Matsushita, B., Chen, J., Fukushima, T., & Ma, R., 2010. An enhanced three-band index for estimating chlorophyll-a in turbid case-II waters: Case studies of Lake Kasumigaura, Japan and Lake Dianchi, China. *IEEE Geoscience and Remote Sensing Letters*, 7, 655–659.

Yang, W., Matsushita, B., Chen, J., Yoshimura, K., & Fukushima, T., 2013. Retrieval of inherent optical properties for turbid inland waters from remote-sensing reflectance. *IEEE Transactions on Geoscience and Remote Sensing*, 51, 3761–3773.

Yang, W., Matsushita, B., Chen, J., Yoshimura, K., & Fukushima, T., 2014. Application of a semianalytical algorithm to remotely estimate diffuse attenuation coefficient in turbid inland waters. *IEEE Geoscience and Remote Sensing Letters*, 11, 1046–1050.

Yang, W., Matsushita, B., Yoshimura, K., Chen, J., & Fukushima, T., 2015. A modified semianalytical algorithm for remotely estimating euphotic zone depth in turbid inland waters. *IEEE Journal of Selected topics in Applied Earth Observations and Remote Sensing*, 8, 1545–1554.

Section III

Remote Sensing for Sustainable Environmental Systems

10 Urban Air Quality Studies Using EO Data

Xuefei Hu

CONTENTS

10.1 PM$_{2.5}$ AND ITS HEALTH EFFECTS

PM$_{2.5}$ refers to particles with an aerodynamic diameter of less than 2.5 μm and is also known as fine particulate matter, including dust, dirt, soot, smoke, and liquid droplets. PM$_{2.5}$ originates from both natural and anthropogenic sources (Alves et al. 2000). Some particulates are emitted naturally from volcano eruptions, dust storms, and forest and grassland fires. However, natural sources only make a small contribution to the total concentration, and anthropogenic sources are more important. For instance, human activities, including the burning of fossil fuels in road vehicles, power plants, and certain industrial processes, contribute a significant amount of fine particles. In addition, secondary aerosols derived from precursors emitted from various sources such as cars, trucks, power plants, and industrial facilities are also a major contributor. Secondary aerosol formation occurs because of chemical reaction in the atmosphere generally downwind a distance from the original emission sources (Hodan and Barnard 2004). Besides outdoor sources, PM$_{2.5}$ is also produced by indoor activities such as cooking (e.g., frying). The composition of particles varies and depends on their sources. For example, mineral dust is generally made of mineral oxides. Primary aerosols that are directly emitted from emission sources into the atmosphere may include sulfur dioxide (SO$_2$), nitrogen oxides (NOx), volatile organic compounds (VOCs), ammonia (NH_3), elemental carbon (EC), and crustal materials such as soil and ash. Secondary aerosols may include ammonium sulfate and ammonium nitrate, which are typically formed through chemical reaction of precursors such as ammonia, SO_2, and NOx (Squizzato et al. 2013). Organic carbon (OC) is one of a large group of compounds (VOCs) and can be either primary or secondary. It is formed by a variety of processes, including combustion and secondary organic aerosol formation. In general, coarse particles are mainly made up of primary aerosols, and fine particles are heavily contributed by secondary aerosols (Deng et al. 2013).

Numerous epidemiological and experimental studies have shown that $PM_{2.5}$, one of the criterion air pollutants regulated by the US National Ambient Air Quality Standard, is associated with various adverse health outcomes including cardiovascular and respiratory mobility and mortality (Dominici et al. 2006). Peters et al. (2001) suggested that the elevated concentrations of fine particles may increase the risk of myocardial infarction after a few hours or 1 day of exposure to those air pollutants. Riediker et al. (2004) found that in-vehicle exposure to $PM_{2.5}$ is associated with cardiovascular effects and may cause pathophysiologic changes including inflammation, coagulation, and cardiac rhythm. Miller et al. (2007) revealed that long-term exposure to $PM_{2.5}$ is associated with increased risk of cardiovascular disease and death for postmenopausal women. Puett et al. (2009) further demonstrated that chronic exposure to $PM_{2.5}$ tends to increase the risk of all-cause and cardiovascular mortality. Madrigano et al. (2013) found that long-term exposure to area $PM_{2.5}$ was associated with the occurrence of acute myocardial infarction. Neophytou et al. (2014) revealed that occupational $PM_{2.5}$ exposure increases the risk of incident ischemic heart disease in both aluminum smelting and fabrication facilities. Sunyer and Basagaña (2001) reported that fine particles are associated with the risk of death in patients with chronic obstructive pulmonary disease (COPD), and their findings pointed out the adverse impact of fine particles on the trigger of death in COPD patients. Bose et al. (2015) demonstrated that even low indoor $PM_{2.5}$ levels may cause a systematic inflammatory response in COPD. Norris et al. (1999) found a significant association between emergency department visit for asthma in children and fine particulate matter. Lin et al. (2002) reported that exposure to fine particles emitted from heavy traffic contributes to childhood asthma hospitalization. Brauer et al. (2002) also revealed a positive association between respiratory diseases (e.g., respiratory infection and asthma) and fine particles in children. Spira-Cohen et al. (2011) suggested that pollution-related asthma exacerbations in children living near roadways are mainly attributed to the diesel soot fraction of $PM_{2.5}$. Habre et al. (2014) reported that exposure to indoor and outdoor $PM_{2.5}$ may exacerbate cough and wheeze symptoms in asthmatic children. Laden et al. (2006) found an increase in overall mortality associated with each 10 $\mu g/m^3$ increase in $PM_{2.5}$, indicating that cardiovascular mortality and lung cancer mortality were positively associated with ambient $PM_{2.5}$ concentrations. Kioumourtzoglou et al. (2016) found strong evidence showing potentially harmful effects of long-term exposure to $PM_{2.5}$ on neurodegeneration. Chung et al. (2015) pointed out that long-term exposure to $PM_{2.5}$ was associated with mortality in the elderly population of the eastern United States.

10.2 REMOTE SENSING APPLICATIONS FOR PM$_{2.5}$ CONCENTRATION PREDICTION

Most studies of the associations between PM$_{2.5}$ and health effects have relied on ground measurements typically from regulatory monitoring systems, such as United States Environmental Protection Agency (USEPA) Air Quality System to estimate PM$_{2.5}$ exposure. Although the EPA measurements are considered ground truth and the most accurate, the spatial distribution of those monitoring stations is sparse and uneven. For example, the United States has an extensive monitoring network and has approximately 1200 stations. However, those stations only cover ~30% of ~3000 counties in the conterminous United States. In addition, many of those stations are located in or near urban areas, which leaves ~30% of the US population living in suburban and rural areas without any PM$_{2.5}$ monitors. Thus, the limited coverage of ground monitoring stations limited the epidemiological studies to areas near monitoring sites. In addition, collecting PM$_{2.5}$ measurements from ground monitoring station is also expensive and time-consuming.

Satellite remote sensing data, given its comprehensive spatiotemporal coverage, have the potential to expand ground network by estimating PM$_{2.5}$ concentrations using aerosol optical depth (AOD) in areas where monitoring stations are not available or too sparse. AOD measures light extinction by aerosol scattering and absorption in an atmospheric column, and AOD retrieved from visible channels is most sensitive to particles with size from 0.1 to 2 μm (Kahn et al. 1998). Thus, it can be considered a representation of the loadings of fine particles in an atmospheric column. In addition, Wang and Christopher (2003) reported a good correlation between the satellite-derived AOD and PM$_{2.5}$. Using satellite-derived AOD to estimate PM$_{2.5}$ exposure is not only time and cost efficient, but could substantially improve estimates of population exposure to PM$_{2.5}$ (van Donkelaar et al. 2010). In contrast, using ground measurements inevitably introduces measurement errors to estimation of population exposure (Zeger et al. 2000).

10.2.1 DATA

10.2.1.1 AOD Data

To date, AOD derived from a number of satellite sensors, including MODerate resolution Imaging Spectroradiometer (MODIS) (Hu et al. 2013; Liu et al. 2007a; Zhang et al. 2009), Multiangle Imaging SpectroRadiometer (MISR) (Liu et al. 2007a), Geostationary Operational Environmental Satellite Aerosol/Smoke Product (GASP) (Liu et al. 2009; Paciorek et al. 2008), and Multi-Angle Implementation of Atmospheric Correction (MAIAC) (Hu et al. 2014a,b,c), has been used in previous studies to estimate PM$_{2.5}$ concentrations. In addition, the new Visible Infrared Imaging Radiometer Suite (VIIRS) AOD product will potentially have wide applications for PM$_{2.5}$ concentration prediction when the data quality becomes acceptable. Table 10.1 lists all available AOD products, their characteristics, and the studies in which these AOD products were used for PM$_{2.5}$ concentration estimation.

TABLE 10.1

Available AOD Products, Their Characteristics, and Related Studies

AOD Products	Spatial Resolutions	Studies
GASP	4 km	Chudnovsky et al. 2012b; Green et al. 2009; Liu et al. 2009; Paciorek et al. 2008 etc.
MODIS	10 km	Hu et al. 2013; Lee et al. 2011; Lin et al. 2015; Ma et al. 2015; Wang and Christopher 2003 etc.
MISR	17.6 km	Liu et al. 2004, 2007b; Ma et al. 2014; van Donkelaar et al. 2006 etc.
MAIAC	1 km	Chudnovsky et al. 2012a, 2013; Hu et al. 2014a,b; Just et al. 2015; Lee et al. 2015 etc.
VIIRS	6 km	Schliep et al. 2015 etc.

10.2.1.1.1 Geostationary Operational Environmental Satellite Aerosol/Smoke Product

GASP retrieves AOD over the contiguous United States from the GOES-East and GOES-West visible imagery with a temporal resolution of 30 min (up to 15 min) and a spatial resolution of 4 km. Unlike instruments onboard polar-orbiting satellites that provide only one daily AOD retrieval, the GOES satellite's geostationary orbit allows AOD retrievals with higher frequency. As a result, the GASP AOD product with high temporal resolution has the ability to capture rapidly changing aerosol conditions attributed to dust storms and fire events.

The GASP AOD has been used as an important proxy for $PM_{2.5}$. Paciorek et al. (2008) examined the association between GASP AOD and ground-level $PM_{2.5}$ concentrations and found that GASP AOD has the potential to improve exposure estimates for epidemiological studies, considering its higher temporal coverage than MODIS and MISR. Green et al. (2009) evaluated the quality of GASP AOD and its potential to predict surface $PM_{2.5}$. The results point out that using satellite AOD to predict surface $PM_{2.5}$ needs to take several factors into account, including seasonal and diurnal variations in particle size distribution, relative humidity, and seasonal change in boundary layer height. Liu et al. (2009) developed a two-stage generalized addictive model (GAM) with GASP AOD, meteorological fields, and land use variables as predictors to estimate ground-level $PM_{2.5}$ concentrations in a domain centered in Massachusetts. The results show that the model with GASP AOD outperforms the non-AOD model and has a greater predicting power. Chudnovsky et al. (2012b) used a mixed-effects model with the control for the day-to-day variability in the $PM_{2.5}$–AOD relationship to predict $PM_{2.5}$ concentrations in the New England region from GASP AOD. The predicted $PM_{2.5}$ concentrations have a good agreement with observations, which indicates that accounting for daily variability in the $PM_{2.5}$–AOD relationship is essential to obtain spatiotemporally resolved $PM_{2.5}$ exposure.

10.2.1.1.2 Moderate Resolution Imaging Spectroradiometer

The MODIS aerosol product monitors AOD over both ocean and land and provides daily near-global observations at a spatial resolution of 10 km. Two types of MODIS

AOD products are available, including MOD04 and MYD04. MOD04 contains data collected from the TERRA satellite since 2000, while MYD04 includes data acquired from the AQUA platform from 2002 onward. Compared to previous efforts such as GOES and AVHRR, MODIS expands spectral channels for aerosol measurement, and as a result, MODIS not only can retrieve AOD over dark vegetated/soiled land by the Dark-Target aerosol retrieval approach but also derive AOD over some brighter surfaces such as deserts using the Deep-Blue algorithm through the blue band. In addition, MODIS is also the first to provide a real solution for AOD retrievals over land on a global scale.

MODIS AOD has been widely used in $PM_{2.5}$ concentration predictions. For instance, Wang and Christopher (2003) examined the relationship between MODIS AOD and hourly surface $PM_{2.5}$ at seven locations in Jefferson county, Alabama. Zhang et al. (2009) examined the relation between MODIS AOD and $PM_{2.5}$ over the 10 USEPA-defined geographic regions in the United States, and the results show good correlations over the eastern United States in summer and fall. Hu (2009) used geographically weighted regression (GWR) to derive a spatially complete $PM_{2.5}$ surface covering the conterminous United States from MODIS AOD. Lee et al. (2011) developed a mixed-effects model that allows day-to-day variability in daily $PM_{2.5}$–AOD relationships to predict $PM_{2.5}$ concentrations in the New England region from MODIS AOD. Kloog et al. (2012) developed an advanced three-stage model to predict spatiotemporally resolved $PM_{2.5}$ concentrations using MODIS AOD in the Mid-Atlantic States. The model has the capability to predict $PM_{2.5}$ exposure in areas where AOD is missing. Hu et al. (2013) developed a GWR model with MODIS AOD, meteorological fields, and land use variables as predictors to predict $PM_{2.5}$ concentrations in the southeastern United States by incorporating spatially varying relationships between $PM_{2.5}$ and AOD. Lin et al. (2015) retrieved ground-level $PM_{2.5}$ concentrations from MODIS AOD by conducting both vertical and humidity corrections. Xie et al. (2015) developed a mixed-effects model to derive daily estimations of surface $PM_{2.5}$ in Beijing from 3-km-resolution MODIS AOD. Ma et al. (2015) developed a two-stage statistical model to estimate ambient $PM_{2.5}$ concentrations from 2004 to 2013 in China using MODIS collection 6 AOD data.

10.2.1.1.3 Multiangle Imaging Spectroradiometer

MISR is an instrument onboard the Terra satellite in a sun-synchronous orbit, and its five viewing angles (0°, 26.1°, 45.6°, 60.0°, and 70.5°) allow changes in reflection that provides the means to distinguish different types of aerosols. The swath of the MISR instrument is ~400 km, which allows it to view the entire Earth surface every 9 days. The MISR aerosol product is at a spatial resolution of 17.6 km derived by the MISR Standard Aerosol Retrieval Algorithm from MISR top-of-atmosphere radiances. Kahn et al. (2005) compared MISR AOD and Aerosol Robotic Network (AERONET) data on a global scale and obtained high correlation coefficients between them for various sites.

Many studies have examined the relationship between MISR AOD and surface $PM_{2.5}$ concentrations. Liu et al. (2004) used the global chemical transport models output and obtained the local $PM_{2.5}$–AOD conversion factors over the contiguous United States to predict surface $PM_{2.5}$ concentrations from MISR AOD. van

Donkelaar et al. (2006) used the factors affecting the relationship between $PM_{2.5}$ and AOD simulated from GEOS-CHEM to estimate ground-level $PM_{2.5}$ concentrations from both MISR and MODIS AOD, and they found significant spatial variation of the relationship between the annual mean ground-level measurements and $PM_{2.5}$ estimated from MODIS and MISR. Liu et al. (2007b) used the MISR fractional AOD and aerosol transport model constraints to predict ground-level $PM_{2.5}$ concentrations and its major constituents in the continental United States. The results show that using fractional AOD can significantly improve the estimating power of the model, compared with similar models with total-column AOD as the single predictor. Liu et al. (2007a) compared the ability of MISR and MODIS AOD to predict ground-level $PM_{2.5}$ concentrations in St. Louis, MO, and found that MISR achieves higher prediction accuracy, while MODIS provides better spatial coverage. Ma et al. (2014) developed a national-scale GWR model to estimate daily $PM_{2.5}$ concentrations in China with AOD data fused from MISR and MODIS data.

10.2.1.1.4 Multi-Angle Implementation of Atmospheric Correction

The MAIAC AOD product is derived from MODIS radiances using a newly developed MAIAC algorithm. The algorithm has a global scope and can retrieve AOD over both dark and bright surfaces at a spatial resolution of 1 km. Validation over the continental United States shows that MAIAC and operational Collection 5 MODIS Dark Target AOD have a similar accuracy over dark and vegetated surfaces, while MAIAC achieved higher accuracy over brighter surfaces (Lyapustin et al. 2011). In addition, MAIAC provides greater spatial coverage and more AOD retrievals than MODIS (Hu et al. 2014a).

Chudnovsky et al. (2012a) investigated the relationship between MAIAC AOD and $PM_{2.5}$ concentrations measured by EPA ground monitors at a variety of spatial scales and found that the correlation between $PM_{2.5}$ and AOD decreased significantly with the decrease of AOD resolution. Chudnovsky et al. (2013) further compared the relationship between MAIAC AOD and $PM_{2.5}$ measured from 84 EPA ground monitors with the $PM_{2.5}$–AOD relationship using MODIS AOD, and the results show that the correlation coefficient for MAIAC is slightly higher, and MAIAC AOD is more capable of capturing spatial patterns of $PM_{2.5}$. In addition, MAIAC AOD can also help increase the number of days for $PM_{2.5}$ prediction. Hu et al. (2014a) developed an advanced two-stage model with MAIAC AOD as the primary predictor and meteorological fields and land use variables as secondary predictors to estimate spatiotemporally resolved $PM_{2.5}$ concentrations in the southeastern United States at 1 km resolution. The model includes a linear mixed-effects model to account for the day-to-day variability in the $PM_{2.5}$–AOD relationship and a geographical weighted regression model to explain the spatial variability. Hu et al. (2014b) further investigated the 10-year spatiotemporal trend of $PM_{2.5}$ concentrations in the southeastern United States using $PM_{2.5}$ at 1 km resolution estimated from MAIAC AOD. Kloog et al. (2014) predicted daily $PM_{2.5}$ at 1 km resolution across the northeastern United States, including New England, New York, and New Jersey, from 2003 to 2011 using MAIAC AOD. Lee et al. (2015) applied a separate mixed-effects model to predict daily $PM_{2.5}$ concentrations at 1 km resolution in the southeastern United States for the years 2003–2011 using MAIAC AOD. Just et al. (2015) calibrated the relationship between $PM_{2.5}$ and MAIAC AOD using ground monitors, land use, and meteorological features and predicted daily $PM_{2.5}$ concentrations at 1 km resolution across the great Mexico City area for 2004–2014.

10.2.1.1.5   Visible Infrared Imaging Radiometer Suite

The VIIRS instrument is onboard the Suomi National Polar Partnership, which was launched in late 2011 and is a polar-orbiting satellite with afternoon overpass (similar to Aqua, ~1:30 p.m. local time). VIIRS provides two high-resolution AOD products, including Intermediate Product at a spatial resolution of 750 m and Environmental Data Record with a resolution of 6 km. Two types of AOD products make it possible for air quality applications to examine spatial detail on an urban scale and also monitor large-scale events like wildfires.

Because of the recent launch, applications of VIIRS AOD products for $PM_{2.5}$ prediction have been limited but are promising. Schliep et al. (2015) developed a hierarchical autoregressive model with daily spatially varying coefficients to estimate daily average $PM_{2.5}$ across the conterminous United States from VIIRS AOD. Because of the quality assurance protocol of VIIRS AOD, many missing AOD data contribute to the ineffectiveness of AOD in their model.

10.2.1.2   Fire Data

Another remote sensing product that can be used for $PM_{2.5}$ concentration estimation is fire data. Fires, including prescribed burning and wildfires, are important sources of fine particulate matter. Zeng et al. (2008) estimated a maximum increase of 25 $\mu g/m^3$ in $PM_{2.5}$ concentrations within a day owing to prescribed burning emissions. Zhang et al. (2008) conducted a research in Beijing with the time span from July 2002 to July 2003 and found that biomass burning contributes 18%–38% of $PM_{2.5}$ organic carbon. Jaffe et al. (2008) reported that the increase in $PM_{2.5}$ attributed to fires reaches a high fraction of the annual National Ambient Air Quality Standards in summer and has a significant contribution to regional haze in the western United States. Zhang et al. (2010) revealed that 13% of $PM_{2.5}$ concentrations are attributed to biomass burning annually in the southeastern United States. Tian et al. (2009) reported that prescribed burning contributes 55% and 80% of $PM_{2.5}$ mass concentrations in January and March of 2002, respectively. Christopher et al. (2009) found a threefold increase of $PM_{2.5}$ concentrations collected from ground monitors in Birmingham, Illinois, during fire events, compared to background values, although those monitors are hundreds of miles away from the fire sources.

10.2.1.2.1   MODIS Fire Product

The MODIS fire products are produced for the MODIS sensors onboard two Earth Observing System (EOS) satellites: Terra and Aqua. The two satellites are in sun-synchronous orbits with different local overpass times (~1:30 p.m. for Aqua and ~10:30 a.m. for Terra). As a result, MODIS can provide two overpasses over a location in a single day, which is important because the differences in fires between morning and afternoon can be detected. Two products exist. One is the MODIS Active Fire Product, which provides actively burning fire locations at satellite overpass times, and the other is the MODIS Burned Area Product, which provides the burned area. The active fire products are generated daily at 1 km spatial resolution.

The use of MODIS fire in $PM_{2.5}$ concentration prediction has been limited. Hu et al. (2014c) incorporated MODIS fire counts in an advanced two-stage model to examine if remotely sensed fire data can help improve the prediction accuracy of $PM_{2.5}$ concentrations in the southeastern United States and conducted a sensitivity analysis to determine the optimum buffer radius centered around each $PM_{2.5}$ monitoring site, which is crucial to count the fire incidents that may have impact on the corresponding monitors. The results show that when the radius reaches 75 km, fire count data achieve the greatest predictive power of $PM_{2.5}$ concentrations. A comparison between the fire model and the nonfire model shows that the prediction accuracy increases more substantially at the sites with higher fire occurrence when MODIS fire count data are incorporated in the model, indicating that remotely sensed fire count data can provide a measurable improvement in $PM_{2.5}$ concentration prediction, particularly in areas where fires occur frequently.

10.2.1.3   LIDAR Data

Light Detection and Ranging (LIDAR) data have also been used for estimating surface $PM_{2.5}$ concentrations. A LIDAR system can use a laser to measure aerosol scattering as a function of height in the atmosphere, and the advantage of the LIDAR method over satellite instruments is that it can calculate optical depth for selected ranges in the atmosphere. Engel-Cox et al. (2006) pointed out that LIDAR apportionment of the fraction of AOD within the planetary boundary layer (PBL) can achieve better agreement with surface $PM_{2.5}$ than does the total column amount. He et al. (2008) used LIDAR measurements to obtain significant improvements in correlation between AOD with surface extinction and $PM_{2.5}$ by considering aerosol vertical distribution. Chu et al. (2013) demonstrated that surface $PM_{2.5}$ can be better estimated using AOD normalized by haze layer height derived from LIDAR aerosol extinction profiles than that using AOD only.

10.2.2   METHODS

10.2.2.1   Physically Based Methods

Liu et al. (2004) developed a simple approach using outputs from a global chemistry and transport model (CTM) to examine the physical basis for the relationship between $PM_{2.5}$ and AOD. The local $PM_{2.5}$–AOD conversion factors were obtained using simulated $PM_{2.5}$ concentrations and AOD. Simulated $PM_{2.5}$ concentrations include sea salt and dust mass concentrations from the Global Ozone Chemistry Aerosol Radiation and Transport (GOCART) model and mass concentrations for SO_4^{2-}, NO_3^-, NH_4^+, EC, and OC derived from GEOS-CHEM, while simulated AOD is composed of AOD values for sea salt and dust from GOCART and those for other particulate species from GEOS-CHEM. The results show that the predicted $PM_{2.5}$ concentrations from MISR AOD are strongly correlated with EPA $PM_{2.5}$ measurements, and the proposed method has the capability to reduce the uncertainty in estimated $PM_{2.5}$ concentrations owing to the discrepancy of correlations between lower and upper tropospheric aerosols. van Donkelaar et al. (2006) extended this method to estimate $PM_{2.5}$ concentrations from both MISR and MODIS and examined the factors affecting the $PM_{2.5}$–AOD relationship using the GEOS-CHEM simulation. The GEOS-CHEM aerosol simulation was conducted for each day from January 2001 to October 2002 at the MODIS and MISR overpass times to obtain local values for each parameter required in the $PM_{2.5}$ prediction model. The results show that the relative vertical profile is the most influential factor affecting the spatial relationship between predicted and measured $PM_{2.5}$ concentrations, while the temporal variation in AOD is the most important factor that affects the temporal relationship between predicted and measured concentrations. Lin et al. (2015) also used a physically based method to retrieve ground-level $PM_{2.5}$ concentrations from MODIS AOD by taking account of the effect of the main aerosol characteristics. The effects on hygroscopic growth, particle mass extinction efficiency, and size distribution are estimated and incorporated in the $PM_{2.5}$–AOD relationship.

10.2.2.2   Statistical Methods

Statistical models have also been used to predict $PM_{2.5}$ concentrations from remotely sensed AOD, and the prediction models have been evolving from using AOD as the only predictor to the incorporation of multiple predictors including meteorological and land use variables and from one-stage models to advanced multiple-stage models. Wang and Christopher (2003) explored the relationship between MODIS AOD and hourly $PM_{2.5}$ mass concentrations at seven locations in Jefferson County, Alabama, and used an empirical linear regression model to derive $PM_{2.5}$ concentrations from MODIS AOD. The results indicate that MODIS AOD has great potential for air quality applications. However, this model assumes that the relationships between $PM_{2.5}$ and AOD are constant over space and time, and a single linear $PM_{2.5}$–AOD relationship was established for all sampling sites and days, which may lead to bias for $PM_{2.5}$ prediction. Hu (2009) found that the $PM_{2.5}$–AOD relationships are spatially inconsistent across the conterminous United States and fitted a GWR model with MODIS AOD as the independent variable to investigate the $PM_{2.5}$–AOD relationship using the 2-year average $PM_{2.5}$ and AOD data. GWR is a linear regression model that can model spatial varying relationship and reveal how the $PM_{2.5}$–AOD relationship changes over space by generating a continuous surface of estimates for each predictor at each local location instead of a universal value. The results show that it is appropriate to estimate $PM_{2.5}$ surface from AOD using GWR, and the estimated $PM_{2.5}$ reaches an accuracy of 84%. Ma et al. (2014) developed a national-scale GWR model to estimate daily $PM_{2.5}$ concentrations in China from fused satellite AOD, and their results confirm the satisfactory performance of the GWR model. Lee et al. (2011) developed a mixed-effects model with MODIS AOD as the predictor to predict daily $PM_{2.5}$ concentrations in the New England region. The mixed-effects model with random intercept and slopes allows day-to-day variability in the $PM_{2.5}$–AOD relationship and, as a result, can calculate a $PM_{2.5}$–AOD slope separately for each day. The results suggest that the proposed method makes it possible to determine spatial and temporal patterns of $PM_{2.5}$ concentrations in a relatively large study domain from remotely sensed AOD. Xie et al. (2015) also developed a mixed-effects model to derive daily estimations of surface $PM_{2.5}$ in Beijing from 3-km-resolution

MODIS AOD. The model can account for daily variations of the $PM_{2.5}$–AOD relationship and shows good performance in model predictions. In addition to AOD values, many other parameters have been demonstrated as influential factors that can affect the relationship between $PM_{2.5}$ and AOD and, as a result, have the potential to be used as predictors for $PM_{2.5}$ concentration estimation. Liu et al. (2005) incorporated meteorological fields, including PBL height and relative humidity, as covariates in linear regression models to predict $PM_{2.5}$ concentrations and found that they are highly significant predictors of $PM_{2.5}$. Furthermore, Liu et al. (2007a) introduced other meteorological parameters such as air temperature and wind speed to be included in a general linear regression model for $PM_{2.5}$ concentration prediction, and the results show that both wind speed and air temperature are significant predictors of $PM_{2.5}$. Moving one step further, Liu et al. (2009) used generalized additive models incorporated with land use information, including road length and population density, to predict daily $PM_{2.5}$ concentrations on a regional scale and demonstrate that land use variables are effective predictors of $PM_{2.5}$ concentrations. Kloog et al. (2011) extended previous work by introducing numerous land use variables that can be potentially used in statistical models for $PM_{2.5}$ concentration prediction, including percent of open spaces, elevation, major roads, $PM_{2.5}$ point emissions, and area-source point emissions. The results indicate that by including land use variables, the proposed model outperforms previous AOD-$PM_{2.5}$ models. Hu et al. (2013) incorporated forest cover in the GWR model for $PM_{2.5}$ concentration estimation in the southeastern United States and found that forest cover is a statistically significant predictor of $PM_{2.5}$. Hu et al. (2014c) further tested the inclusion of MODIS fire counts in the linear mixed-effects model to examine if fire count data are an effective predictor of $PM_{2.5}$. The results show that the prediction accuracy improved more from the nonfire model to the fire model at sites with higher fire occurrence, and the inclusion of fire count data can provide a measurable improvement in $PM_{2.5}$ concentration estimation, particularly in areas and seasons prone to fire events. In addition, to account for both spatial and temporal variability in the $PM_{2.5}$–AOD relationship and predict $PM_{2.5}$ concentrations in areas and days where and when AOD is not available, multistage statistical models are becoming more and more popular. Liu et al. (2009) proposed a two-stage GAM model to examine the spatial and temporal variability in $PM_{2.5}$ concentrations separately. The first stage aims to account for the temporal variability. The dependent variable is the daily $PM_{2.5}$ concentrations, and all the covariates are averaged spatially and therefore only vary temporally. The second stage is to explain the spatial variability. All the covariates in this stage are averaged over the entire period and therefore only vary spatially. The results show that the first stage model contributes more to the overall model performance because temporal variability dominates the overall $PM_{2.5}$ variability. Hu et al. (2014a) developed a two-stage model incorporated with meteorological fields and land use variables to predict 1-km $PM_{2.5}$ surface from MAIAC AOD in the southeastern United States. The model includes a first-stage linear mixed-effects model to account for the day-to-day variability in the relationship between $PM_{2.5}$ and AOD and a second-stage GWR model to explain the spatial variability in the $PM_{2.5}$–AOD relationship. Hence, the model has the capability to predict spatiotemporally resolved $PM_{2.5}$ concentrations with high accuracy. Ma et al. (2015) also developed a two-stage spatial statistical model to estimate ambient $PM_{2.5}$ concentrations from 2004 to 2013 in China. The model includes a linear mixed-effects model as the first stage and a GAM model as the second stage to calibrate the spatiotemporal relationship between $PM_{2.5}$ and AOD. Kloog et al. (2011) and Kloog et al. (2012) introduced a third-stage GAM model to estimate daily $PM_{2.5}$ concentrations in the study domain for days when AOD data were unavailable. The model includes a smooth function for the coordinates of each grid cell centroid and a random intercept for each grid cell, uses predicted $PM_{2.5}$ concentrations from previous stages as the dependent variable and the mean of $PM_{2.5}$ measurements collected from monitors on that day as the predictor, and is similar to universal kriging. In addition, to account for temporal variation, the model was fit for each 2-month period of each year to generate separate spatial surfaces for those periods.

10.3 CASE STUDY

A case study was conducted to estimate ground-level $PM_{2.5}$ concentrations from MODIS AOD in Atlanta, Georgia, in 2011using statistical methods. The study area was approximately 130 km × 130 km, covering the Atlanta Metro area. The domain included a large urban center and surrounding suburban areas.

10.3.1 DATA

The 24-h averaged $PM_{2.5}$ concentrations in 2011 were collected from 13 EPA Federal Reference Monitors and downloaded from the EPA's Air Quality System Technology Transfer Network (http://www.epa.gov/ttn/airs/airsaqs/). The 2011 MODIS AOD data at a spatial resolution of 10 km were downloaded from the Earth Observing System Data Gateway at the Goddard Space Flight Center (https://ladsweb.nascom .nasa.gov/data/search.html). Hourly meteorological parameters were obtained from the North American Land Data Assimilation System Phase 2 (http://ldas.gsfc.nasa .gov/nldas/) at a spatial resolution of ~13 km. Elevation data were obtained from the National Elevation Dataset (http://ned.usgs.gov) at a spatial resolution of 1 arc-sec (~30 m). Road data were extracted from ESRI StreetMap USA (Environmental Systems Research Institute, Inc., Redland, California). Forest cover data were generated using 2011 land cover maps downloaded from the National Land Cover Database (http://www.mrlc.gov). Primary $PM_{2.5}$ emissions were obtained from the 2011 EPA National Emission Inventory facility emissions reports. All data were integrated for model fitting and $PM_{2.5}$ prediction, a 1 km × 1 km square buffer was generated for each $PM_{2.5}$ monitoring site and MODIS pixel centroid, and meteorological fields and AOD values were assigned using the nearest-neighbor approach. Forest cover and elevation were averaged, while road length and point emissions were summed over the 1 km × 1 km square buffer. A fitting data set with 1320 records was obtained.

10.3.2 METHODOLOGY

A linear mixed-effects model (Kloog et al. 2011) with 24-h averaged $PM_{2.5}$ measurements as the dependent variable, MODIS AOD as the primary predictor, and meteorological and land use variables as secondary predictors is developed to estimate ground-level $PM_{2.5}$ concentrations. The model explains the temporal variability in the $PM_{2.5}$–AOD relationship by including day-specific random intercepts and slopes for AOD and meteorological variables. The model structure can be expressed as

$$PM_{2.5,st} = (b_0 + b_{0,t}) + (b_1 + b_{1,t})AOD_{st} + (b_2 + b_{2,t})\text{Relative Humidity}_{st}$$
$$+ (b_3 + b_{3,t})\text{Wind Speed}_{st} + b_4\text{Elevation}_s$$
$$+ b_5\text{Major Roads}_s + b_6\text{Forest Cover 2011}_s + b_7\text{Point Emissions 2011}_s$$
$$+ \varepsilon_{st}(b_{0,t}b_{1,t}b_{2,t}b_{3,t}) \sim N[(0,0,0,0),\Psi],$$

where $PM_{2.5,st}$ is the measured ground-level $PM_{2.5}$ concentration ($\mu g/m^3$) at site s on day t; b_0 and $b_{0,t}$ (day-specific) are the fixed and random intercept, respectively; AOD_{st} is the MODIS AOD value (unitless) at site s on day t; b_1 and $b_{1,t}$ (day-specific) are the fixed and random slopes for AOD, respectively; Relative Humidity$_{st}$ is the relative humidity at site s on day t; b_2 and $b_{2,t}$ (day-specific) are the fixed and random slopes for relative humidity, respectively; Wind Speed$_{st}$ is the 2-m wind speed (m/s) at site s on day t; b_3 and $b_{3,t}$ (day-specific) are the fixed and random slopes for wind speed, respectively; Elevation$_s$ is elevation values (m) at site s; Major Roads$_s$ is road length values (m) at site s; Forest Cover 2011$_s$ is forest cover values (unitless) at site s; Point Emissions 2011$_s$ is point emissions (tons per year) at site s; and Ψ is an unstructured variance–covariance matrix for the random effects.

A 10-fold cross-validation (CV) is conducted to assess the model performance, and statistical indicators, including the coefficient of determination (R^2) and the square root of the mean squared prediction errors (RMSPE), are calculated.

10.3.3 RESULTS

The 2011 annual mean $PM_{2.5}$ surface for the Atlanta region is illustrated in Figure 10.1. The mean $PM_{2.5}$ concentration for the entire study domain in 2011 is 10.06 $\mu g/m^3$.

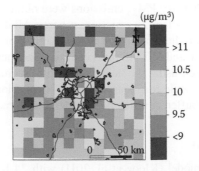

FIGURE 10.1 The 2011 annual mean $PM_{2.5}$ surface in the Atlanta region.

The results show that high $PM_{2.5}$ concentrations appear in urban areas, while low concentrations occur in rural and suburban areas. In addition, the results also show that model fitting generates an R^2 of 0.7 and an RMSPE of 3.15 $\mu g/m^3$, while CV generates an R^2 of 0.6 and an RMSPE of 3.45 $\mu g/m^3$, indicating a good fit of the model.

REFERENCES

Alves, C.A., Pio, C.A., & Duarte, A.C. (2000). Particulate size distributed organic: Compounds in a forest atmosphere. *Environmental Science & Technology, 34*, 4287–4293.

Bose, S., Hansel, N., Tonorezos, E., Williams, D., Bilderback, A., Breysse, P., Diette, G., & McCormack, M.C. (2015). Indoor particulate matter associated with systemic inflammation in COPD. *Journal of Environmental Protection, 6*, 566–572.

Brauer, M., Hoek, G., Van Vliet, P., Meliefste, K., Fischer, P.H., Wijga, A., Koopman, L.P., Neijens, H.J., Gerritsen, J., Kerkhof, M., Heinrich, J., Bellander, T., & Brunekreef, B. (2002). Air pollution from traffic and the development of respiratory infections and asthmatic and allergic symptoms in children. *American Journal of Respiratory and Critical Care Medicine, 166*, 1092–1098.

Christopher, S.A., Gupta, P., Nair, U., Jones, T.A., Kondragunta, S., Yu-Ling, W., Hand, J., & Xiaoyang, Z. (2009). Satellite remote sensing and mesoscale modeling of the 2007 Georgia/Florida Fires. *IEEE Journal of Selected Topics in Applied Earth Observations and Remote Sensing, 2*, 163–175.

Chu, D.A., Tsai, T.-C., Chen, J.-P., Chang, S.-C., Jeng, Y.-J., Chiang, W.-L., & Lin, N.-H. (2013). Interpreting aerosol lidar profiles to better estimate surface PM2.5 for columnar AOD measurements. *Atmospheric Environment, 79*, 172–187.

Chudnovsky, A.A., Kostinski, A., Lyapustin, A., & Koutrakis, P. (2012a). Spatial scales of pollution from variable resolution satellite imaging. *Environmental Pollution, 172*, 131–138.

Chudnovsky, A.A., Lee, H.J., Kostinski, A., Kotlov, T., & Koutrakis, P. (2012b). Prediction of daily fine particulate matter concentrations using aerosol optical depth retrievals from the Geostationary Operational Environmental Satellite (GOES). *Journal of the Air & Waste Management Association, 62*, 1022–1031.

Chudnovsky, A.A., Tang, C., Lyapustin, A., Wang, Y., Schwartz, J., & Koutrakis, P. (2013). A critical assessment of high resolution aerosol optical depth (AOD) retrievals for fine particulate matter (PM) predictions. *Atmospheric Chemistry and Physics Discussions, 13*, 14581–14611.

Chung, Y., Dominici, F., Wang, Y., Coull, B.A., & Bell, M.L. (2015). Associations between long-term exposure to chemical constituents of fine particulate matter (PM(2.5)) and mortality in Medicare enrollees in the eastern United States. *Environmental Health Perspectives, 123*, 467–474.

Deng, X.J., Wu, D., Yu, J.Z., Lau, A.K.H., Li, F., Tan, H.B., Yuan, Z.B., Ng, W.M., Deng, T., Wu, C., & Zhou, X.J. (2013). Characterization of secondary aerosol and its extinction effects on visibility over the Pearl River Delta Region, China. *Journal of the Air & Waste Management Association, 63*, 1012–1021.

Dominici, F., Peng, R.D., Bell, M.L., Pham, L., McDermott, A., Zeger, S.L., & Samet, J.M. (2006). Fine particulate air pollution and hospital admission for cardiovascular and respiratory diseases. *Jama—Journal of the American Medical Association, 295*, 1127–1134.

Engel-Cox, J.A., Hoff, R.M., Rogers, R., Dimmick, F., Rush, A.C., Szykman, J.J., Al-Saadi, J., Chu, D.A., & Zell, E.R. (2006). Integrating lidar and satellite optical depth with ambient monitoring for 3-dimensional particulate characterization. *Atmospheric Environment, 40*, 8056–8067.

Green, M., Kondragunta, S., Ciren, P., & Xu, C. (2009). Comparison of GOES and MODIS Aerosol Optical Depth (AOD) to Aerosol Robotic Network (AERONET) AOD and IMPROVE PM2.5 Mass at Bondville, Illinois. *Journal of the Air & Waste Management Association, 59*, 1082–1091.

Habre, R., Moshier, E., Castro, W., Nath, A., Grunin, A., Rohr, A., Godbold, J., Schachter, N., Kattan, M., & Coull, B. (2014). The effects of PM2.5 and its components from indoor and outdoor sources on cough and wheeze symptoms in asthmatic children. *Journal of Exposure Science and Environmental Epidemiology, 24*, 380–387.

He, Q.S., Li, C.C., Mao, J.T., Lau, A.K.H., & Chu, D.A. (2008). Analysis of aerosol vertical distribution and variability in Hong Kong. *Journal of Geophysical Research-Atmospheres, 113*.

Hodan, W.M., & Barnard, W.R. (2004). Evaluating the Contribution of PM2.5 Precursor Gases and Re-entrained Road Emissions to Mobile Source PM2.5 Particulate Matter Emissions. MACTEC Under Contract to the Federal Highway Administration. MACTEC Federal Programs, Research Triangle Park, NC.

Hu, X., Waller, L.A., Al-Hamdan, M.Z., Crosson, W.L., Estes Jr, M.G., Estes, S.M., Quattrochi, D.A., Sarnat, J.A., & Liu, Y. (2013). Estimating ground-level PM2.5 concentrations in the southeastern U.S. using geographically weighted regression. *Environmental Research, 121*, 1–10.

Hu, X., Waller, L.A., Lyapustin, A., Wang, Y., Al-Hamdan, M.Z., Crosson, W.L., Estes Jr, M.G., Estes, S.M., Quattrochi, D.A., Puttaswamy, S.J., & Liu, Y. (2014a). Estimating ground-level PM2.5 concentrations in the Southeastern United States using MAIAC AOD retrievals and a two-stage model. *Remote Sensing of Environment, 140*, 220–232.

Hu, X., Waller, L.A., Lyapustin, A., Wang, Y., & Liu, Y. (2014b). 10-year spatial and temporal trends of PM2.5 concentrations in the southeastern US estimated using high-resolution satellite data. *Atmospheric Chemistry and Physics, 14*, 6301–6314.

Hu, X., Waller, L.A., Lyapustin, A., Wang, Y., & Liu, Y. (2014c). Improving satellite-driven PM2.5 models with Moderate Resolution Imaging Spectroradiometer fire counts in the southeastern U.S. *Journal of Geophysical Research: Atmospheres, 119*, 2014JD021920.

Hu, Z.Y. (2009). Spatial analysis of MODIS aerosol optical depth, PM2.5, and chronic coronary heart disease. *International Journal of Health Geographics, 8*.

Jaffe, D., Hafner, W., Chand, D., Westerling, A., & Spracklen, D. (2008). Interannual variations in PM2.5 due to wildfires in the western United States. *Environmental Science & Technology, 42*, 2812–2818.

Just, A.C., Wright, R.O., Schwartz, J., Coull, B.A., Baccarelli, A.A., Tellez-Rojo, M.M., Moody, E., Wang, Y., Lyapustin, A., & Kloog, I. (2015). Using high-resolution satellite aerosol optical depth to estimate daily PM2.5 geographical distribution in Mexico City. *Environmental Science & Technology, 49*, 8576–8584.

Kahn, R., Banerjee, P., McDonald, D., & Diner, D.J. (1998). Sensitivity of multiangle imaging to aerosol optical depth and to pure-particle size distribution and composition over ocean. *Journal of Geophysical Research—Atmospheres, 103*, 32195–32213.

Kahn, R.A., Gaitley, B.J., Martonchik, J.V., Diner, D.J., Crean, K.A., & Holben, B. (2005). Multiangle Imaging Spectroradiometer (MISR) global aerosol optical depth validation based on 2 years of coincident Aerosol Robotic Network (AERONET) observations. *Journal of Geophysical Research—Atmospheres, 110*.

Kioumourtzoglou, M.-A., Schwartz, J.D., Weisskopf, M.G., Melly, S.J., Wang, Y., Dominici, F., & Zanobetti, A. (2016). Long-term PM(2.5) exposure and neurological hospital admissions in the northeastern United States. *Environmental Health Perspectives, 124*, 23–29.

Kloog, I., Chudnovsky, A.A., Just, A.C., Nordio, F., Koutrakis, P., Coull, B.A., Lyapustin, A., Wang, Y., & Schwartz, J. (2014). A new hybrid spatio-temporal model for estimating daily multi-year PM2.5 concentrations across northeastern USA using high resolution aerosol optical depth data. *Atmospheric Environment, 95*, 581–590.

Kloog, I., Koutrakis, P., Coull, B.A., Lee, H.J., & Schwartz, J. (2011). Assessing temporally and spatially resolved PM2.5 exposures for epidemiological studies using satellite aerosol optical depth measurements. *Atmospheric Environment, 45*, 6267–6275.

Kloog, I., Nordio, F., Coull, B.A., & Schwartz, J. (2012). Incorporating local land use regression and satellite aerosol optical depth in a hybrid model of spatiotemporal PM2.5 exposures in the mid-Atlantic States. *Environmental Science & Technology, 46*, 11913–11921.

Laden, F., Schwartz, J., Speizer, F.E., & Dockery, D.W. (2006). Reduction in fine particulate air pollution and mortality—Extended follow-up of the Harvard six cities study. *American Journal of Respiratory and Critical Care Medicine, 173*, 667–672.

Lee, H.J., Liu, Y., Coull, B.A., Schwartz, J., & Koutrakis, P. (2011). A novel calibration approach of MODIS AOD data to predict PM2.5 concentrations. *Atmospheric Chemistry and Physics, 11*, 7991–8002.

Lee, M., Kloog, I., Chudnovsky, A., Lyapustin, A., Wang, Y., Melly, S., Coull, B., Koutrakis, P., & Schwartz, J. (2015). Spatiotemporal prediction of fine particulate matter using high-resolution satellite images in the Southeastern US 2003–2011. *Journal of Exposure Science and Environmental Epidemiology, 26*, 377–384.

Lin, C., Li, Y., Yuan, Z., Lau, A.K.H., Li, C., & Fung, J.C.H. (2015). Using satellite remote sensing data to estimate the high-resolution distribution of ground-level PM2.5. *Remote Sensing of Environment, 156*, 117–128.

Lin, S., Munsie, J.P., Hwang, S.A., Fitzgerald, E., & Cayo, M.R. (2002). Childhood asthma hospitalization and residential exposure to state route traffic. *Environmental Research, 88*, 73–81.

Liu, Y., Franklin, M., Kahn, R., & Koutrakis, P. (2007a). Using aerosol optical thickness to predict ground-level PM2.5 concentrations in the St. Louis area: A comparison between MISR and MODIS. *Remote Sensing of Environment, 107*, 33–44.

Liu, Y., Koutrakis, P., Kahn, R., Turquety, S., & Yantosca, R.M. (2007b). Estimating fine particulate matter component concentrations and size distributions using satellite-retrieved fractional aerosol optical depth: Part 2—A case study. *Journal of the Air & Waste Management Association, 57*, 1360–1369.

Liu, Y., Paciorek, C.J., & Koutrakis, P. (2009). Estimating regional spatial and temporal variability of PM2.5 concentrations using satellite data, meteorology, and land use information. *Environmental Health Perspectives, 117*, 886–892.

Liu, Y., Park, R.J., Jacob, D.J., Li, Q.B., Kilaru, V., & Sarnat, J.A. (2004). Mapping annual mean ground-level PM2.5 concentrations using Multiangle Imaging Spectroradiometer aerosol optical thickness over the contiguous United States. *Journal of Geophysical Research—Atmospheres, 109*.

Liu, Y., Sarnat, J.A., Kilaru, A., Jacob, D.J., & Koutrakis, P. (2005). Estimating ground-level PM2.5 in the eastern United States using satellite remote sensing. *Environmental Science & Technology, 39*, 3269–3278.

Lyapustin, A., Wang, Y., Laszlo, I., Kahn, R., Korkin, S., Remer, L., Levy, R., & Reid, J.S. (2011). Multiangle implementation of atmospheric correction (MAIAC): 2. Aerosol algorithm. *Journal of Geophysical Research—Atmospheres, 116*.

Ma, Z., Hu, X., Huang, L., Bi, J., & Liu, Y. (2014). Estimating ground-level PM2. 5 in China using satellite remote sensing. *Environmental Science & Technology, 48*, 7436–7444.

Ma, Z., Hu, X., Sayer, A.M., Levy, R., Zhang, Q., Xue, Y., Tong, S., Bi, J., Huang, L., & Liu, Y. (2015). Satellite-based spatiotemporal trends in PM2.5 concentrations: China, 2004–2013. *Environmental Health Perspectives, 124*, 184–192.

Madrigano, J., Kloog, I., Goldberg, R., Coull, B.A., Mittleman, M.A., & Schwartz, J. (2013). Long-term exposure to PM2.5 and incidence of acute myocardial infarction. *Environmental Health Perspectives, 121*, 192–196.

Miller, K.A., Siscovick, D.S., Sheppard, L., Shepherd, K., Sullivan, J.H., Anderson, G.L., & Kaufman, J.D. (2007). Long-term exposure to air pollution and incidence of cardiovascular events in women. *New England Journal of Medicine, 356*, 447–458.

Neophytou, A.M., Costello, S., Brown, D.M., Picciotto, S., Noth, E.M., Hammond, S.K., Cullen, M.R., & Eisen, E.A. (2014). Marginal structural models in occupational epidemiology: Application in a study of ischemic heart disease incidence and PM2.5 in the US aluminum industry. *American Journal of Epidemiology*.

Norris, G., YoungPong, S.N., Koenig, J.Q., Larson, T.V., Sheppard, L., & Stout, J.W. (1999). An association between fine particles and asthma emergency department visits for children in Seattle. *Environmental Health Perspectives, 107*, 489–493.

Paciorek, C.J., Liu, Y., Moreno-Macias, H., & Kondragunta, S. (2008). Spatiotemporal associations between GOES aerosol optical depth retrievals and ground-level PM2.5. *Environmental Science & Technology, 42*, 5800–5806.

Peters, A., Dockery, D.W., Muller, J.E., & Mittleman, M.A. (2001). Increased particulate air pollution and the triggering of myocardial infarction. *Circulation, 103*, 2810–2815.

Puett, R.C., Hart, J.E., Yanosky, J.D., Paciorek, C., Schwartz, J., Suh, H., Speizer, F.E., & Laden, F. (2009). Chronic fine and coarse particulate exposure, mortality, and coronary heart disease in the Nurses' Health Study. *Environmental Health Perspectives, 117*, 1697–1701.

Riediker, M., Cascio, W.E., Griggs, T.R., Herbst, M.C., Bromberg, P.A., Neas, L., Williams, R.W., & Devlin, R.B. (2004). Particulate matter exposure in cars is associated with cardiovascular effects in healthy young men. *American Journal of Respiratory and Critical Care Medicine, 169*, 934–940.

Schliep, E., Gelfand, A., & Holland, D. (2015). Autoregressive spatially varying coefficients model for predicting daily PM 2.5 using VIIRS satellite AOT. *Advances in Statistical Climatology, Meteorology and Oceanography, 1*, 59–74.

Spira-Cohen, A., Chen, L.C., Kendall, M., Lall, R., & Thurston, G.D. (2011). Personal exposures to traffic-related air pollution and acute respiratory health among Bronx schoolchildren with asthma. *Environmental Health Perspectives, 119*, 559–565.

Squizzato, S., Masiol, M., Brunelli, A., Pistollato, S., Tarabotti, E., Rampazzo, G., & Pavoni, B. (2013). Factors determining the formation of secondary inorganic aerosol: A case study in the Po Valley (Italy). *Atmospheric Chemistry and Physics, 13*, 1927–1939.

Sunyer, J., & Basagaña, X. (2001). Particles, and not gases, are associated with the risk of death in patients with chronic obstructive pulmonary disease. *International Journal of Epidemiology, 30*, 1138–1140.

Tian, D., Hu, Y., Wang, Y., Boylan, J.W., Zheng, M., & Russell, A.G. (2009). Assessment of biomass burning emissions and their impacts on urban and regional PM2.5: A Georgia case study. *Environmental Science & Technology, 43*, 299–305.

van Donkelaar, A., Martin, R.V., Brauer, M., Kahn, R., Levy, R., Verduzco, C., & Villeneuve, P.J. (2010). Global estimates of ambient fine particulate matter concentrations from satellite-based aerosol optical depth: Development and application. *Environmental Health Perspectives, 118*, 847–855.

van Donkelaar, A., Martin, R.V., & Park, R.J. (2006). Estimating ground-level PM2.5 using aerosol optical depth determined from satellite remote sensing. *Journal of Geophysical Research-Atmospheres, 111*.

Wang, J., & Christopher, S.A. (2003). Intercomparison between satellite-derived aerosol optical thickness and PM2.5 mass: Implications for air quality studies. *Geophysical Research Letters, 30*, 2095.

Xie, Y., Wang, Y., Zhang, K., Dong, W., Lv, B., & Bai, Y. (2015). Daily estimation of ground-level PM2.5 concentrations over Beijing using 3 km resolution MODIS AOD. *Environmental Science & Technology, 49*, 12280–12288.

Zeger, S.L., Thomas, D., Dominici, F., Samet, J.M., Schwartz, J., Dockery, D., & Cohen, A. (2000). Exposure measurement error in time-series studies of air pollution: Concepts and consequences. *Environmental Health Perspectives, 108*, 419–426.

Zeng, T., Wang, Y., Yoshida, Y., Tian, D., Russell, A.G., & Barnard, W.R. (2008). Impacts of prescribed fires on air quality over the southeastern United States in spring based on modeling and ground/satellite measurements. *Environmental Science & Technology, 42,* 8401–8406.

Zhang, H., Hoff, R.M., & Engel-Cox, J.A. (2009). The relation between Moderate Resolution Imaging Spectroradiometer (MODIS) Aerosol Optical Depth and PM2.5 over the United States: A geographical comparison by US Environmental Protection Agency regions. *Journal of the Air & Waste Management Association, 59,* 1358–1369.

Zhang, T., Claeys, M., Cachier, H., Dong, S., Wang, W., Maenhaut, W., & Liu, X. (2008). Identification and estimation of the biomass burning contribution to Beijing aerosol using levoglucosan as a molecular marker. *Atmospheric Environment, 42,* 7013–7021.

Zhang, X., Hecobian, A., Zheng, M., Frank, N.H., & Weber, R.J. (2010). Biomass burning impact on PM2.5 over the southeastern US during 2007: Integrating chemically speciated FRM filter measurements, MODIS fire counts and PMF analysis. *Atmospheric Chemistry and Physics, 10,* 6839–6853.

11 Heat Hazard Monitoring with Satellite-Derived Land Surface Temperature

Yitong Jiang and Qihao Weng

CONTENTS

11.1 INTRODUCTION

Land surface temperature (LST) is a key parameter in heat hazard monitoring. The trade-off between spatial and temporal resolutions of currently available thermal infrared (TIR) images and the need for obtaining TIR have been discussed (Agam et al. 2007; Kustas et al. 2003; Schmugge et al. 1998; Weng 2009). The sensors on polar orbiting satellites, such as Landsat sensors and the Advanced Spaceborne Thermal Emission and Reflectance Radiometer (ASTER), can provide TIR data with relatively high spatial resolution. However, their low temporal resolutions are not sufficient for monitoring the diurnal change of LST. Although the Moderate-Resolution Imaging Spectroradiometer (MODIS) and the Advanced Very High Resolution Radiometer (AVHRR) produce one to two images per day for the same area, cloud coverage reduces the usage of the image data and thus increases the time between two image acquisitions. The Geostationary Operational Environmental Satellite (GOES) imager on the geostationary satellite has a much higher frequency of observation, which is every 15 min, but with a much coarser spatial resolution of 4 km. Therefore, a common solution for characterizing heat waves is to downscale GOES images from 4 to 1 km while keeping its temporal resolution.

Thermal sharpening, also called downscaling, is the way to increase the spatial and temporal resolution of TIR images (Weng et al. 2014). The spatial sharpening techniques focus on downscaling the surface temperature of a sensor to higher resolution with its visible and near-infrared bands (Weng and Fu 2014). The downscaling of Landsat LST based on its normalized difference vegetation index (NDVI) is an example of spatial sharpening. Downscaling temporal sharpening techniques are to downscale surface temperature from a coarser spatial resolution but higher temporal resolution sensor to generate high temporal resolution images (Weng and Fu 2014). To downscale GOES or SEVIRI LST to generate hourly or every 15 min LST are examples of temporal downscaling. There are two major ways to increase the spatial resolution of TIR images: the emissivity-based method (Nichol 2009) and the vegetation cover–based method (Agam et al. 2007; Jiang and Weng 2013; Kustas et al. 2003). The vegetation cover–based method was developed upon the assumption that vegetation cover is the primary driver of LST variation. Besides vegetation cover, researchers also discovered other factors that affect LST. Such auxiliary data include solar irradiation, albedo, topography, thermal inertia, and surface moisture (Zakšek and Oštir 2012).

The knowledge of surface moisture in urban areas is significant because it links LST to air temperature. A combination of air temperature, wind speed, humidity, and short- and long-wave radiation exposure determines a person's thermal comfort during urban heat island effects and heat hazard. An accurate estimation of surface moisture would aid the public health reaction during extreme weather events. Evapotranspiration (ET), the sum of evaporation and plant transpiration, is one of the important measures of the surface moisture, and it has profound impact on a person's thermal comfort. Therefore, the interactions between ET and LST are worth studying. Transpiration is the process in which water is absorbed by plants' roots and is moved to pores on the underside of leaves, where it changes to vapor and is released to the atmosphere. It is controlled by several atmospheric factors: temperature, relative humidity, topographic condition, wind and air movement, and soil and moisture availability (Christensen et al. 2008; Strelcova et al. 2013). In agricultural and forest areas, the studies on transpiration were usually conducted on individual or several species (Strelcova et al. 2013). However, in urban areas, urban forests and urban street trees contain a diverse mix of species from many regions worldwide. Therefore, the studies of transpiration in urban areas specifically considered the diversity of the species and hypothesized that species played a more important role than meteorological variables (Pataki et al. 2011). Urban built environments increase urban runoff and slow down the infiltration rate, which reduces the transpiration rate and limits the rooting depth of urban trees (Barthens et al. 2009). Plant density varies in urban areas as well. Hagishima et al. (2007) suggested that small-size plants and low spatial density vegetation may result in an increase in ET and latent heat flux in urban areas. The land cover of urban surfaces affects urban tree transpiration because energy fluxes and ambient humidity may influence the leaf-to-air vapor pressure, consequently prolonging or shortening the stomatal closure (Kjelgren and Montague 1998).

Evaporation over impervious surfaces is one of the least studied topics in the fields of urban hydrology and microclimatology (Ramamurthy and Bou-Zeid 2014). The urban area plays an important role in global climate change because of the urban heat island effects, heat wave hazards, and urban runoff. To measure the impact of urban climate in regional- or global-scale models, one of the areas that has not been studied thoroughly is urban surface moisture. Impervious surfaces contribute to urban evaporation (Ramamurthy and Bou-Zeid 2014). With the LST derived from satellite images in various combinations of temporal and spatial resolutions, and the maturity of the techniques of surface temperature derivation, the LST is becoming more and more popular for heat-related hazard monitoring. However, no single currently available TIR satellite system is capable of capturing all features of ET dynamics of agricultural fields, such as weekly or monthly trends, day-to-day fluctuations, and peaks (Cammalleri et al. 2014). Therefore, thermal downscaling and fusion seem promising to combine the advantages of multiple data sets, although the wide range of spectral, spatial, and temporal resolutions may affect the consistency of ET estimation (Cammalleri et al. 2014).

In urban areas, the land surface can be depicted by three major components: vegetation cover, impervious surfaces, and soil (Ridd 1995; Weng and Lu 2009). In remote sensing, the land cover surface in each pixel can be described by the percentage of vegetation cover, impervious surfaces, and soil. The surface moisture on pure vegetation cover, impervious surfaces, and soil can be measured by vegetation transpiration and evaporation over impervious surfaces and soil. The objectives of this study are (1) to produce the instantaneous LST with hourly and daily temporal resolution by downscaling the GOES LST from 4 to 1 km and MODIS LST from 960 to 30 m; and (2) to estimate the instantaneous latent flux by the downscaled LST, and compare the variation of LST and latent flux over vegetation, impervious surfaces, and soil before and after precipitation.

11.2   METHODOLOGY

11.2.1   STUDY AREA

The study area, Marion County, Indiana, is located in the Midwestern United States (Figure 11.1). It is the county seat of Indianapolis, the capital and largest city in the State of Indiana. Marion County is part of the Indianapolis–Carmel–Anderson Metropolitan Statistical Area. It covers both the urban and suburban areas of the city of Indianapolis. The county is located on a flat plain, which makes it possible for the city to expand in all directions. Indianapolis was the 14th largest city in the United States by population in 2014. The area lies between latitudes 39°55′37.75″ to 39°37′54.28″ and longitudes −86°19′33.65″ to −85°56′10.11″. According to the 2010 Census, the total area of Marion County is 403.01 square miles. 98.34% is land, and 1.66% is water. According to Census 2000, Indianapolis is the most populous city in

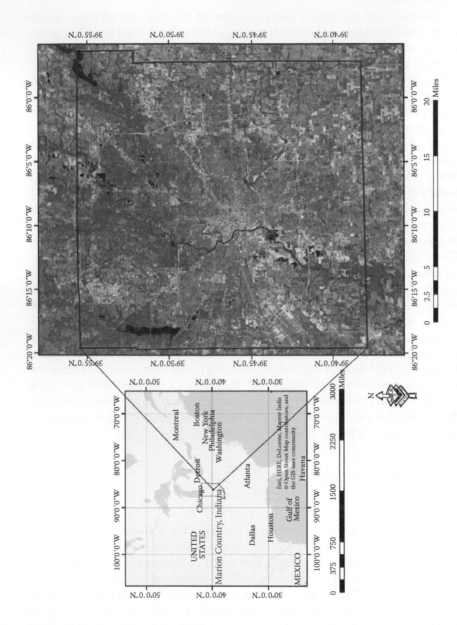

FIGURE 11.1   The location of the study area, Marion County, Indiana.

the State of Indiana and the 12th largest city in the United States. From 2000 to 2010, the population continued to increase from 781,926 to 820,445. The rate of increase is 4.9%. The estimated population in 2015 is 853,173.

Indianapolis has undergone land use/land cover (LULC) change because of the urbanization process. According to the National Land Cover Database (NLCD) 2006, the conversion to developed land from 2001 to 2006 mainly took place in the suburban areas, specifically between the circle of Interstate Highway 465 and the county boundary, with the largest changes in the Southern and Eastern fringes. Sparse land cover change in terms of density took place in the urban areas within the circle of Highway 465. According to the NLCD 2001–2006 land cover change data, 3.65% of the total land cover was changed from 2001 to 2006 in Indianapolis, Indiana, among which, 51% were changed from cultivated land to developed land, 31% were changed from a lower level to a higher level of developed land (e.g., open space to low density developed, medium density developed to high density developed), and 3.6% were changed from forest to developed.

From 2006 to 2011, more changes from lower intensity to higher intensity of developed land than from cultivated crop to developed land were found in Indianapolis. Among the total changed land covers, the first three biggest amounts of land cover change were from a lower level of developed land cover to a higher level of developed land cover: 21.7% were changed from Developed Open Space, to Developed, Medium Intensity. 12.1% were changed from Developed, Lower Intensity to Medium Intensity, and 11.2% were changed from Developed Open Space to High Intensity. Changes from cultivated crops to open space and low-intensity developed land were the next two categories; they were 9.6% and 8.1%, respectively. Changes from deciduous forest to developed open space, low intensity, and medium intensity were relatively less; they were 2.9% in total. Changes from pasture/hay to developed open space and low intensity were 3.0%.

11.2.2 DATA SETS

11.2.2.1 Data for Linear Spectral Mixture Analysis

Landsat TM images that overpass path 21 row 32 on June 17, 2001, and July 1, 2006, were used for applying Linear Spectral Mixture Analysis (LSMA) and producing images of the percentage of vegetation in each pixel. The percentage of impervious surfaces is from NLCD 2001 and 2006.

11.2.2.2 Data for Downscaling TIR Images

GOES-8 band 4 and band 5 images from June 15 to 17, 2001, were selected for downscaling. These images were chosen because they were acquired within 48 h after a precipitation, because according to Ramamurthy and Bou-Zeid (2014), the highest evaporation from impervious surface were found 48 h after precipitations. The auxiliary data included MODIS NDVI and enhanced vegetation index products; albedo in visible, NIR, and short-wave bands; emissivity in MODIS band 31 and band 32; Shuttle Radar Topography Mission (SRTM) digital elevation data; and percentage of imperviousness from NLCD 2001.

MODIS LST products acquired from DOY 179 to 200 in 2006 were used to downscale to Landsat resolution. The auxiliary data included NDVI, fractional vegetation cover, emissivity, broadband albedo from the Landsat image collected on July 1, 2006, as well as the SRTM elevation data.

11.2.2.3 Data for ET Estimation

Hourly solar radiation data were acquired from the National Solar Radiation Database (n.d.). Daily solar radiation was configured by averaging hourly data. Hourly wind speed, atmospheric temperature, air pressure, and relative humidity were acquired from the Indiana State Climate office (iClimate.org). The data were collected in the International Airport of Indianapolis. Percent of Imperviousness was downloaded from NLCD 2011. Downscaled LSTs were produced by GOES and MODIS from Section 11.2.2.2, and fractional vegetation cover data were configured by LSMA in Section 11.2.2.1.

11.2.3 Data Processing

The flowchart of methodology is shown in Figure 11.2. It is a combination of LST downscaling, LSMA, and the Two-Source Energy Balance (TSEB) model. In the LST downscaling, the hourly LST was produced by downscaling GOES LST from 4 to 1 km; the daily LST was produced by downscaling MODIS LST from 960 to 30 m. In the LSMA, the percentage of soil, vegetation, and impervious surfaces in each pixel was calculated. The TSEB model generated the instantaneous latent flux, and latent flux over pure soil, vegetation, and impervious surfaces was compared on an hourly and daily bases.

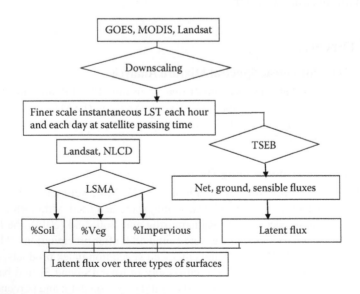

FIGURE 11.2 Flowchart of methodology.

11.2.3.1   Data Preparation and Preprocessing

The atmospheric effect of Landsat TM images was corrected using the simplified dark object subtraction method (Song et al. 2001), and the reference values rescaling gain factor and bias factor values were from Chander et al. (2009). Landsat LSTs were computed according to Coll et al. (2010). GOES LSTs were computed by split-window algorithm according to Jiménez-Muñoz and Sobrino (2008).

11.2.3.2   Vegetation–Impervious Surface–Soil Model

LSMAs were applied to derive fractional vegetation cover. First, principal component analysis (PCA) was applied to all the spectral bands of Landsat images, and the highest ranked components were selected and plotted. Second, the endmembers were selected. According to Johnson et al. (1992), the potential endmembers lay at the vertices of these PCA bands' scatterplots. The three selected endmembers were vegetation, soil, and impervious surfaces. Details about the selection of endmembers and the estimation of the fraction were discussed in Weng et al. (2008). To estimate the fraction of impervious surface, different combinations of three or four endmembers were compared. The endmembers included high albedo, low albedo, vegetation, and soil. Since this study was conducted in an urban area, the criteria for selecting the best-suited fraction images included high-quality fraction images for urban landscape, low error, and the distinction among typical LULC types. The fractional soil cover was produced based on the percentage of impervious surface and vegetation cover—besides the vegetation and impervious surfaces, the rest of each pixel was assumed to be soil. Water bodies were masked out.

11.2.3.3   Downscaling LST

The GOES-8 Imager possesses one visible band and four infrared bands. The data were downloaded from the National Oceanic and Atmospheric Administration (NOAA) comprehensive large array-data stewardship system (http://www.nsof .class.noaa.gov/saa/products/search?datatype_family=GVAR_IMG). The NOAA Weather and Climate Toolkit was used to export the AREA files to TIFF format. AREA files are count data with calibration coefficients. The toolkit uses the calibration information contained in the calibration block of the AREA file and converts the raw counts (10 bit precision) to brightness temperatures for the IR channels. Then, GOES LSTs were retrieved using the split-window algorithm and coefficients from Jiménez-Muñoz and Sobrino (2008). Since LST is strongly influenced by parameters such as solar irradiation, albedo, topography, thermal inertia, and vegetation cover (Weng et al. 2004), the corresponding auxiliary data were obtained for downscaling. The downscaling method followed Zakšek and Oštir (2012). First, PCA was applied on auxiliary data in finer spatial resolution; second, the highest ranked principal components were upscaled to coarser resolution; third, regressions between LST and upscaled components were formed; last, the LSTs at finer spatial resolution were estimated by finer resolution components and the regressions (Figure 11.3). The GOES LSTs were downscaled from 4 to 1 km every hour; the MODIS LSTs are downscaled from 960 to 240 m, and then from 240 to 30 m each day at the satellite passing time.

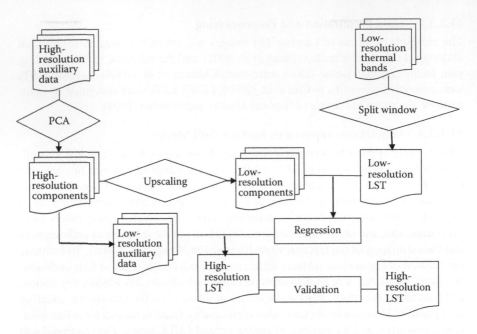

FIGURE 11.3 The flowchart for LST downscaling.

11.2.3.4 The Land Surface Moisture Model

The method of estimating LE followed Weng et al. (2013). Net radiation (R_n) was calculated as

$$R_n = (1-\alpha)R_{short} + \varepsilon\varepsilon_a\sigma T_a^4 - \varepsilon\sigma T_s^4, \qquad (11.1)$$

where R_n is the net radiation, σ is the Stefan–Boltzmann constant (5.67×10^{-8}W/m²K⁴), ε_a is atmospheric emissivity, α is broadband albedo, R_{short} is short-wave radiation, ε is surface broadband emissivity, T_a is atmospheric temperature, and T_s is surface temperature. Broadband albedo was calculated according to Liang (2000), and broadband emissivity was calculated according to Ogawa et al. (2003).

Atmospheric emissivity was calculated as

$$\varepsilon_a = 1.24\left(\frac{e_a}{T_a}\right)^{1/7}, \qquad (11.2)$$

where e_a is atmospheric water vapor pressure, which was estimated based on saturation water vapor pressure and relative humidity (Weng et al. 2013).

Ground heat flux was calculated based on net radiation and a coefficient (c_g) that describes the influence of surface cover material, seasonality, and diurnal change.

The land cover was classified into water, bare soils, grass, forest, urban, and agriculture, and c_g values vary among different land cover types:

$$G = c_g \times R_n. \tag{11.3}$$

Since the study area is the same, the c_g values were adopted from Weng et al. (2013) in Table 11.1.

Sensible heat flux (H) was calculated separately for non-vegetated areas and vegetated areas. The non-vegetated areas and vegetated areas are differentiated by the amount of vegetation cover, and the value of 0.5 was used as the threshold:

$$H = f_{\text{non-veg}} H_{\text{non-veg}} + f_{\text{veg}} H_{\text{veg}} \tag{11.4}$$

$$H_{\text{non-veg}} = \rho_a c_p \frac{T_s - T_a}{R_{AH} + R_s} \tag{11.5}$$

$$H_{\text{veg}} = \rho_a c_p \frac{T_c - T_a}{R_{AH}}, \tag{11.6}$$

where ρ_a is the air density in kg/m³, c_p is the specific heat of air at constant pressure in J/(kg·K), T_s and T_c are surface temperature for non-vegetated and vegetated areas, and T_a is air temperature. R_{AH} is the aerodynamic resistance in s/m, and R_s is the resistance to heat flow in the boundary layer immediately above soil surface. R_{AH} was calculated as

$$R_{AH} = \frac{\left[\ln\left(\frac{z_u - d_0}{z_{0M}} \right) - \Psi_M \right]\left[\ln\left(\frac{z_t - d_0}{z_{0H}} \right) - \Psi_H \right]}{k^2 u}. \tag{11.7}$$

TABLE 11.1
c_g Values According to LULC Types for Ground Heat Flux

LULC Types	c_g
Water	0.35
Bare soils	0.30
Grass	0.30
Forest	0.15
Urban	0.40
Agriculture	0.30

According to Weng et al. (2013), z_u and z_t are the heights at which the wind speed u and atmospheric temperature are measured, which are 10 and 2 m, respectively. d_0 is the displacement height, and z_{0M} and z_{0H} are the roughness lengths for momentum and heat transport, respectively. Ψ_M and Ψ_H are stability correction functions for momentum and heat, respectively, and k is von Karman's constant, which is equal to 0.4 (Weng et al. 2013). This equation may be simplified by removing Ψ_M and Ψ_H.

R_s can be calculated as

$$R_s = \frac{1}{a + bu_s},$$ (11.8)

where a is the free convective velocity, which is equal to 0.04 m/s, b is a coefficient to represent the typical soil surface roughness, and u_s is the wind speed over soil surface at the height of 0.05–0.2 m.

LE was also calculated for vegetated and non-vegetated areas separately:

$$LE = f_{non\text{-}veg} LE_{non\text{-}veg} + f_{veg} LE_{veg}$$ (11.9)

$$LE_{non\text{-}veg} = R_{n,non\text{-}veg} - G - H_{non\text{-}veg}$$ (11.10)

$$LE_{veg} = \alpha_{PT} f_G \frac{\Delta}{\Delta + \gamma} R_{n,veg},$$ (11.11)

where α_{PT} is the Priestley–Taylor parameter, which is equal to 1.26, γ is the psychrometric constant, Δ is the slope of saturation vapor pressure–temperature curve. f_G is the fraction of the leaf area index that is green, and it is equal to unity when it is not available. If $LE_{non\text{-}veg}$ is negative, it was set to zero, $H_{non\text{-}veg}$ was recomputed as the residual of Equation 11.10.

11.3 RESULTS

11.3.1 PURE IMPERVIOUS SURFACES, SOIL, AND VEGETATION COVER PIXELS

The spatial distribution of vegetation, soil, and impervious surface fractions, as well as the pure pixels of the three land cover types (red dots), are shown in Figure 11.4. Figure 11.4a through c are in 30 m resolution, and Figure 11.4d through f are in 1 km resolution. A resolution of 30 m shows the distribution of the three land types much clearer than the 1-km resolution. The high-percentage impervious surfaces were distributed in the central urban areas, along the major roads and highways, and in the commercial sites along the highway. The vegetation cover is mainly located in the urban forest along the White River, beside Eagle Creek Reservoir and Geist Reservoir, in low-density residential land in the north and in agricultural fields in the south. The distribution of pure soil pixels was mainly in the agriculture fields, open space, and airport in the south.

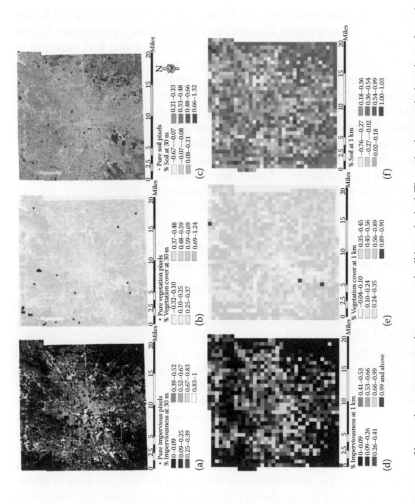

FIGURE 11.4 The percentage of imperviousness, vegetation cover, and soil in each pixel. The red points are the locations of pure impervious surface, vegetation cover, and soil pixels. (a) Percentage of imperviousness at 30 m resolution; (b) percentage of vegetation cover at 30 m resolution; (c) percentage of soil at 30 m resolution; (d) percentage of imperviousness at 1 km resolution; (e) percentage of vegetation at 1 km resolution; and (f) percentage of soil at 1 km resolution.

In the 1-km-resolution images, there are no 100% pure impervious surface and vegetation cover pixels. Therefore, the highest concentration pixels with 99% of impervious surface and 89% of vegetation cover were used as the pure pixels. The pure pixels of the three land cover types were selected to monitor the characteristics of the daily and hourly LST and Latent flux changes over the tree land cover types.

11.3.2 The Changes of Spatial Distribution of Daily and Hourly LST

The spatial and temporal variability of LST at 1 km resolution is shown in Figure 11.5. It was produced using GOES 4 km LST and the downscaling technique. Taking June 16, 2001, for example, from 9:00 to 15:00, the downscaled LST was able to record the changes in spatial distribution over time. At 9:00, the built-up areas were 300–305 K, and the vegetated areas were 295–300 K; at 10:00, the LST for the whole area increased 5–10 K; at 11:00, the LST of central urban area, airport, and commercial land along the Interstate Highway reached 315 K; at noon, the built-up area remained above 310 K; the high LST shrank at 14:00, and the LST of the whole area decreased below 310 K at 15:00. The highest LST was found at the center of Indianapolis, Indiana, and it lasted longer than surrounding areas. Compared to 4-km-resolution images, the heat hazard can be located more accurately and efficiently with the increased spatial and temporal resolution. With exposure and vulnerability information, the heat hazard risk areas can be decided (Jiang et al. 2015).

11.3.3 Instantaneous LST and Latent Flux over Pure Impervious Surfaces, Soil, and Vegetation Cover with an Hour or Day Interval

The instantaneous LST and latent flux over vegetation, soil, and impervious surface show various responses to rainfall during multiple precipitations in July 2006 (Figure 11.6). The three land cover types had a similar pattern of LST change. In particular, the LSTs of soil and impervious surfaces were parallel to each other. This is because the impervious surface usually has the highest sensible heat. Soil moisture and evaporation from soil may lower down the surface temperature. The pattern of LST of vegetation cover is a little different: it tended to have smoother change of LST during rainfalls. It may attributed to evaporation and transpiration, which adjusted the direct impact from solar radiation and reflectance from nearby objects. The impervious surfaces had the highest LST before and after the precipitations, it was 1–2 K higher than soil, and 8–10 K higher than vegetation cover. Precipitation tended to make the LST difference between the three land surfaces smaller. After the 6.32-cm precipitation on 193 days of the year, the LST of impervious surfaces dropped 2 K, and the LST of soil surfaces dropped less than 2 K in 2 days. Precipitation has much smaller impact on LST over vegetation cover. There is no apparent temperature drop after precipitation over vegetation cover.

Figure 11.6b explains the LE changes over three land cover types over time. Vegetation cover and soil surface share the similar pattern of latent flux change during multiple precipitations. The latent flux of vegetation cover and soil was high not long after the 0.64-cm precipitation on DOY 179, and it gradually dropped until it received the next rainfall. After the 6.32-cm precipitation, the latent flux of

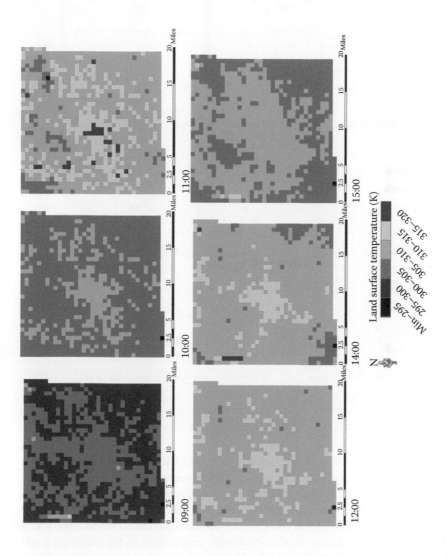

FIGURE 11.5   Selected hourly LSTs at 1 km resolution on June 16, 2001.

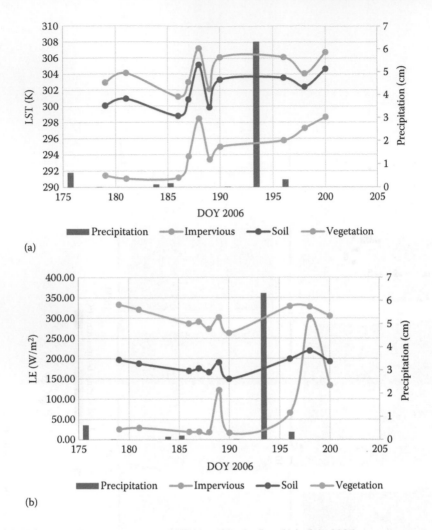

FIGURE 11.6   The instantaneous LST (a) and latent flux (b) in July 2006.

vegetation cover and soil increased and then dropped within a week. The pattern of latent flux over impervious surfaces is markedly different from the natural land covers such as vegetation and soils. The LE over impervious surfaces generated sharp increase and decrease within a week after the rainfall, and it was higher than soil surfaces and close to the amount from vegetation. Lacking the ability to store water, the impervious surfaces contribute considerable evaporation in the urban area and thus alter the natural energy and water cycle.

Figure 11.7a shows the changes of LST before and after a 2.54-cm precipitation. During and shortly after the precipitation, there is a drop of LST from three types of land cover types, among which the LST over soil decreased more rapidly, and LST of impervious surfaces remained higher compared to vegetation and soil. There was a sudden drop of LST around midnight. The LST of the three land cover types in the second day was relatively smooth. The highest LST from impervious surface and the LST of soil and impervious surface were almost the same. The figure only shows 48 h after the precipitation.

Figure 11.7b shows the change of latent flux over the three land cover types after the 2.54-cm precipitation. During the rainfall, the latent flux from soil is much higher than that from vegetation and impervious surfaces. However, after the precipitation, the latent flux over impervious surface is more than 100 W/m^2 higher than soil and more than 250 W/m^2 higher than vegetation cover at noon. This result indicates that the evaporation over impervious surfaces may remain high for more than 48 h in the Midwest.

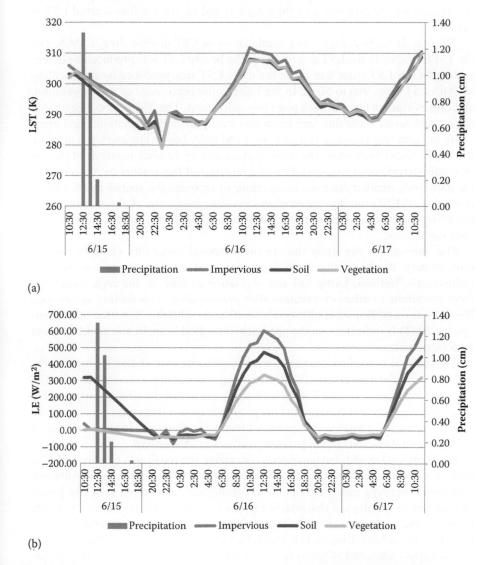

(a)

(b)

FIGURE 11.7   The hourly LST (a) and latent flux (b) in June 2001 after precipitation.

11.4 CONCLUSIONS

This chapter introduced a case study of using the satellite-derived LST to monitor the heat hazard. It focuses on two aspects: first, using the thermal downscaling technique to increase the spatial and temporal resolution; second, applying the TSEB model in urban areas, categorizing the urban land cover into impervious surfaces, soil, and vegetation, and discussing the characteristics of LST, evaporation, and transpiration over three land cover types. This research is meaningful because the surface moisture in urban areas is one of the least studied field. To monitor the heat hazard accurately, the impacts from surface moisture cannot be ignored, and an efficient way to estimate air temperature based on the satellite-derived LST is desirable.

The overall methodology was a combination of LST downscaling, LSMA, and the TSEB model. In the LST downscaling, the hourly LST was produced by downscaling GOES LST from 4 to 1 km; the daily LST was produced by downscaling MODIS LST from 960 to 30 m. In the LSMA, the percentage of soil, vegetation, and impervious surfaces in each pixel was calculated. The TSEB model generated the instantaneous latent flux, and latent flux over pure soil, vegetation, and impervious surfaces was compared on an hourly and daily basis. The approach of using the TSEB model to estimate the urban surface energy balance maximized the use of multispectral remote sensing data and minimized the amount of meteorological data from weather stations. In addition, to increase the spatial and temporal resolution of LST, the downscaling process also made use of the satellite data, including the thermal bands from GOES and MODIS and the auxiliary data at finer resolutions.

The findings for the study site are the temporal variability of surface moisture, namely, the latent flux from vegetation, soil, and impervious surfaces in Indianapolis, Indiana. Using soil and vegetation as control, the impervious surfaces contribute to urban evaporation after precipitation. This finding agrees with Ramamurthy and Bou-Zeid (2014) in a small scale with the Princeton urban canopy model. This indicates that the approached method is applicable in urban areas with large scales.

The study succeeds in accounting for urban moisture in the following aspects: the LSMA picked the pure pixels, LST downscaling successfully generated finer spatial and temporal scale LST, and TSEB successfully estimated the similar pattern of ET over vegetation and soil surfaces and the prompt response from impervious surfaces after precipitation. Therefore, the proposed method is promising in the estimation of ET over vegetation, soil, and impervious surfaces in urban areas.

In the original model, the parameters describing the roughness condition of land surface, namely, d_0, z_{0M}, and z_{0H}, were fixed according to the LULC types. Although each land cover type may yield similar morphological parameters, each city has its own unique morphological characteristics, which relate to history, culture, tradition, planning, and topography; it varies over regions, countries, and continents. Contexts that affect the urban setting include hydraulic factors, transportation hubs, trade centers, defensive sites, and religious factors. Therefore, in future research, the estimation of morphological parameters from LiDAR data is highly desirable.

REFERENCES

Agam, N., Kustas, W. P., Anderson, M. C., Li, F., and Neale, C. M. U. (2007). A vegetation index based technique for spatial sharpening of thermal imagery. *Remote Sensing of Environment, 107*(2007), 545–558.

Barthens, J., Day, S. D., Harris, J. R., Wynn, T. M., and Dove, J. E. (2009). Transpiration and root development of urban trees in structural soil stormwater reservoirs. *Environmental Management, 44*(4), 646–657.

Cammalleri, C., Anderson, M. C., Gao, F., Hain, C. R., and Kustas, W. P. (2014). Mapping daily evapotranspiration at field scales over rainfed and irrigated agricultural areas using remote sensing data fusion. *Agricultural and Forest Meteorology, 186*(2014), 1–11.

Chander, G., Markham, B. L., & Helder, D. L. (2009). Summary of current radiometric calibration coefficients for Landsat MSS, TM, ETM+, and EO-1 ALI sensors. *Remote Sensing of Environment, 113*, 893–903.

Christensen. L., Tague, C. L., and Baron, J. S. (2008). Spatial patterns of simulated transpiration response to climate variability in a snow dominated mountain ecosystem. *Hydrological Processes, 22*(18), 3576–3588.

Coll, C., Galve, J. M., Sanchez, J. M., and Caselles, V. (2010). Validation of Landsat-7/ETM+ thermal-band calibration and atmospheric correction with ground-based measurements. *IEEE Transactions on Geoscience and Remote Sensing, 48*, 547–555.

Hagishima, A., Narita, K. I., and Tanimoto, J. (2007). Field experiment on transpiration from isolated urban plants. *Hydrological Processes, 21*(9), 1217–1222.

Jiang, Y., Fu, P., and Weng, Q. (2015). Downscaling GOES land surface temperature for assessing heat wave health risks. *IEEE Geoscience and Remote Sensing Letters, 12*(8), 1605–1609.

Jiang, Y. and Weng, Q. (2013). Estimating LST using a vegetation-cover-based thermal sharpening technique. *IEEE Geoscience and Remote Sensing Letters, 10*(5), 1249–1252.

Jiménez-Muñoz, J. and Sobrino, J. A. (2008). Split-window coefficients for land surface temperature retrieval from low-resolution thermal infrared sensors. *IEEE Geoscience and Remote Sensing, 5*(4), 806–809.

Johnson, P. E., Smith, M. O., and Adams, J. B. (1992). Simple algorithms for remote determination for mineral abundances and particles sizes from reflectance spectra. *Journal of Geophysical Research, 97*, 2649–2657.

Kjelgren, R., and Montague, T. (1998). Urban tree transpiration over turf and asphalt surfaces. *Atmospheric Environment, 32*(1), 35–41.

Kustas, W. P., Norman, J. M., Anderson, M. C., and French, A. N. (2003). Estimating subpixel surface temperatures and energy fluxes from the vegetation index–radiometric temperature relationship. *Remote Sensing of Environment, 85*(2003), 429–440.

Liang, S. (2000). Narrowband to broadband conversions of land surface albedo I: Algorithms. *Remote Sensing of Environment, 76*, 213–238.

National Solar Radiation Database (n.d.). Retrieved from http://rredc.nrel.gov/solar/old_data /nsrdb/1991-2010/targzs/targzs_by_state.html.

Nichol, J. E. (2009). An emissivity modulation method for spatial enhancement of thermal satellite images in urban heat island analysis. *Photogrammetric Engineering & Remote Sensing, 60*(5), 547–556.

Ogawa, K., Schmugge, T., Jacob, F., and French, A. (2003). Estimation of land surface window (8–12 µm) emissivity from multispectral thermal infrared remote sensing—A case study in a part of Sahara Desert. *Geophysical Research Letters, 30*(2), 1067.

Pataki, D. E., McCarthy, H. R., Litvak, E., and Pincetl, S. (2011). Transpiration of urban forests in the Los Angeles metropolitan area. *Ecological Applications, 21*(3), 661–677.

Ramamurthy, P. and Bou-Zeid, E. (2014). Contribution of impervious surfaces to urban evaporation. *Water Resources Research, 50*, 2889–2902.

Ridd, M. K. (1995). Exploring a V–I–S (Vegetation–Impervious Surface–Soil) model for urban ecosystem analysis through remote sensing: Comparative anatomy for cities. *International Journal of Remote Sensing, 16,* 2165–2185.

Schmugge, T. J., Kustas, W. P., and Humes, K. S. (1998). Monitoring land surface fluxes using ASTER observations. *IEEE Transactions on Geoscience and Remote Sensing, 36*(5), 1421–1430.

Song, C., Woodcock, C. E., Seto, K. C., Lenney, M. P. and Macomber, S. A. (2001). Classification and change detection using Landsat TM data: When and how to correct atmospheric effects? *Remote Sensing of Environment, 75,* 230–244.

Strelcova, K., Kurjak, D., Lestianska, A., Kovalcikova, D., Ditmarova, L., Skvarenina, J., and Ahmed, Y. A. R. (2013). Differences in transpiration of Norway spruce drought stressed trees and trees well supplied with water. *Biologia, 68*(6), 1118–1122.

Weng, Q. (2009). Thermal infrared remote sensing for urban climate and environmental studies: Methods, applications, and trends. *ISPRS Journal of Photogrammetry and Remote Sensing, 64*(4), 335–344.

Weng, Q. and Fu, P. (2014). Modeling diurnal land temperature cycles over Los Angeles using downscaled GOES imagery. *ISPRS Journal of Photogrammetry and Remote Sensing, 97,* 78–88.

Weng, Q., Fu, P., and Gao, F. (2014). Generating daily land surface temperature at Landsat resolution by fusing Landsat and MODIS data. *Remote Sensing of Environment, 145,* 55–67.

Weng, Q., Hu, X., and Lu, D. (2008). Extracting impervious surfaces from medium spatial resolution multispectral and hyperspectral imagery: A comparison. *International Journal of Remote Sensing, 29,* 3209–3232.

Weng, Q., Hu, X., Quattrochi, D. A., and Liu, H. (2013). Assessing intra-urban surface energy fluxes using remotely sensed ASTER imagery and routine meteorological data: A case study in Indianapolis, U.S.A. *IEEE Journal of Selected Topics in Applied Earth Observations and Remote Sensing,* doi: 10.1109/JSTARS.2013.2281776.

Weng, Q. and Lu D. (2009). Landscape as a continuum: An examination of the urban landscape structures and dynamics of Indianapolis city, 1991–2000, by using satellite images. *International Journal of Remote Sensing, 30*(10), 2547–2577.

Weng, Q., Lu, D., and Schubring J. (2004). Estimation of land surface temperature-vegetation abundance relationship for urban heat island studies. *Remote Sensing of Environment, 89*(2004), 467–483.

Zakšek, K. and Oštir, K. (2012). Downscaling land surface temperature for urban heat island diurnal cycle analysis, *Remote Sensing of Environment, 117,* 114–124.

12 Remote Sensing Identification of Threshold Zones along a Mediterranean to Arid Climatic Gradient

Maxim Shoshany

CONTENTS

12.1 INTRODUCTION

Debates regarding desertification have been heated especially since Thomas and Middleton (1994) challenged the *marching desert myth* in response to earlier suggestions by Lamprey (1975) that the Sahara desert boundary in Sudan is advancing by 5.5 km every year. Controversies in interpreting spatial and temporal information regarding shifts in desert threshold zones are not limited to the Sahel region (e.g., Nicholson 2011) but also characterize other semiarid regions (Veron et al. 2006) and the semiarid margins of the Mediterranean in particular (Safriel 2009; Thornes 2000). Although there is a consensus regarding the important role of remote sensing in monitoring desertification, there are disagreements regarding the interpretation of satellite imagery for this purpose. This is well exemplified by the disagreement of Hein and De Ridder (2006) with claims regarding greening of the Sahel made by Herrmann et al. (2005). According to Veron et al. (2006), controversies regarding desertification emerge mainly from methodological and terminological differences. They suggest that improvements in exploring desertification over large regions may be obtained by strengthening the synergy between ecology and

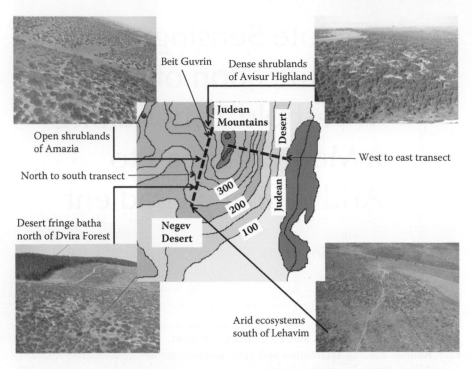

FIGURE 12.1 Map of the climatic gradient of the Judean Mountains with photographs of the four main ecosystems along the gradient and location of transects.

remote sensing. Within such an approach, this remote sensing study implements vegetation phenomenologies in investigating ecosystems' transition between semiarid Mediterranean to arid region. Three modes of vegetation change are analyzed: green vegetation cover (GVC) change between years of extreme high and low rainfall levels, life-forms' composition change, and variations in spatial erosion versus recovery potentials. Landsat images of the climatic gradient between the Judean Mountains in central Israel and the Judean Desert in the east and the Negev Desert in the south (Figure 12.1) facilitate this eco-exploration.

12.2 ECOSYSTEM TRANSITIONS ALONG A CLIMATIC GRADIENT

Mediterranean to arid climatic gradients characterize wide regions at the western sides of South and North America, South Africa, and Australia, and in the southern and eastern margins of the Mediterranean Basin. Ecosystems along these gradients vary between woodlands, dense shrublands, open shrublands, dwarf-shrublands (batha), and arid vegetation (herbaceous species with scattered patches of dwarf-shrubs). These life-form transitions are governed by rainfall, lithology, and soil, together with strong anthropogenic influences attributed to fire, grazing, and wood-cutting. Desertification representing long-term (possibly irreversible) soil erosion and loss of vegetation productivity was reported in several well-known studies (e.g., Cerda and Lavee 1999; Geeson et al. 2002; Hill et al. 2008; Lavee et al. 1998) for the margins of the Mediterranean regions. The interplay between land use change and climate parameters was hypothesized by Milan et al. (2005) and Puigdefabregas and Mendizabal (1998) and is thought to be responsible for the desert-like conditions found in the vast Mediterranean regions of eastern Tunisia, Libya, Egypt, and in Almeria in Spain. Spatiotemporal changes in shrubs, dwarf-shrubs, and grass cover characterize this land degradation (e.g., Maestre et al. 2009; Papanastasis et al. 2003; Peinado et al. 1995). Identification of threshold zones between these vegetation cover typologies is instrumental for studying desertification in semiarid Mediterranean landscapes. According to Fagre and Charles (2009), thresholds can be identified "where small changes in an environmental driver produce large, persistent responses in an ecosystem." In the Mediterranean to arid transition context, this refers to discontinuity among vegetation and soil conditions along gradual climate change as a result of ecosystems' hysteresis (Meron et al. 2004), or from the loss of recovery potential.

The climatic gradients between the Judean Mountains in central Israel and their eastern and southern desert margins represent a wide range of differences in natural settings (e.g., rainfall and lithology) and anthropogenic influences. Implementation of remote sensing on west to east versus north to south gradients is expected to allow one to explore the formation of sharp transition zones between ecosystems in general and of desert thresholds in particular.

12.3 REMOTE SENSING TECHNIQUES AND THEIR IMPLEMENTATION

In this chapter, we will integrate three techniques, developed and implemented in earlier studies for assessing modes of vegetation change along the climatic gradients of the Judean Mountains. The first method is based on a multitemporal technique developed by Shoshany et al. (1994, 1995, 1996), the second method presents an algorithm for life-form decomposition as implemented by Shoshany and Svoray (2002), and the third method concerning estimation of spatial potentials for erosion versus recovery adopts the work reported by Shoshany (2012).

12.3.1 GREEN VEGETATION COVER

Vegetation cover is a primary indicator for the location of desert threshold zones. A study of vegetation cover changes between years of extreme rainfall levels was found to be most informative regarding the type of response of different ecosystems along a climatic gradient to rainfall changes. This was conducted for the years 1991 and 1992, which had a unique combination of rainfall regimes in the Eastern Mediterranean; specifically, 1991 was the 60-year record for low precipitation and 1992 was the year with the highest rainfall during this time frame (Kutiel et al. 1995; Shoshany et al. 1996). Here, we conducted a similar study while extending the investigation from 2 to 16 years, between 1996 and 2011. For this purpose, we used the same empirical multiple regression vegetation reflectance (VR) model, which was developed by Shoshany et al. (1994) specifically for estimating GVC in our study area:

$$\text{GVC} (\%) = 1.48 + 2.5R[1] - 25.5R[2] - 266.97R[3] + 275.5R[4]. \qquad (12.1)$$

For constructing this model, the reflectance data for small plot areas at Landsat TM channels (R[1]...R[4]) were acquired using the NASA Radiometer, while vegetation cover was calculated by classifying close-range photographs of these plot areas into green vegetation and non-vegetation areas. The VR model was tested for 26 plot areas at four sites along the climatic gradient between the Judean Mountains and the Judean Desert; it was found to be highly significant with an R^2 value of 0.88 and a P value of 0.001.

This model was reapplied here to five radiometrically corrected and geometrically co-registered images of the end of the summer, for 3 years of extreme high and low rainfall (Table 12.1) in between 2 years with close to average rainfall (1996 and 2011).

Two transects were selected along the climatic gradients representing the less-disturbed landscape sequences from the Judean Mountains to the Judean Desert in the east and the Negev Desert in the south. The high spatial and temporal variability of GVC values represents both the high GVC characteristic of protected shrublands and forest plantations versus the low green values in areas undergoing high grazing pressures and frequent fires.

A step-like form of change can be used for generalizing GVC changes in both directions (Table 12.1 and Figure 12.2):

- North to south transect (Figure 12.2b): Two *steps* of a sharp GVC decrease characterize the transition from dense shrublands to the desert fringe batha (dwarf-shrubs). The first step occurs where dense shrublands, which recovered (Shoshany 2002) in areas with rainfall above 400 mm/year (section I with GVC values between 22 and 37%), are rapidly transformed into composites of highly disturbed shrublands dissected by stripes of fields and other land uses. Further south, the landscape is characterized by open shrublands of moderate density (section II with GVC values between 17% and 23%) in areas of approximately 350 mm/year average rainfall. The transition of this area into desert fringe Batha (section III with GVC values between 8% and 12%) of *Sarcopoterium spinosum* spread over large extents occurs within less than 2 km. Such a sharp transition is the result of historic high grazing pressures followed by frequent fires ignited by army training in this area. The two sharp transition zones of this transect are thus primarily anthropogenic, where heavy soil erosion in the northern parts of section III would severely slow down any future recovery trends. The transition between the desert fringe batha and the arid unit (section IV with GVC between 6% and 13%) does not form a sharp threshold, as can be inferred from their overlap in GVC values.
- West to east transect (Figure 12.2a): The GVC change along this transect can be well described by both a step-like form and a convex transition from the

TABLE 12.1

GVC and Life-Form Composition in Sections along the Climatic Gradient between the Judean Mountains and the Judean Desert in the East and the Negev Desert in the South (Figure 12.1)

Year	1996	1999	2003	2008	2011	Life-Form Composition		
Percentage of the MAP	102%	66%	143%	70%	94%			
West–east section I	21.08	17.46	25.07	23.74	22.72			
	(5.38)	(4.12)	(5.63)	(4.35)	(5.18)			
West–east section II	15.06	11.05	16.71	16.04	14.15			
	(3.65)	(3.12)	(5.51)	(5.18)	(4.70)			
West–east section III	5.29	5.37	10.19	6.04	6.04			
	(2.55)	(2.67)	(2.71)	(2.06)	(2.62)			
						Shrubs	Dwarf-shrubs	Soil and rocks
North–south section I	34.41	22.61	36.70	30.50	25.80	0.73	0.04	0.23
	(2.06)	(2.49)	(2.75)	(2.67)	(2.79)			
North–south section II	21.78	17.39	24.94	22.76	17.50	0.32	0.23	0.45
	(3.09)	(1.86)	(2.89)	(2.72)	(2.46)			
North–south section III	10.95	8.94	11.95	11.47	8.45	0.065	0.585	0.35
	(3.10)	(1.85)	(2.83)	(2.68)	(2.40)			
North–south section IV	7.67	6.16	12.56	8.64	6.71	0.05	0.31	0.64
	(2.08)	(2.50)	(2.90)	(2.64)	(2.78)			

Source: *Summary of rainy season for 2011/2012 and the main hydrological characteristics.* Reported by the Israel Water Authority.
Note: Rainfall data represent the percentage of the mean annual precipitation (MAP).

crests of the Judean Mountains in the west to the desert margins in the east. Two sharp transition zones can be detected. The western one has GVC values between 17% and 25% of the open shrubland declining toward the highly disturbed zone with GVC levels between 11% and 17%, representing built-up, rural, and agricultural land uses with scattered islands of natural vegetation (Sharakas et al. 2003; Shoshany et al. 1995, 1996). The eastern one represents a drop-down within a distance of 5 km from an average GVC of 15% as obtained for the seminatural desert fringe batha mixed with tree plantations to an average GVC of 6% characterizing the arid ecosystem. The sharp transition between the desert fringe batha and the arid ecosystem corresponds to the boundary between the desert and brown lithosols and the Rendzina soil (Dan and Raz 1970). The rate of GVC that increases west from this boundary is jointly governed by rainfall and both positive and negative human impacts: tree planting, on the one hand, and grazing and fires, on the other.

Generally, profiles from all years fit well and describe similar spatial forms of change beyond significant temporal differences owing to changes in precipitation

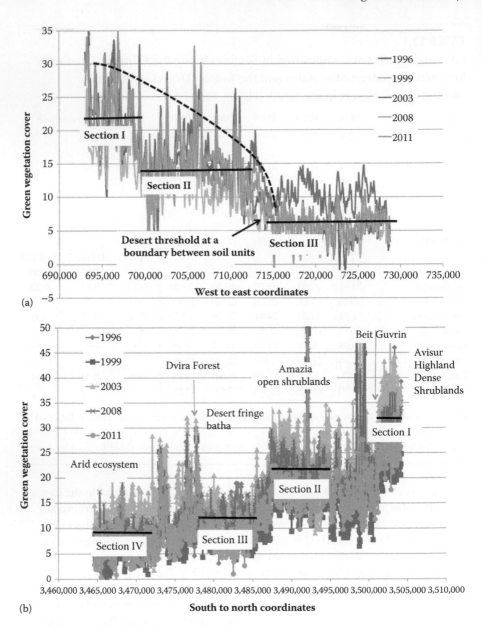

FIGURE 12.2 GVC along transects. (a) GVC change along the west to east transect. (b) GVC change along the north to south transect.

between years. This result is most meaningful, considering the fact that the rainfall fluctuated between 66% and 143% of the mean annual precipitation (MAP) between 1996 and 2011. Within the framework of temporal changes in threshold zones, it is most important to examine the GVC variations in locations of highest transitions in both transects.

The boundary between the desert fringe batha section and the arid section in the west to east transect, which is attributed to soil transition, seems to be maintained even during years of extremely high rainfall (such as 2003). However, when the profiles are examined in detail, there seems to be short extents in the desert fringe batha section in which their GVC values are only slightly higher (~2%) than those of the arid section. During the years with extremely low rainfall, the GVC of these desert fringe batha section drops to its level in the arid section. Permanent desertification may take place in this section with repeated fires, wood-cutting, and gazing, followed by soil erosion.

The boundary between the open shrublands (section II) and the desert fringe batha (section III) in the north to south transect shows a catastrophic shift owing to overgrazing and fire. The GVC changed by only 3% in section III between 1999 and 2003, suggesting that it lost some of its recovery potential. Section II shows an increase in GVC that is twice as high between these years (from 17% to 25%). The open shrublands thus maintain some level of resilience to dry periods. However, since the lithology and rainfall are similar on both sides of this partition line, frequent fires may transform it relatively quickly into a degraded ecosystem similar to Unit III.

12.3.2 LIFE-FORM COMPOSITIONS

The spatial heterogeneity of life-form compositions at the subpixel level results in the non-uniqueness of spectral and phenological information at moderate and low spatial resolutions. Although the mapping of trees, shrubs, and herbaceous growth in large homogenous areas was reported in a number of studies, an estimation of life-form compositions in general, and of dwarf-shrubs in particular, in Mediterranean and semiarid environments having high spatial heterogeneity have received very limited attention (e.g., Hamada et al. 2013; Shoshany and Svoray 2002). An adaptive phenological (seasonal) and zonal unmixing technique was developed by Shoshany and Svoray (2002) in order to estimate the area fraction of these life-forms for each pixel. The fundamental principles underlying this technique are as follows (Figure 12.3):

a. Selecting vegetation, soil, and rock endmembers and performing the unmixing separately for the semiarid and arid regions for each season (end of winter, beginning of summer, and end of summer).
b. Estimating the fraction of each life-form based on subtracting the corresponding seasonal vegetation fractions. For example, fractional cover of herbaceous growth is estimated according to the difference between the end of the winter (end of March) vegetation fraction and that of the beginning of the summer (beginning of June).

A north to south transect (following the same line as in Section 12.3.1) representing changes in life-form compositions allowed us to assess the types of life-form transitions along the climatic gradient (Figure 12.4) for 1995, a year having 113% of the MAP. The spatial structure of life-form compositions along the transects fits well the structure inferred based on the GVC. However, life-form compositions provide

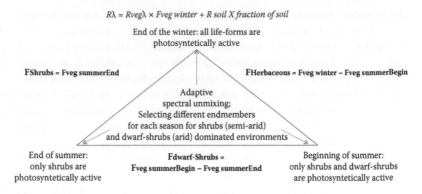

$R\lambda = Rveg\lambda \times Fveg\ winter + R\ soil\ X\ fraction\ of\ soil$

End of the winter: all life-forms are
photosyntetically active

FShrubs = Fveg summerEnd   FHerbaceous = Fveg winter − Fveg summerBegin

Adaptive
spectral unmixing;
Selecting different endmembers
for each season for shrubs (semi-arid)
and dwarf-shrubs (arid) dominated environments

End of summer: **Fdwarf-Shrubs =**   Beginning of summer:
only shrubs are   **Fveg summerBegin − Fveg summerEnd**   only shrubs and dwarf-shrubs
photosyntetically active   are photosyntetically active

$R\lambda = Rveg\lambda \times Fveg\ summerEnd + R\ soil\ X\ fraction\ of\ soil$ $R\lambda = Rveg\lambda \times Fveg\ summerBegin + R\ soil\ X\ fraction\ of\ soil$

FIGURE 12.3   Schematic description of the algorithm for life-forms' unmixing.

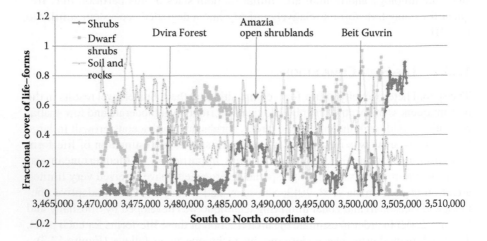

FIGURE 12.4   Life-forms' composition along the north to south transect.

much more explicit information regarding the nature of the transitions within and
between threshold zones:

- The sharp southern boundary of the Dense Shrublands of the Avisur
  Highland (shrub fraction = 0.76) at Beit Guvrin was caused by the dissec-
  tion of the natural vegetation by fields, grazing areas, archeological sites,
  and army training areas.
- The Open Shrublands' recovery boundary: approximately 3 km south of
  Beit Guvrin; a sharp boundary marks the northern edge of the Amazia
  open shrublands (shrub fraction = 0.32; dwarf-shrub fraction = 0.23; soil
  and rock fraction = 0.45) where the landscape is recovering under controlled grazing intensities, with fields extending only in the lower parts of
  the Wadis.

- The southern boundary of the Open Shrublands of Amazia that emerged through frequent fires and historically high grazing intensities. The area south of this boundary consists the desert fringe batha that is dominated by large extents of dwarf-shrubs and bare soil patches (shrub fraction = 0.065; dwarf-shrub fraction = 0.585; soil and rock fraction = 0.35).
- Transition from dwarf-shrub–dominated areas to an arid ecosystem (shrub fraction = 0.05; dwarf-shrub fraction = 0.31; soil and rock fraction = 0.64) of scattered patches of dwarf-shrubs with large extents of bare ground and rocks, typical of the Northern Negev (Desert).

In these two latter desert fringe zones, grazing and fires are balanced by the relatively quick colonization (and recolonization) of the *S. spinosum*, forming a heterogeneous patchy pattern of bare soil and dwarf-shrubs. In the arid zone, dwarf-shrub patches are smaller than those found in the desert fringe batha. The separation line between these two units corresponds to a rainfall level of approximately 270 mm/year. In our study area, it passes at the vicinity of the Dvira Forest.

12.3.3 INVERSE EROSION AND RECOVERY POTENTIALS

Thresholds in the approach presented by Shoshany (2012) are boundaries of geodiversity rather than lines separating homogenous units. Geodiversity in this context concerns the spatial heterogeneity of vegetation and soil patch patterns. Degradation and recovery cycles in desert fringe ecosystems are characterized by dynamic changes in the spatial heterogeneity with mutual processes taking place between the vegetation and soil patterns (Forman 1995; Shoshany and Kelman 2006). During degradation, vegetation continuum is perforated by soil patches, dissected, fragmented, isolated, contracted, and dissolved. In parallel, soil patches grow, aggregate, connect, and expand. Inverse processes are taking place during recovery. As described in Section 12.3.2, along the north to south transect, the shift between the desert fringe batha and the arid ecosystem is characterized by a change from 2/3 dwarf-shrubs' cover and 1/3 bare soil to 2/3 bare soil and 1/3 dwarf-shrubs. Along the west to east transect, the shift is from approximately 15% to 6% GVC (note that *S. spinosum* has low greenness at the end of the summer). Spatial variations between these cover distributions are of significant impact on the degradation and recovery of these ecosystems. Large dwarf-shrub patches may shrink and still sustain during long and hot summers and during drought periods, while maintaining their recovery potential when habitat conditions improve. Isolated scattered plants under such conditions may dry out and die. Soil protection by vegetation in this latter case is low and thus increases potential erosion by runoff and wind (e.g., Ludwig et al. 2005; Okin et al. 2009; Puigdefabregas 2008). At the Landsat TM resolution, spatial heterogeneity of plant patterns is expressed by variations in vegetation versus soil/rock fractional cover. For analyzing potential erosion and potential recovery, Shoshany (2012) suggested to employ the following mathematical morphology technique of dilation on F as fractional cover of vegetation or soil:

$$F \oplus K = \text{Max} \{f(xi + q, j + p) + Kq, p \mid -n < q < n; -n < p < n\}, \qquad (12.2)$$

where Kq,p is the structuring element (kernel) of the dimensions $2n + 1 \times 2n + 1$ ($n = 2$ in this study).

Dilation of fractional cover maps would extend areas of maximal cover. The overall areal change between the original and dilated maps highly depends on the spatial distribution of the maximal values. In a scattered pattern, the change will occur around any core of maximal vegetation cover or inversely of high soil cover, thus producing wide areal changes. In clumped areas of high cover, the change will take place only in the margins of these clumps and thus will cover limited area. Clumped patches of high vegetation fractional cover associated with scattered pattern of low soil fractional cover would represent areas of potential recovery. Clumped patterns of high soil fractional cover associated with scattered pattern of low vegetation cover fraction would represent erosion potential. Within this description, the determination of threshold zone is expected to be configured according to two phenomenologies: first, by sharp increase in soil fractional cover reaching almost complete cover (Threshold Type I), and second, by the shift from higher fractional vegetation cover to higher fractional soil cover (Threshold Type II).

Figure 12.5 presents the methodological outline. In the first stage, spectral unmixing facilitates estimating the subpixel fractions of vegetation, soil, and rocks. In the second stage, the cover proportion map of vegetation and soil surface types was dilated separately using Equation 12.2, indicating the vegetation recovery and soil erosion potentials. In the third stage, recovery versus erosion potentials are determined by subtracting their dilated levels.

The Landsat image of October 1987 was used for detecting erosion and recovery potentials as a reference point before the high rainfall fluctuations that took place in 1991, 1992, 1999, 2003, and 2008. The position of the two threshold zones, as inferred by the new technique (Figure 12.6a and b), corresponds well with the

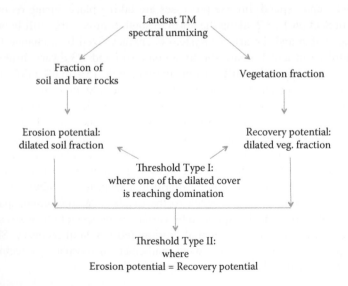

FIGURE 12.5   Schematic description of the erosion and recovery potentials' algorithm.

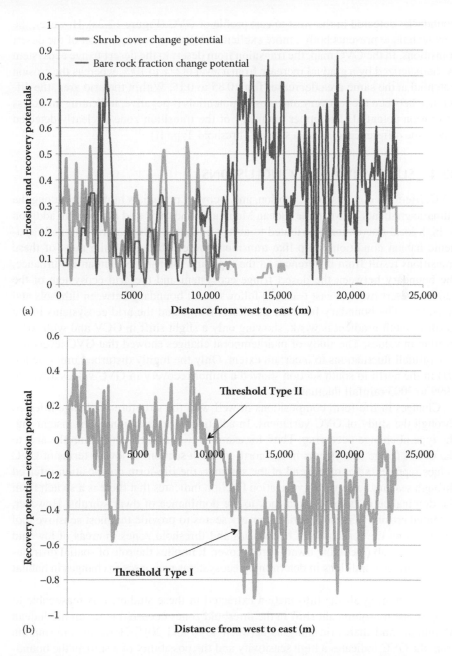

(a)

(b)

FIGURE 12.6 Potential erosion and potential recovery (a) and their subtraction (b).

boundaries detected in the west to east profile of GVC (Figure 12.2a). However, the new technique presents both a more explicit and distinctive expression of the desert transitions. In the GVC map, the transition from desert to the desert fringe ecosystem is characterized by a gradual increase from a level of 6% to 10%, whereas the erosion potential at the same area decreases from 0.85 to 0.15. Within the arid area, the difference between the recovery and erosion potentials is negative, indicating an excess of erosion potential. The upper boundary of the transition zone is clearly detected where the difference approaches zero (Threshold Type II).

12.4   SUMMARY AND CONCLUSIONS

GVC, life-form compositions, erosion, and recovery potentials highly vary along the climatic gradients between the Judean Mountains and their arid margins. In addition to high local fluctuations attributed to abrupt changes in the natural and anthropogenic habitat conditions, step-like transitions were detected. While most of these transitions result from differences in the type and intensity of human disturbance, the boundary between the desert fringe ecosystem and the arid ecosystem of the Judean Desert (west to east transect) follows a soil boundary between lithosols and Rendzina. The boundary between the desert fringe and the arid ecosystems in the north to south gradient is weak, showing only a slight shift in GCV and with some overlap in values. The study of multitemporal changes showed that GVC levels follow rainfall fluctuations to a certain extent. Only the highly disturbed unit (section III) in the north to south section showed a minor recovery in GVC values after the 1999 to 2003 rainfall fluctuations.

Changes in life-form compositions fit well with the structure of change inferred through the study of GVC variations. In addition, it adds a significant insight into the type of change occurring. Thus, for example, whereas the shift from the arid to the desert fringe ecosystems in the north to south sections is low in terms of GVC values representative of the end of the summer, the life-form composition inferred through seasonal changes in vegetation fraction indicates that there is a switch from the dominance of bare soil and rock to the dominance of dwarf-shrubs. However, potential erosion versus potential recovery seems to provide the most sensitive tool for detecting the formation of boundaries and threshold zones in areas of low and moderate shrub (including dwarf-shrub) cover. It implies the role of spatial distribution of surface conditions in determining ecosystems responses to changes in habitat conditions.

By combining all the information extracted in these studies, it is reasonable to suggest that no significant shift in the threshold zones occurred between the Judean Mountains and their arid margins between 1996 and 2011. However, information from the GVC indicates a high sensitivity and the possibility of a shift in the boundary between the desert fringe Batha and the arid zone in both transects. Recent evidence regarding the accumulation of water deficits and the drying of planted forests (Dorman et al. 2012) and of shrublands (Shoshany and Karnibad 2015) in this region may enhance the effect of human disturbance, reduce recovery potential, and result in a significant desertification of the Mediterranean to arid transition zones.

REFERENCES

Cerda, A., and Lavee, H. (1999) The effect of grazing on soil and water losses under arid and Mediterranean climates. Implications for desertification. *Pirineos*, 153–154, 159–174.

Dan, Y., and Raz, Z. (1970) The soil formations map of Israel. Beit Dagan, Israel: The Volcani Centre.

Dorman, M., Svoray, T., and Prevolozki, A. (2012) Drying trees in the Jerusalem Pine forests of Israel—From a high altitude point of view. *Ecol. Environ.*, 3, 230–237. (In Hebrew.)

Fagre, D.B., and Charles, C.W. (2009) Thresholds of climate change in ecosystems: Final Report, Synthesis and Assessment Product 4.2 U.S. Climate Change Science Program and the Subcommittee on Global Change Research.

Forman, R.T.T. (1995) *Land Mosaics: The Ecology of Landscapes and Regions*. Cambridge University Press. 632 pp.

Geeson, N., Brandt, J., and Thornes, J.B. (2002) *Mediterranean Desertification: A Mosaic of Processes and Responses*. John Wiley & Sons.

Hamada, Y., Stow, D.A., Roberts, D.A., Franklin, J., and Kyriakidis, P.C. (2013) Assessing and monitoring semi-arid shrublands using object-based image analysis and multiple endmember spectral mixture analysis. *Environ. Monit. Assess.*, 185, 3173–3190.

Hein, L., and De Ridder, N. (2006) Desertification in the Sahel: A reinterpretation. *Glob. Chang. Biol.*, 12, 751–758.

Herrmann, S.M., Anyamba, A., and Tucker, C.J. (2005) Recent trends in vegetation dynamics in the African Sahel and their relationship to climate. *Glob. Environ. Chang.*, 15, 394–404.

Hill, J., Stellmes, M., Udelhoven, Th., Röder, A., and Sommer, S. (2008) Mediterranean desertification and land degradation Mapping related land use change syndromes based on satellite observations. *Glob. Planet. Change*, 64, 146.

Kutiel, P., Lavee, H., and Shoshany, M. (1995) The influence of a climatic gradient upon vegetation dynamics along a Mediterranean arid transect. *J. Biogeogr.*, 22, 1065–1071.

Lamprey, H.F. (1975) Report on the desert encroachment reconnaissance in northern Sudan. UNESCO/UNEP. *Desertification Control Bull.*, 17, 1–7.

Lavee, H., Imeson, A.C., and Sarah, P. (1998) The impact of climate change on geomorphology and desertification along a Mediterranean–arid transect. *Land Degrad. Dev.*, 9, 407–422.

Ludwig, J.A., Wilcox, B.P., Breshears, D.D., Tongway, D.J., and Imeson, A.C. (2005) Vegetation patches and runoff-erosion as interacting ecohydrological processes in semi-arid landscapes. *Ecology*, 86(2), 288–297.

Maestre, F.T., Bowker, M.A., Puche, M.D., Belen Hinnojosa, M., Martinez, I., Garcia-Palacios, P., Castillo, A.P., Soliveres, S., Luzuriaga, A.L., Sanchez, A.M., Carreira, J.A., Gullardo, A., and Escudero, A. (2009) Shrub encroachment can reverse desertification in semi-arid Mediterranean grasslands. *Ecol. Lett.*, 12, 930–941.

Meron, E., Gilad, E., von Hardenberg, J., Shachak, M., and Zarmi, Y. (2004) Vegetation patterns along a rainfall gradient. *Chaos, Solitons Fractals*, 19, 367–376.

Milan, M.M., Estrela, M.J., Sanz, M.J., Mantilla, E., Martin, M., Pastor, F., Salvador, R., Valledo, R., Alonso, L., Gangoitti, G., Ilardia, J.L., Navazo, M., Albizuri, A., Artinano, B., Ciccioli, P., Kallos, G., Carvalho, R.A., Andres, D., Hoff, A., Werhahn, J., Seufert, G., and Versino, B. (2005) Climatic feedbacks and desertification: The Mediterranean model. *J. Clim.*, 18, 684–701.

Nicholson, S.E. (2011) *Dryland Climatology, Desertification Chapter*, Cambridge University Press, 431–447.

Okin, G.S., Parsons, A.J., Wainwright, J., Herrick, J.E., Bestelmeyer, B.T., Peters, D.C., and Fredrickson, E.L. (2009) Do changes in connectivity explain desertification? *Bioscience*, 59, 237–244.

Papanastasis, V.P., Kyriakakis, S., Kazakis, G., Abid, M., and Doulis, A. (2003) Plant cover as a tool for monitoring desertification in mountain Mediterranean rangelands. *Manage. Environ. Qual.*, 14(1), 69–81.

Peinado, M., Alcaraz, F., Aguirre, J.L., Delgadillo, J., and Aguado, I. (1995) Shrubland formations and association in Mediterranean–desert transitional zones of northwestern Baja California. *Vegetatio*, 117, 165–179.

Puigdefabregas, J. (2008) The role of vegetation patterns in structuring runoff and sediment fluxes in drylands. *Earth Surf. Processes Landforms.* 30, 133–147.

Puigdefabregas, J., and Mendizabal, T. (1998) Perspectives on desertification: Western Mediterranean. *J. Arid Environ.*, 39, 209–224.

Safriel U. (2009) Status of desertification in the Mediterranean region. In: *Water Security, Land Degradation and Desertification in the Mediterranean Region*. Edited by Rubio, J.L., Rubio, J.L., Daussa, R., Blum, W., and Pedrazzini, F. Springer: Berlin, Germany; pp. 33–43.

Sharakas, O., Hammad, A.A., Nubani, A., and Abdullah, A. (2003) Land Degradation Risk Assessment in the Palestinian Central Mountains Utilizing Remote Sensing and GIS Technique; Geography Department, Birzeit University: Birzeit, Palestine, 97 pp.

Shoshany, M. (2002) Landscape fragmentation and soil erodibility in south and north facing slopes during ecosystems recovery: An analysis from multi-date air photographs. *Geomorphology*, 45, 3–20.

Shoshany, M. (2012) Desert threshold identification by maximizing potential instability in soil and shrubs' patterns: A remote sensing study. *Land Degrad. Dev.*, 23, 331–338.

Shoshany, M., and Karnibad, L. (2015) Relative water use efficiency and shrublands biomass along a semi-arid climatic gradient: A remote sensing study. *Remote Sens. (online)*, 7, 2283–2301.

Shoshany, M., and Kelman, E. (2006) Mutuality in soil and vegetation pattern characteristics: Modeling with cellular automata. *Geomorphology*, 77(1–2), 35–46.

Shoshany, M., Kutiel, P., and Lavee, H. (1995) Seasonal vegetation cover changes as indicators of soil types along a climatological gradient: A mutual study of environmental patterns and controls using remote sensing. *Int. J. Remote Sens.*, 16, 2137–2151.

Shoshany, M., Kutiel, P., and Lavee, H. (1996) Monitoring temporal vegetation cover changes in Mediterranean and arid ecosystem using a remote sensing technique: Case study of the Judean Mountain and the Judean Desert. *J. Arid Environ.*, 32, 1–13.

Shoshany, M., Kutiel, P., Lavee, H., and Eichler, M. (1994) Remote sensing of vegetation cover along a climatological gradient. *ISPRS J. Photogramm. Remote Sens.* 49, 1–8.

Shoshany, M., and Svoray, T. (2002) Multi-date adaptive spectral unmixing and its application for the analysis of ecosystems' transition along a climatic gradient. *Remote Sens. Environ.*, 81, 1–16.

Thomas, D.S.G., and Middleton, N.J. (1994) *Desertification: Exploding the Myth*. John Wiley & Sons Ltd., Chichester, UK. xiv + 194 pages. ISBN 0-471-94815-2.

Thornes, J.B. (2000) Mediterranean desertification: The issues. In: *Mediterranean Desertification: Research Results and Policy Implications*. Edited by Balabanis, P., Peter, D., Ghazi, A., and Tsogas, M. European Commission, EUR 19303, Brussels; pp. 9–17.

Veron, S.R., Paruelo, J.M., and Oesterheld, M. (2006) Assessing desertification. *J. Arid Environ.*, 66, 751–763.

13 Soil Moisture Using Optical Remote Sensing and Ground Measurements
A Case Study from Pakistan

Mudassar Umar, Siraj Munir, Iftikhar Ali,
Salman Qureshi, Claudia Notarnicola,
Said Rahman, and Qihao Weng

CONTENTS

13.1 INTRODUCTION

Soil moisture is an important component of the terrestrial environment that significantly regulates water circulation and surface energy exchanges between land surface and the atmosphere (Jackson 1993; Vereecken et al. 2014). It is an important factor affecting the budget of the hydrological cycle, specifically by separating rainfall into runoff, surface infiltration, and evapotranspiration (Jawson and Niemann 2007). Many scientific applications require soil moisture data to represent the initial state of soil moisture such as numerical weather prediction and climate projections because they play a key role in hydro-meteorological processes (Gao et al. 2014). A precise estimation of soil moisture can help in improving the forecasting of precipitation, temperature, droughts, and floods (Albergel et al. 2013). There are generally four methods to retrieve soil moisture: (i) ground-based soil sampling or sensors, (ii) land surface model using meteorological data, (iii) remote sensing from airborne or satellite data, and (iv) data assimilation techniques that integrate remote sensing signals into land surface models (Qin et al. 2013).

The spatial variability of soil moisture is high owing to the significant heterogeneity of soil, topography, vegetation, and precipitation (Crow et al. 2012), making field campaigns challenging and time-intensive. Furthermore, the gravimetric measurements (i.e., ground measurements) are simply representative over a small spatial scale. In data assimilation and land surface models, the model grid-box size is greater than the gravimetric measurements (Qin et al. 2013). Remote sensing provides effective methods for collecting information from a range of samples over a large area in relatively shorter and repeated intervals of time (Nichols et al. 2011).

Microwave remote sensing offers a great potential to measure soil moisture across varying spatial and temporal scales (Singh et al. 2005). The poor spatial resolution of passive microwave and the strong sensitivity of active microwave over vegetation cover and in surface roughness limit the capability of microwave remote sensing to infer soil moisture (Srivastava et al. 2003). The weakness of microwave sensing soil moisture is the impact of ground vegetation cover. Only longer-wavelength microwave could penetrate thick vegetation cover and can appropriately sense soil moisture. To overcome this problem, a number of approaches using visible, near-infrared (NIR), and thermal infrared wavelengths were evaluated for soil moisture assessment.

Remote sensing data have been effectively used for the estimation of soil moisture near surface, but this information is limited to a few centimeters below the surface (Santos et al. 2014). Remote sensing does not allow access to the whole root zone where water can be absorbed by the roots (Liu et al. 2012). Soil moisture has varying spectral patterns across different wavelengths as remote sensing data from visible to microwave lengths can effectively be used for monitoring the variations in soil moisture (Wang and Qu 2009). Vegetation indices can be used to relate soil moisture in the root zone and soil moisture influences vegetation growth and thus changes the spectral characteristics of vegetation (Wang et al. 2010). The spectral reflectance decreases with increasing soil moisture in the visible and NIR range. Some studies have shown soil moisture modeling with visible and NIR data such as the vegetation anomaly index combining NIR–red data with land surface temperature (LST) for studying land surface parameters (vegetation anomaly, soil moisture), and vegetation indices serve as indicators of drought as well (Huete et al. 1992). The soil moisture reflectance method is dependent on the relationship between soil moisture and vegetation (Zhan et al. 2007). Zhan et al. (2007) validated the NIR–red observed soil moisture with gravimetric measurements, and the effective soil depth for moisture estimation was 5 cm and was found to be significant. Nevertheless, the major limitations in case of optical remote sensing of soil moisture relate to surface roughness, soil structure, and the presence of organic matters (Wang and Qu 2009).

The use of remote sensing for the estimation of surface energy fluxes and surface soil water dates back to 1970. For instance, it was used by meteorologists to estimate the surface energy fluxes and soil water (Carlson et al. 1984; Price 1980, 1982). Several other studies have approached them with similar methods (Kustas and Norman 1996; Sandholt et al. 2002). Information about surface energy and water status has been determined by many researchers by developing the relationship between remotely sensed LST and normalized difference vegetation index (NDVI) (Amiri et al. 2009; Carlson et al. 1994; Rajasekar and Weng 2009). The correlation between thermal and visible/NIR wavelengths has proven to be useful for appropriate monitoring of vegetation and water stress. A method was developed for mapping the surface moisture and land surface energy fluxes in the 1990s (Price 1990), which was referred to as the triangle method. This method lies on the interpretation of the pixel distribution in T_s–NDVI space. If an image contains a wide range of soil water content and vegetation cover, the space presents a triangle. This triangle is formed because surface temperature decreases as vegetation cover increases. The concept of the triangle was first given by Price (1990) and later further elaborated by several researchers (Gillies et al. 1997) while Sandholt et al. (2002) and others adopted and applied the triangle method (Chauhan et al. 2003; Stisen et al. 2008; Wang et al. 2006). Wang et al. (2007) stated that MODIS land parameters (LST, NDVI) are significantly correlated with ground soil moisture. Chauhan et al. (2003) applied this approach to obtain nearly accurate, high soil moisture by linking microwave-derived soil moisture with optical parameters.

Li and Dong (1996) stated that a relationship can be developed using satellite-derived NDVI, surface temperature, and soil moisture to estimate moisture contents in deep soil. Guo et al. (1997) estimated soil moisture at 20 cm depth using vegetation and surface temperature information and reported that satellite data have a significant relationship to the mentioned depth. Li et al. (2008) analyzed the spatiotemporal variability of land surface moisture based on vegetation and temperature using the triangle method. Wang et al. (2010) examined a relationship between TVDI and soil moisture to estimate soil moisture from 10- to 20-cm depths. They stated that satellite data have a significant relationship only to the upper layers (10 and 20 cm). The relationship was not significant for deep soil layers (20–50 cm). The developed estimation model provides reliable estimates of soil moisture at depths of 20 cm but cannot estimate moisture levels in deeper soil layers (Wang et al. 2010). It was suggested that the combination of LST and NDVI could provide better estimates of vegetation and soil moisture on the land surface. T_s–NDVI space is related to surface evapotranspiration rate and has been used for the assessment of temperature and soil moisture condition (Goetz 1997; Prihodko and Goward 1997).

Sandholt et al. (2002) proposed a moisture index called temperature vegetation dryness index (TVDI) based on the simplified interpretation of T_s–NDVI space. Xin et al. (2006) evaluated the potential of TVDI by using Pathfinder data (8 km) to assess the soil moisture status by using routine level measurements of soil moisture at the station level. The potential of the moderate resolution imaging spectroradiometer (MODIS) is investigated to capture spatial variability in soil moisture with intensive measurements (Patel et al. 2009). Less focus has been given in investigating the potential of TVDI from Landsat TM data for soil moisture assessment.

The main aim of this study was to evaluate soil moisture through remotely sensed data and its comparison using field measurements in the Umer Kot, Sindh province of Pakistan. To achieve this goal, our objectives were to develop a remote sensing soil moisture model (RSSMM) using NIR–red spectral reflectance space and to develop TVDI by combining LST and NDVI for estimating soil moisture. However, despite numerous work on soil moisture estimation through TVDI, a comparison between RSSMM and TVDI has not been carried out.

13.2   DATA AND METHODOLOGY

13.2.1   STUDY AREA

The Umer Kot district located in the Sindh province of Pakistan was selected as the study area. This region is geographically situated between 24°10′ to 25°45′ north latitude and 69°04′ to 71°06′ east longitude. The total area of the district is 5608 km². The study area comprises two distinct land covers, the irrigated portion in the northwest and *barony/desert in the* north-southeast. The agricultural land of the study area is very fertile. The main crops grown in the irrigated belt of the study area are wheat and chili. The other crops such as sugarcane, mustard, sunflower, ispaghol, and saunf are also successfully grown in the area. *Bajra, guwar*, and *till* are grown in the rainfed belt (desert) of the area. The general classification of the land cover of this study area is shown in Figure 13.1. The study area bears distinct climatic conditions; the irrigated land within the study area has temperate conditions, that is, neither extremely hot nor very cold in winter as compared to the eastern portion of the study area. The summer heat is reduced by constant blowing of the southwest breeze from the sea. The climatic pattern of the eastern (desert) portion is tropical, hot, and dry. In this area, air temperature can rise above 40°C during May to August and drops down to 6°C in December and January. The average annual rainfall is approximately 222 mm, which mostly occurs during the monsoon.

The Sindh province was severely affected by drought in past years (FAO 2001), and the study area is part of the drought-affected region. It is a well-established fact that soil moisture is an essential parameter for early drought prediction. The main reason behind the selection of this region was that it is a mixture of semiarid and desert land. The total population of the district is 2,065,590 in 2010 and is heavily dependent on agricultural production. Hence, the assessment of moisture conditions using advanced remote sensing techniques will help in the effective handling of drought and other hydrological conditions.

13.2.2   REMOTE SENSING DATA

Optical and thermal infrared band data of the Landsat 5 Thematic Mapper (TM) were used in this study. The scene was acquired on May 3, 2011. The data acquisition

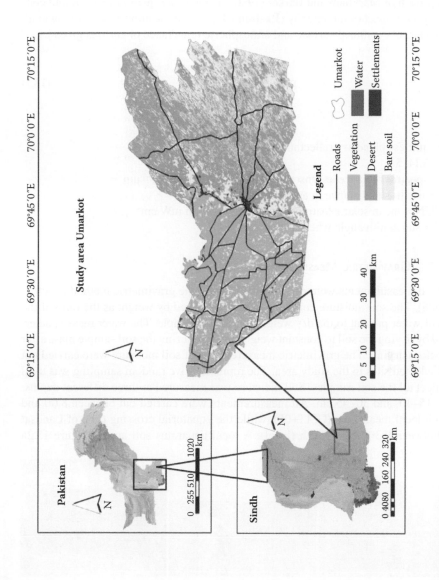

FIGURE 13.1 Study area and its land cover.

date has clear atmospheric condition and images were acquired through the USGS Earth Resource Observation Systems Data Center, which has corrected radiometric and geometrical distortions of the images to a quality level of 1G before delivery. The image covering the Umer Kot district was a subset from the geometrically corrected image (Path 152, Row 41). The image was atmospherically corrected by using the approach of Markham and Barker (1986). They gave Equation 13.1 without considering the impact of topography (Ekstrand 1996) and the atmosphere (Liang et al. 1997). Hence, combined surface and atmospheric reflectance of earth is given by

$$\rho = \frac{\pi L d^2}{\text{ESUN}\cos(\text{SZ})},$$

(13.1)

where

ρ = unitless planetary reflectance at the satellite
π = 3.1415
L = spectral radiance at sensor aperture in mW cm^{-2} ster^{-1} μm^{-1}
d^2 = earth–sun distance in astronomical units
ESUN = mean solar exoatmospheric irradiance in mW cm^{-2} μm^{-1}
SZ = sun zenith angle when scene is recorded

13.2.3 Gravimetric Measurements

The field measurements were carried out by using the gravimetric method (Evett et al. 2008). The soil moisture content can be determined by weight as the ratio of the mass of water present to the dry weight of the soil sample. The water mass is determined by drying the soil to constant weight and measuring the soil sample mass after and before drying. The gravimetric measurements of soil moisture were carried out at 24 selected sites in the study area. The representative random sampling was used to collect all the soil samples. Soil moisture was measured at three different depths: 0–15, 15–30, and 30–45 cm. All measurements were carried out between 8:00 and 16:00 h local mean time (GMT+05), while the equatorial crossing time of Landsat was between 9:30 and 10:30 h (local). A local apparatus soil auger (Figure 13.2a

(a)   (b)   (c)

FIGURE 13.2   (a and b) Field sample collection using soil auger. (c) Samples in the laboratory oven.

and b) was used for the measurement. The collected samples were analyzed in the laboratory (Figure 13.2c); the weight of each sample was 50 g. The samples were heated in an oven for 18 h at 105°C. The dry weight of each sample was noted and moisture was calculated by the following formula:

$$\text{Moisture}\,(\%) = \frac{\text{Wt.Soil}\,(g) - \text{Dry soil}\,(g)}{\text{Dry Soil}\,(g)} * 100. \tag{13.2}$$

13.2.4 THE NIR–RED SPECTRAL REFLECTANCE SPACE

Vegetation strongly absorbs blue and red light and reflects green in the NIR spectrum. The reflectance of bare soil is high in the red to NIR spectral region. Densely vegetated area shows smaller reflectance in red and higher reflectance in NIR. This concept lays the theory of NDVI, which helps assess the changes in vegetation fraction, leaf area index, and chlorophyll content. Richardson and Weigand (1977) developed a NIR–red spectral space with MSS and a perpendicular vegetation index for distribution of vegetation in space. The NIR–red spectral reflectance space was constructed by using atmospherically corrected TM band 3 and band 4. The NIR–red spectral scatterplot constructs a triangular shape (Figure 13.3). The NIR–red space developed by TM band 3 and band 4 validates the distribution of vegetation in space (Richardson and Weigand 1977).

In Figure 13.3, the change in surface vegetation cover is represented by line AD from the densely vegetated area (A) and sparse vegetation (E) to bare soil (D). Line BC refers to soil moisture status, extending from the wet area (B) to semiarid to dry soil (C). Line BC, which shows the direction of dryness, helps in inferring the drought severity. The scatterplot of the NIR–red reflectance space demonstrates the triangular shape. Surface coverage and surface condition can be described significantly in this space. Relationships exist among surface spectrum, land cover types, and dryness condition. This prompts us to build a NIR–red spectral space model for soil moisture monitoring.

13.2.5 DEVELOPMENT OF A RSSM MODEL

Figure 13.3 shows that dryness gradually rises from B to C and climaxes at C. Line BC is the soil line and is mathematically expressed by the following equation:

$$R_{\text{NIR}} = MR_{\text{red}} + I, \tag{13.3}$$

where R_{NIR} is the reflectance of the NIR band, R_{red} is the reflectance of the red band, M is the slope of the soil line, and I is the intercept.

According to the perpendicular vegetation index (PVI) developed by Richardson and Weigand (1977), any point in the NIR–red reflectance space to the line vertical to the soil line represents surface soil moisture. Line L dissects the coordinate origin

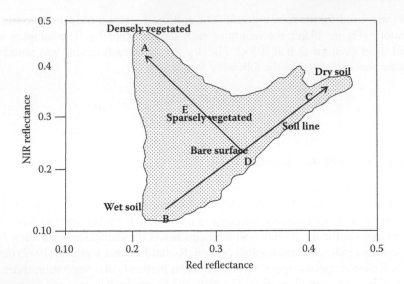

FIGURE 13.3   Conceptual diagram of NIR–red reflectance space.

and is vertical to the soil line (Figure 13.4). Hence, the normal function of line L can be formulated from the soil line expression

$$R_{\mathrm{NIR}} = \left(-\frac{1}{M} R_{\mathrm{red}} \right). \tag{13.4}$$

The distance from any points in the NIR–red reflectance space to line L represents surface soil moisture conditions. The farther the distance, the stronger the drought and the less soil moisture and vice versa. Let there be a random point K (R_{red}, R_{NIR}); then, the vertical distance from K (R_{red}, R_{NIR}) to line L can calculated as follows:

$$KM = \frac{1}{\sqrt{M^2 + 1}} (R_{\mathrm{NIR}} + MR_{\mathrm{red}}). \tag{13.5}$$

The objects with some reflectance near the wet area of Figure 13.3 have a moisture content close to 1, while the objects with high reflectance are in the direction of C showing dryness and have a moisture content close to 0. The following model has been established by subtracting the normalized values of Equation 13.5 from 1:

$$\mathrm{RSSMM} = 1 - \frac{1}{\sqrt{M^2 + 1}} (R_{\mathrm{NIR}} + MR_{\mathrm{red}}). \tag{13.6}$$

The soil line extracted from NIR and red band reflectance of the observed area in Figure 13.5 is expressed as follows:

$$R_{\mathrm{NIR}} = 0.8745 R_{\mathrm{red}} + 0.0745. \tag{13.7}$$

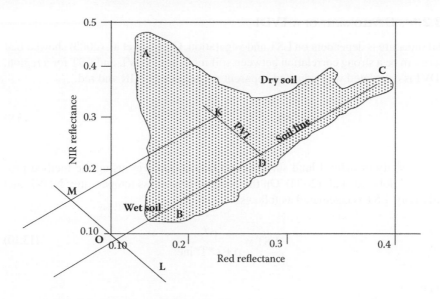

FIGURE 13.4 Sketch of soil moisture modeling.

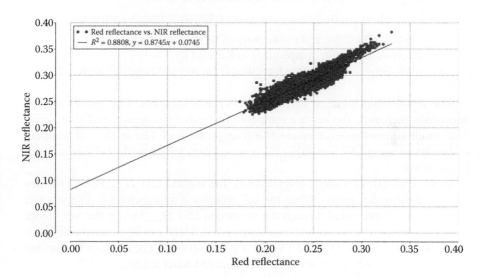

FIGURE 13.5 Soil line extraction.

It can be seen from Equation 13.7 that the slope $(M) = 0.8745$ and the intercept $(I) = 0.0745$; introducing into Equation 13.5, soil moisture can be obtained over the study area:

$$RSSMM = 1 - \frac{1}{\sqrt{0.8745^2 + 1}}(R_{NIR} + 0.8745 R_{red}). \qquad (13.8)$$

13.2.6   Development of a TVDI

Soil moisture is dependent on LST and vegetation. Chauhan et al. (2003) showed that there can be a strong correlation between soil moisture, NDVI, and LST for a region. NDVI is calculated with adjusted apparent reflectance of NIR and red:

$$NDVI = \frac{\rho(band\,4) - \rho(band\,3)}{\rho(band\,4) + \rho(band\,3)}. \tag{13.9}$$

Emissivity of natural land surfaces can be measured by using the method proposed by Sobrino et al. (2004). On the basis of brightness temperature, NDVI, and emissivity, LST is calculated as follows:

$$LST = \frac{\rho T}{\rho + \lambda * T * \ln\varepsilon}, \tag{13.10}$$

where
   $\lambda =$ wavelength of emitted radiance for which the average of limiting wavelengths ($\lambda = 11.5$ μm) is used (Markham et al. 1985)
   $\rho = \dfrac{h * c}{\sigma}$, where $\rho =$ Boltzmann constant ($1.38 * 10^{-23}$ J/K), $h =$ Planck's constant (6.626×10^{-34} Js), and $c =$ velocity of light (3×10^{8} m/s)
   $T =$ effective at-satellite temperature in Kelvin (K)
   $\varepsilon =$ emissivity of land surface (obtained from thermal band of Landsat TM)

The relationship between LST and NDVI was negative. This proves that as vegetation increases, the surface temperature tends to decrease, and vice versa. This result allows the development of a triangle method. The method of mapping soil moisture using LST and NDVI is known as the triangle method (Carlson 2007; Price 1990).

In the LST–NDVI triangle (Figure 13.6), the highest LST (T_{smax}) along the dry edge establishes dry surface when the soil wetness is approaching 0. The wet soil conditions are represented through the minimum LST (T_{smin}) along the wet border when the soil wetness is high and approaching 1. A TVDI can be established by having TVDI = 1 at the dry edge and TVDI = 0 at the wet edge. The dry edge has limited water availability, while the wet edge has unlimited water access.

The mathematical relation is as follows:

$$TVDI = \frac{T_s - T_{smin}}{T_{smax} - T_{smin}}, \tag{13.11}$$

where T_s is the LST, T_{smin} is the minimum LST in the triangle and represents the wet edge, and T_{smax} is the maximum LST observed for a given NDVI and signifies the dry edge in the triangle.

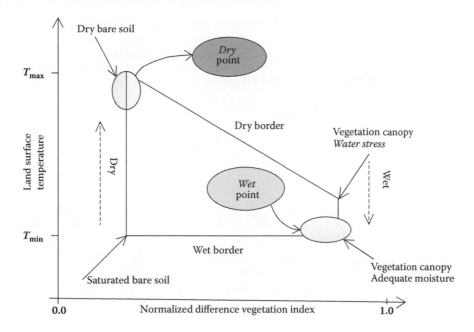

FIGURE 13.6   TVDI representation.

$T_{smin} = a_1 + b_1\text{NDVI}$, where a_1 and b_1 are coefficients of the regression equation for the wet edge of the moisture condition and $T_{smax} = a_2 + b_2\text{NDVI}$, where a_2 and b_2 are coefficients of the regression equation for the dry edge. The TVDI broadly considers the changes between vegetation and surface temperature.

Four important parameters, dryness, wetness, NDVI, and LST, can be perceived from the triangle. A dry edge will develop when the LST tends to increase against minimum NDVI. Similarly, wet edge develops when the LST tends to decrease against maximum NDVI (Figure 13.6). As the LST decreases, the vegetation increases, the higher the NDVI value becomes, and the lower the LST will be with the exception of small variations. The small changes in the LST with higher NDVI values point to the wetness of the soil in vegetation. According to Carlson (2007), variations in LST only reveal the dryness and wetness. The drier conditions can easily be indicated by steeper T_s–NDVI slopes (Goetz 1997). The scatterplot of T_s–NDVI has a negative relation for most of the pixels found in the triangle. The dry edge and the wet edge were estimated on the basis of pixel information from the area that is large enough to represent the range of surface moisture condition and from sandy soil to a fully vegetated surface. A plot of an NDVI pixel against a corresponding T_s is shown in Figure 13.7 for wet and dry edges. T_{smax} (dry edge) and T_{smin} (wet edge) were observed for intervals of NDVI extracted from T_s–NDVI space as shown in Figure 13.7.

The performance of satellite-based soil moisture retrieved through both RSSMM and TVDI was assessed using root mean square error (RSME) and index of agreement (Willmott 1981) d:

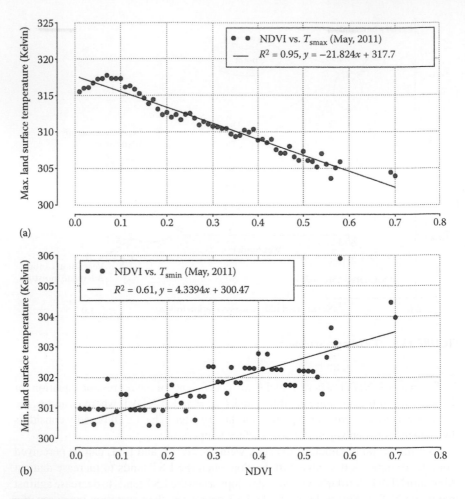

FIGURE 13.7 Relationship between (a) maximum T_s and NDVI (T_{smax}—dry edge) and (b) minimum T_s and NDVI (T_{smin}—wet edge).

$$\text{RSME} = \sqrt{\frac{\sum_{i=1}^{N}(O_i - E_i)^2}{N}} \tag{13.12}$$

$$d = 1 - \frac{\sum_{i=1}^{N}(O_i - E_i)^2}{\sum_{i=1}^{N}\left(\left|E_i - \bar{O}\right| + \left|O_i - \bar{O}\right|\right)^2}, \tag{13.13}$$

where N is the number of observations and O_i and E_i are the observed and estimated values, respectively. The d index values vary between 0 and 1, indicating the low and high relationship between observed and estimated values, respectively. Willmott

(1981) developed a d index, known as Willmott's index of agreement, which is used for validation in regression and prediction models. The model is perfect when $d = 1$ and RSME = 0.

13.3   RESULTS AND DISCUSSION

The RSSMM was used to analyze the soil moisture at the pixel level. The RSSMM-derived results were then correlated with soil moisture level calculated by gravimetric measurements. It is clear that results derived from RSSMM and ground measurement have a significant correlation (Figure 13.8). The relationship between RSSMM and gravimetric (ground) measurements is statistically significant ($P <$ 0.05) and correlation coefficients $\left(R^2_{\text{land surface depth (RSSMM)}} \right)$ were $R^2_{0-15 \text{ cm (RSSMM)}} = 0.51$, $R^2_{15-30 \text{ cm (RSSMM)}} = 0.60$, and $R^2_{30-45 \text{ cm (RSSMM)}} = 0.57$.

The effective soil depth for soil moisture measurement has been the subject of debate in scientific literature. For example, Ghulam et al. (2004) proves that at a 10-cm depth, the visible and NIR spectral space have a close relationship, whereas Guo et al. (1997) reported that satellite data have a correlation with soil moisture at a depth of 20 cm. Liu et al. (1997) stated that the effective depth for estimation of soil moisture through visible NIR is 10 cm. Carlson et al. (1995) calculated soil water content at two depths: 0–15 and 15–30 cm. On the basis of these contradictory arguments, field measurements were conducted at depths of 0–15, 15–30, and 30–45 cm to aid our methods based on NIR–red reflectance data and TVDI.

It is evident from the results that the satellite estimates (i.e., RSSMM) and field-measured (i.e., gravimetric soil moisture) data are in agreement (Figure 13.8 and Table 13.1). The impact of vegetation on soil moisture was not considered in RSSMM; hence, the results present mixed information on soil and vegetation. This poses a limitation on the precision of estimated results together with the uncertainty of field measurements attributed to sample transportation as wind and other metrological factors have a strong effect on soil moisture conditions. The results are acceptable and the developed method can be applied to soil moisture monitoring.

13.3.1 T_s–NDVI SPACE (COMBINATION OF LST AND NDVI)

A plot of NDVI pixels against the corresponding surface temperature (T_s) presents a triangular shape. A linear regression was applied to obtain the dry edge (T_{smax}) and wet edge (T_{smin}) (Figure 13.7). A strong negative and positive relationship was found in T_{smax} and T_{smin} observations, respectively. The dry edge has a negative relationship, indicating the decrease in $\text{LST}_{(\text{max})}$ when the NDVI increases. The wet edge has a positive relationship, signifying the decrease in $\text{LST}_{(\text{min})}$ with NDVI. The AT was developed using T_s–NDVI space. These wet and dry edges were used to calculate TVDI.

The analysis of the scatterplot (Figure 13.7) shows a stronger correlation with R^2 between 0.61 and 0.95, indicating that dry and wet edges are adequately represented by a linear equation (Figure 13.7). The results are in agreement with previous work of Holzman et al. (2014), Patel et al. (2009), and Sandholt et al. (2002).

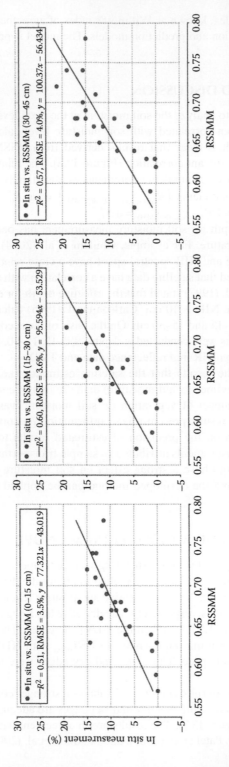

FIGURE 13.8　Linear relation between gravimetric and satellite-based soil moisture measurements at different depths (0–15, 15–30, and 30–45 cm).

TABLE 13.1

Error Statistic of RSSMM and TVDI in Comparison with Gravimetric Soil Moisture Measurements

RSSMM Relation with Gravimetric Measurements	R^2	RSME	Willmott's Index (d)
0–15 cm	0.51	3.5%	0.93
15–30 cm	0.60	3.6%	0.95
30–45 cm	0.57	4.0%	0.94
TVDI Relation with Gravimetric Measurements	**R^2**	**RSME**	**Willmott's Index (d)**
0–15 cm	0.75	2.4%	0.97
15–30 cm	0.70	3.0%	0.96
30–45 cm	0.63	3.8%	0.95

Figure 13.7 illustrates that the slope (–21.82) of the dry edge tends toward the negative side as compared to the slope (4.33) of the wet edge. This difference is attributed to the fraction of vegetation cover and LST as the dry edge was determined by observing maximum LST against an interval on NDVI values from T_s–NDVI space. The negative slope associated with the dry edge may be the result of evapotranspiration and bare soil surface with confining water condition.

Hence, the dry and wet edges have a high correlation ($R^2 = 0.95$ and $R^2 = 0.61$, respectively). The relationship showed that the LST was higher in the area with low NDVI. The relationship with the dry edge shows a decrease in maximum LST with increasing NDVI. The relationship with the wet edge is positive, showing an increase in NDVI with an increase in minimum LST. The variation shows the rise and decline in both the LST and NDVI. This shows that when soil moisture is high, the absorption of solar energy is used for evaporation. When soil moisture is low, the bare soil becomes drier rapidly and there is less evaporation, and as a result, absorption is consumed in LST. Hence, when surface temperature is high, the moisture in the root layer is consumed to maintain a high transpiration rate.

13.3.2 Spatial Variation of TVDI

With dry and wet edges, the TVDI image was obtained to infer the pattern of wetness and dryness. By applying T_{smin} and T_{smax} (extracted from T_s–NDVI) in Equation 13.11, the TVDI was obtained. Dry and very dry (TVDI > 0.6) conditions are located in bare areas with more soil mainly in the eastern region (Figure 13.9). Moist areas having low and very low (TVDI < 0.5) values are noticeable in the western region (Figure 13.9). The influence of moisture was clearly visible in the study area. Moist areas had low TVDI values where the vegetation cover is robust (Figure 13.1).

Dry condition was present in the eastward region of the study area owing to the high TVDI values. These high TVDI values in the eastern region show a drier condition as compared to the western region of the study area with predominantly irrigated cultivation. On the whole, the western region of the study area has low TVDI values (0.1–0.5), which allow surface wetness as a result of the high vegetation cover

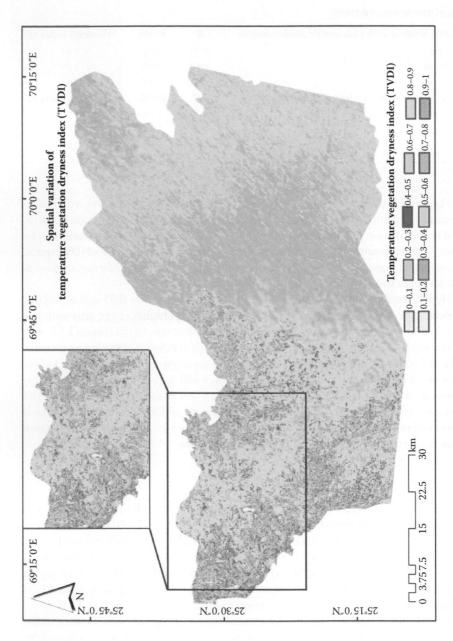

FIGURE 13.9 Spatial pattern of TVDI in the study area.

and moisture retention capacity of soil. The extremely low TVDI values (<0.2) were observed along the water bodies/channels. TVDI was found very effective for determining the surface wetness (moisture) and dryness.

13.3.3 Comparison of TVDI with Gravimetric Measurements

To prove the effectiveness of the methodology, TVDI and gravimetric measurements were analyzed in the study area. The efficiency of TVDI as an index for assessing soil moisture was compared with gravimetric measurements (Figure 13.10). This illustrates the relationship between TVDI and gravimetric measurements at various depths (0–15, 15–30, and 30–45 cm) for a field campaign (May 2011). The result shows a significantly negative correlation between TVDI and gravimetric measurements across the respective depths during the field campaign. The correlation $\left(R^2_{\text{land surface depth (TVDI)}} \right)$ was significant ($P < 0.05$) at all three depths $(R^2_{0-15\,\text{cm (TVDI)}} = 0.75,\ R^2_{15-30\,\text{cm (TVDI)}} = 0.70,\ \text{and}\ R^2_{30-45\,\text{cm (TVDI)}} = 0.63)$. A significant relationship was observed at 0–15 cm. This relationship was better than those for other depths.

Previous studies have compared the TVDI and direct estimation of soil moisture content. According to Sandholt et al. (2002), the relationship between TVDI and soil moisture yields a coefficient of determination ranging from 0.23 to 0.81. Wang et al. (2004) have found a significant negative correlation ($R^2 = 0.35 - 0.68$) between TVDI from NOAA-AVHRR and surface soil moisture. Patel et al. (2009) have shown that TVDI has a significant correlation with gravimetric soil moisture at a depths of 15, 30, and 45 cm. Chen et al. (2011) also describe that a significantly negative relationship exists between the TVDI and gravimetric soil moisture measurement at different soil depths. However, the relationship at 10–20 cm depth ($R^2 = 0.43$) was closest. Holzman et al. (2014) reported that the TVDI showed a strong negative correlation with soil moisture measurement with R^2 values ranging from 0.61 to 0.83. In this study, the relationship is significant for depths of 0–15 and 15–30 cm, which is also evident for some other studies (Chen et al. 2011; Holzman et al. 2014). Hence, these results reveal the ability of the TVDI to reflect spatial variation of soil moisture (Figure 13.9). It is observed that the TVDI better explains the variance in soil moisture in upper soil layers as compared to the deeper layers. Moreover, the TVDI results were found satisfactory for estimation of soil moisture.

The results show that the soil moisture (%) can be estimated by using linear regression models. The linear regression models developed for the soil moisture (%) are as follows:

$$0 \text{ to } 15 \text{ cm: Moisture (\%)} = -31.04\text{TVDI} + 24.21 \tag{13.14}$$

$$15 \text{ to } 30 \text{ cm: Moisture (\%)} = -36.58\text{TVDI} + 29.09 \tag{13.15}$$

$$30 \text{ to } 45 \text{ cm: Moisture (\%)} = -38.94\text{TVDI} + 30.98. \tag{13.16}$$

The spatial variation of soil moisture (%) at three different depths is shown in Figure 13.11.

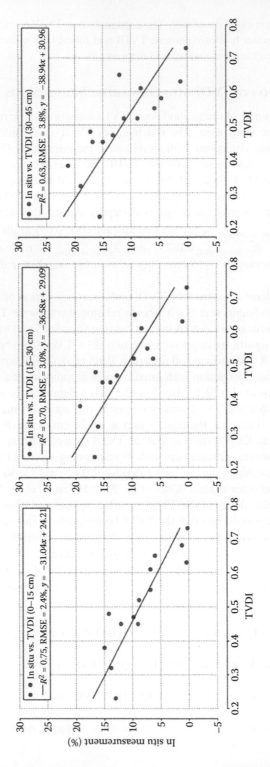

FIGURE 13.10 Linear regression between TVDI and gravimetric soil moisture at different depths (0–15, 15–30, and 30–45 cm).

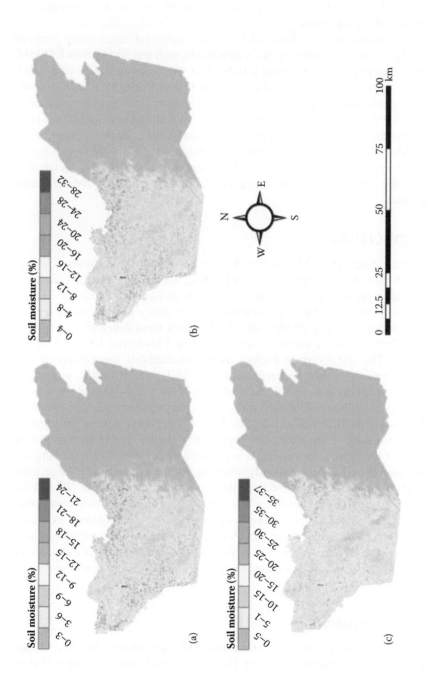

FIGURE 13.11 Spatial distribution of soil moisture (%) at soil depths (a) 0–15 cm, (b) 15–30 cm, and (c) 30–45 cm using TVDI and regression equations.

RSSMM and TVDI were found to be appropriate methods for estimating soil moisture. Willmott's Index (Willmott 1981) and RMSE were used to determine the performance of both methods.

It can be observed from Table 13.1 that the statistical relationship between TVDI and gravimetric measurements is significant, with less RSME than RSSMM. Moreover, it is clear from Willmott's Index that validation of TVDI with gravimetric measurements is more satisfactory than that of RSSMM. RSSMM shows potential for estimating surface soil moisture. However, TVDI is a combination of LST and NDVI (T_s–NDVI space), and it has been noted that LST and NDVI present considerable change in land surface characteristics between wet edge (T_{smin}) and dry edge (T_{smax}). Hence, TVDI has an advantage over RSSMM in that it considers both vegetation (NDVI) and LST. Comparing RSSMM and TVDI, it is found that the latter is an effective and a more appropriate method in determining soil moisture.

13.4 CONCLUSION

In this study, both RSSMM and TVDI have been used for estimation of soil moisture. It was observed that the TVDI method, which considers LST and vegetation index relationship, can be used effectively and appropriately for the estimation of soil moisture. The main objective of this study was to investigate soil moisture estimation using RSSM and TVDI, which were developed using NIR and red reflectance and a combination of Landsat TM–driven LST and NDVI, respectively. The soil moisture derived from both methods was compared with gravimetric measurements. The relationship of TVDI and RSSMM with gravimetric measurements was significant at depths of 0–15, 15–30, and 30–45 cm. The two models developed in this study delivered relatively reliable estimates of soil moisture at different soil depths. RSSMM presents diverse information of soil and vegetation; therefore, disintegration of the diverse pixel information may produce more accurate results. It is also concluded that TVDI extracts soil moisture and is useful for frequently studying soil moisture over a large area. The TVDI has a good correlation with field-collected soil moisture ($R^2 > 0.70$). The results of the study demonstrate that the TVDI is a more effective and comprehensive method for soil moisture estimation. Moreover, rigorous examination with concurring field data to the TVDI with temporal and spatial variation may improve the present approach. The inferring of surface moisture status on a regular basis can help in the monitoring of early drought, agriculture, and water management.

ACKNOWLEDGMENTS

We are thankful to glovis.usgs.gov for creating and disseminating the data. We are thankful to Sindh Agriculture University Tandojam for assistance with the field campaign and laboratory provisions. We also express our sincere thanks to the Institute of Space Technology (Department of Remote Sensing and Geoinformation Science) for funding this research study.

REFERENCES

Albergel, C., Dorigo, W., Balsamo, G., Muñoz-Sabater, J., de Rosnay, P., Isaksen, L., & Wagner, W. (2013). Monitoring multi-decadal satellite earth observation of soil moisture products through land surface reanalyses. *Remote Sensing of Environment, 138*, 77–89.

Amiri, R., Weng, Q., Alimohammadi, A., & Alavipanah, S. K. (2009). Spatial–temporal dynamics of land surface temperature in relation to fractional vegetation cover and land use/cover in the Tabriz urban area, Iran. *Remote Sensing of Environment, 113*(12), 2606–2617.

Carlson, T. (2007). An overview of the "triangle method" for estimating surface evapotranspiration and soil moisture from satellite imagery. *Sensors, 7*(8), 1612.

Carlson, T. N., Gillies, R. R., & Perry, E. M. (1994). A method to make use of thermal infrared temperature and NDVI measurements to infer surface soil water content and fractional vegetation cover. *Remote Sensing Reviews, 9*(1–2), 161–173.

Carlson, T. N., Gillies, R. R., & Schmugge, T. J. (1995). An interpretation of methodologies for indirect measurement of soil water content. *Agricultural and Forest Meteorology, 77*(3), 191–205.

Carlson, T. N., Rose, F. G., & Perry, E. M. (1984). Regional-scale estimates of surface moisture availability from GOES infrared satellite measurements. *Agronomy Journal, 76*(6), 972–979.

Chauhan, N., Miller, S., & Ardanuy, P. (2003). Spaceborne soil moisture estimation at high resolution: A microwave-optical/IR synergistic approach. *International Journal of Remote Sensing, 24*(22), 4599–4622.

Chen, J., Wang, C., Jiang, H., Mao, L., & Yu, Z. (2011). Estimating soil moisture using Temperature–Vegetation Dryness Index (TVDI) in the Huang-huai-hai (HHH) plain. *International Journal of Remote Sensing, 32*(4), 1165–1177.

Crow, W. T., Berg, A. A., Cosh, M. H., Loew, A., Mohanty, B. P., Panciera, R., de Rosnay, P., Ryu, D., & Walker, J. P. (2012). Upscaling sparse ground-based soil moisture observations for the validation of coarse-resolution satellite soil moisture products. *Reviews of Geophysics, 50*(2).

Ekstrand, S. (1996). Landsat TM-based forest damage assessment: Correction for topographic effects. *Photogrammetric Engineering and Remote Sensing, 62*(2), 151–162.

Evett, S., Heng, L., Moutonnet, P., & Nguyen, M. (2008). *Field Estimation of Soil Water Content: A Practical Guide to Methods, Instrumentation, and Sensor Technology.* IAEA: Vienna.

FAO. (2001). Special Report: FAO/WFP Crop and Food Supply Assessment Mission to Pakistan.

Gao, S., Zhu, Z., Liu, S., Jin, R., Yang, G., & Tan, L. (2014). Estimating the spatial distribution of soil moisture based on Bayesian maximum entropy method with auxiliary data from remote sensing. *International Journal of Applied Earth Observation and Geoinformation, 32*, 54–66.

Ghulam, A., Qin, Q., Zhu, L., & Abdrahman, P. (2004). Satellite remote sensing of groundwater: Quantitative modelling and uncertainty reduction using 6S atmospheric simulations. *International Journal of Remote Sensing, 25*(23), 5509–5524.

Gillies, R., Kustas, W., & Humes, K. (1997). A verification of the 'triangle' method for obtaining surface soil water content and energy fluxes from remote measurements of the Normalized Difference Vegetation Index (NDVI) and surface e. *International Journal of Remote Sensing, 18*(15), 3145–3166.

Goetz, S. (1997). Multi-sensor analysis of NDVI, surface temperature and biophysical variables at a mixed grassland site. *International Journal of Remote Sensing, 18*(1), 71–94.

Guo, N., Cheng, T., Lei, J., & Yang, L. (1997). Estimating farmland soil moisture in eastern Gansu Province using NOAA satellite data. *Journal of Applied Meteorological Science (in Chinese), 8*, 212–218.

Holzman, M., Rivas, R., & Piccolo, M. (2014). Estimating soil moisture and the relationship with crop yield using surface temperature and vegetation index. *International Journal of Applied Earth Observation and Geoinformation, 28*, 181–192.

Huete, A., Hua, G., Qi, J., Chehbouni, A., & Van Leeuwen, W. (1992). Normalization of multidirectional red and NIR reflectances with the SAVI. *Remote Sensing of Environment, 41*(2), 143–154.

Jackson, T. J. (1993). III. Measuring surface soil moisture using passive microwave remote sensing. *Hydrological Processes, 7*(2), 139–152.

Jawson, S. D., & Niemann, J. D. (2007). Spatial patterns from EOF analysis of soil moisture at a large scale and their dependence on soil, land-use, and topographic properties. *Advances in Water Resources, 30*(3), 366–381.

Kustas, W., & Norman, J. (1996). Use of remote sensing for evapotranspiration monitoring over land surfaces. *Hydrological Sciences Journal, 41*(4), 495–516.

Li, X., & Dong, W. (1996). Methods research on monitoring drought by using remote sensing and GIS. *Remote Sensing Technology and Application, 11*(3).

Li, Z., Wang, Y., Zhou, Q., Wu, J., Peng, J., & Chang, H. (2008). Spatiotemporal variability of land surface moisture based on vegetation and temperature characteristics in Northern Shaanxi Loess Plateau, China. *Journal of Arid Environments, 72*(6), 974–985. doi: http://dx.doi.org/10.1016/j.jaridenv.2007.11.014

Liang, S., Fallah-Adl, H., Kalluri, S., JáJá, J., Kaufman, Y. J., & Townshend, J. R. (1997). An operational atmospheric correction algorithm for Landsat Thematic Mapper imagery over the land. *Journal of Geophysical Research: Atmospheres (1984–2012), 102*(D14), 17173–17186.

Liu, P., Zhang, L., Kurban, A., Chang, P., Li, L., & Polat Zhao, B. (1997). A method for monitoring soil water contents using satellite remote sensing. *Journal of Remote Sensing, 1*(2), 135–138.

Liu, S., Roberts, D. A., Chadwick, O. A., & Still, C. J. (2012). Spectral responses to plant available soil moisture in a Californian grassland. *International Journal of Applied Earth Observation and Geoinformation, 19*, 31–44.

Markham, B. L., & Barker, J. (1986). Landsat MSS and TM post-calibration dynamic ranges, exoatmospheric reflectances and at-satellite temperatures. *EOSAT Landsat Technical Notes, 1*(1), 3–8.

Nichols, S., Zhang, Y., & Ahmad, A. (2011). Review and evaluation of remote sensing methods for soil-moisture estimation. *Journal of Photonics for Energy*, 028001–028017.

Patel, N., Anapashsha, R., Kumar, S., Saha, S., & Dadhwal, V. (2009). Assessing potential of MODIS derived temperature/vegetation condition index (TVDI) to infer soil moisture status. *International Journal of Remote Sensing, 30*(1), 23–39.

Price, J. C. (1980). The potential of remotely sensed thermal infrared data to infer surface soil moisture and evaporation. *Water Resources Research, 16*(4), 787–795.

Price, J. C. (1982). Estimation of regional scale evapotranspiration through analysis of satellite thermal-infrared data. *IEEE Transactions on Geoscience and Remote Sensing*, (3), 286–292.

Price, J. C. (1990). Using spatial context in satellite data to infer regional scale evapotranspiration. *IEEE Transactions on Geoscience and Remote Sensing, 28*(5), 940–948.

Prihodko, L., & Goward, S. N. (1997). Estimation of air temperature from remotely sensed surface observations. *Remote Sensing of Environment, 60*(3), 335–346.

Qin, J., Yang, K., Lu, N., Chen, Y., Zhao, L., & Han, M. (2013). Spatial upscaling of in-situ soil moisture measurements based on MODIS-derived apparent thermal inertia. *Remote Sensing of Environment, 138*, 1–9. doi: 10.1016/j.rse.2013.07.003

Rajasekar, U., & Weng, Q. (2009). Application of association rule mining for exploring the relationship between urban land surface temperature and biophysical/social parameters. *Photogrammetric Engineering & Remote Sensing, 75*(4), 385–396.

Richardson, A. J., & Weigand, C. (1977). Distinguishing vegetation from soil background information. *Photogrammetric Engineering and Remote Sensing, 43*(12).

Sandholt, I., Rasmussen, K., & Andersen, J. (2002). A simple interpretation of the surface temperature/vegetation index space for assessment of surface moisture status. *Remote Sensing of Environment, 79*(2), 213–224.

Santos, W. J. R., Silva, B. M., Oliveira, G. C., Volpato, M. M. L., Lima, J. M., Curi, N., & Marques, J. J. (2014). Soil moisture in the root zone and its relation to plant vigor assessed by remote sensing at management scale. *Geoderma, 221*, 91–95.

Singh, R., Oza, S., Chaudhari, K., & Dadhwal, V. (2005). Spatial and temporal patterns of surface soil moisture over India estimated using surface wetness index from SSM/I microwave radiometer. *International Journal of Remote Sensing, 26*(6), 1269–1276.

Sobrino, J. A., Jiménez-Muñoz, J. C., & Paolini, L. (2004). Land surface temperature retrieval from LANDSAT TM 5. *Remote Sensing of Environment, 90*(4), 434–440.

Srivastava, H. S., Patel, P., Manchanda, M., & Adiga, S. (2003). Use of multiincidence angle RADARSAT-1 SAR data to incorporate the effect of surface roughness in soil moisture estimation. *IEEE Transactions on Geoscience and Remote Sensing, 41*(7), 1638–1640.

Stisen, S., Sandholt, I., Nørgaard, A., Fensholt, R., & Jensen, K. H. (2008). Combining the triangle method with thermal inertia to estimate regional evapotranspiration—Applied to MSG-SEVIRI data in the Senegal River basin. *Remote Sensing of Environment, 112*(3), 1242–1255.

Vereecken, H., Huisman, J., Pachepsky, Y., Montzka, C., Van Der Kruk, J., Bogena, H., Weihermüller, L., Herbst, M., Martinez, G., & Vanderborght, J. (2014). On the spatiotemporal dynamics of soil moisture at the field scale. *Journal of Hydrology, 516*, 76–96.

Wang, C., Qi, S., Niu, Z., & Wang, J. (2004). Evaluating soil moisture status in China using the temperature–vegetation dryness index (TVDI). *Canadian Journal of Remote Sensing, 30*(5), 671–679.

Wang, H., Li, X., Long, H., Xu, X., & Bao, Y. (2010). Monitoring the effects of land use and cover type changes on soil moisture using remote-sensing data: A case study in China's Yongding River basin. *Catena, 82*(3), 135–145.

Wang, K., Li, Z., & Cribb, M. (2006). Estimation of evaporative fraction from a combination of day and night land surface temperatures and NDVI: A new method to determine the Priestley–Taylor parameter. *Remote Sensing of Environment, 102*(3), 293–305.

Wang, L., Qu, J., Zhang, S., Hao, X., & Dasgupta, S. (2007). Soil moisture estimation using MODIS and ground measurements in eastern China. *International Journal of Remote Sensing, 28*(6), 1413–1418.

Wang, L., & Qu, J. J. (2009). Satellite remote sensing applications for surface soil moisture monitoring: A review. *Frontiers of Earth Science in China, 3*(2), 237–247.

Willmott, C. J. (1981). On the validation of models. *Physical Geography, 2*(2), 184–194.

Xin, J., Tian, G., Liu, Q., & Chen, L. (2006). Combining vegetation index and remotely sensed temperature for estimation of soil moisture in China. *International Journal of Remote Sensing, 27*(10), 2071–2075.

Zhan, Z., Qin, Q., Ghulan, A., & Wang, D. (2007). NIR-red spectral space based new method for soil moisture monitoring. *Science in China Series D: Earth Sciences, 50*(2), 283–289.

Section IV

Remote Sensing for Sustainable Energy

14 Earth Observation and Its Potential to Implement a Sustainable Energy Supply
A German Perspective

Thomas Esch, Markus Tum, and Annekatrin Metz

CONTENTS

14.1 REMOTE SENSING IN SUPPORT OF ENERGY POLICY

Today, satellite sensors provide digital recordings of the earth's surface in a spatial resolution of approximately 1 km (e.g., SPOT-VEGETATION, MODIS, MERIS, Proba-V) to less than 1 m (z. B. WorldView, QuickBird, IKONOS). The temporal repetition cycle varies from daily to monthly surveys. The Earth observation technology has steadily evolved in recent years from a beginning characterized by a rather experimentally imprinted alignment toward operational services, which ensures a long-term, regular, and quality-assured provision of spatial data and higher-value information products. On the one hand, these data and products can be recorded while targeted on request for specific applications, periods, or regions as needed, or on the other hand, these can be implemented in a systematic and comprehensive monitoring of the Earth's surface. Thus, satellite-based Earth observation provides promising applications to implement a sustainable energy supply. Remote sensing–aided applications such as wind field analysis and the levying of irradiation data or the determination of solar surface potentials are commercially exploited and have been available for several years and are firmly established in existing planning processes.

Nevertheless, the need for the provision of missing or supplementation of existing geodata and geoinformation products to support the energy turnaround in Germany has been formulated repeatedly by representatives from policy, management, planning, the private sector, and academia.

14.2 GEODATA FOR THE ENERGY SECTOR— EXAMPLES OF SATELLITE-BASED ANALYSIS

Looking over the formulated demand for spatially and thematically enhanced spatial data in support of sustainable energy supply, the collection of geospatial data from satellite image analysis especially with regard to the potential assessment, site selection, conflict prevention of interest, and monitoring of impacts and trends must be regarded as particularly beneficial. This applies to the comprehensive determination of the biomass volume in the context of site assessment or the parcel-related acquisition of agricultural crops for the estimation of regional straw potentials. Furthermore, the use of remote sensing techniques can also be used for the detection, visualization, and documentation of developments and trends of, for example, the transformation of landscapes in the context of energy policy decisions or the growing energy consumption and increasing use conflicts in the wake of steadily progressing land use by settlements and transport infrastructure. The following three exemplary recent research and development works of the department's land surface of the Earth Observation Center at the German Aerospace Center (DLR) will be presented: the modeling of sustainable bioenergy potentials, the acquisition of agricultural growing patterns, and the accounting of settlement structures with regard to their suitability for the construction of heating networks. All these applications are aimed at supporting land management, which is geared toward the promotion of sustainable energy supply.

14.2.1 POTENTIAL ANALYSIS: BIOENERGY

The Biosphere Energy Transfer Hydrology (BETHY/DLR) model of the DLR is driven by high spatial and temporal resolution remote sensing data and derived parameters such as leaf area index (LAI) and is used for the quantification of biomass and bioenergy potentials. As an input data set for the evaluation described below, global LAI data, which are available in the form of 10-day composite time series as well as information on land cover and land use, derived from the Global Land Cover 2000 (GLC2000) were used. In addition to these remote sensing data, BETHY/DLR also requires diverse meteorological information on air temperature, precipitation, wind speed, and cloud cover. These data are provided by the European Center for Medium-Range Weather Forecast in an appropriate format. Furthermore, BETHY/DLR uses information on the dominant soil type from the Harmonized World Soil Database and a digital elevation model of the Shuttle Radar Topography Mission.

In vegetation models such as BETHY/DLR, plant growth is parameterized in a way that, in a first step, the biochemical processes of photosynthesis are modeled at the leaf level. This is followed by an extrapolation of both the structure of the plant and the interaction (e.g., energy flows, water circulation) included in the calculations between soil, vegetation, and atmosphere from the leaf level to the inventory level. As a result, the model initially provides the carbon amount absorbed by the vegetation per unit area and time out of the atmosphere—the gross primary production. Since each plant emits carbon back to the atmosphere through autotrophic respiration in the form of CO_2, in the balance sheet, less carbon in the form of biomass is bound in the plant than was originally recorded. This output of the model is called Net Primary Productivity (NPP). With the use of conversion factors, NPP can be converted into biomass and energy potentials. For this purpose, conversion factors such as the relationship of grain to straw or root to shoot and the ratio of aboveground to underground biomass are also needed, as well as information on the specific water content of the dry biomass and its energy yield per kilogram (lower heating value). Additionally, estimates of competing uses are likewise required (e.g., soil fertilization by straw, entry into stables, etc.).

In Figure 14.1, agricultural and forestry bioenergy potentials for Germany in 2012 are illustrated. Energy potentials of straw are locally a factor of 10 lower than the potential growth of wood; as for the agricultural potential, because of competing uses, only a small proportion of the straw (20%) is considered to be usable for energy production. In contrast, 80% of forest growth is considered to be theoretically available. For 2012, the overall energy potential was calculated to be 572 petajoules (PJ) for agriculture and 1938 PJ for forestry. A detailed description of all input data and the methodology for modeling of the NPP and conversion of NPP in agricultural and forest energy potentials can be found in Wißkirchen et al. (2013) and Tum et al. (2013).

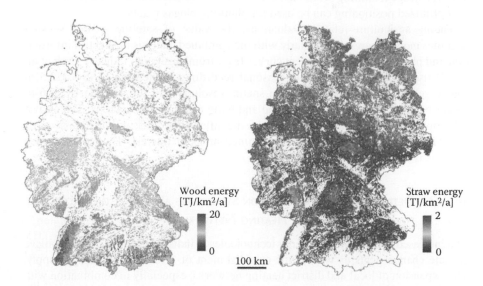

FIGURE 14.1 Energy potential of wood growth (forestry) and straw (agriculture) in TJ/km²/a for 2012.

14.2.2 MAPPING OF AGRICULTURAL CULTIVATION PATTERNS

In connection with the policy on climate change, the efforts to reduce CO_2 consumption and the implementation to a sustainable energy supply have been added as a new component in addition to aspects such as food security, food quality, and sustainable management in view of the agricultural production in Germany and the increased use of bioenergy raw materials. This contributed input to the funding policy with regard to biogas plants and the Renewable Energies Act (EEG). The opportunity to obtain co-financing of biogas power plants through the EEG has greatly affected the cultivation of crops in the past decade, for example, in the form of an increase in the acreage of silage maize (DMK 2013). The new support measures for the cultivation of energy raw materials also drive increased reallocation of grassland in areas used for farming. In addition to the loss of arable land for food production, the environmental impact of large-scale cultivation of maize and other energy crops should be evaluated critically.

The sometimes rapid processes of change in the cultural landscape bear in mind that a continuous and timely monitoring of intensity of use and acreage development is needed to promptly identify trends and unfavorable regional developments in terms of sustainable management to be able to effectively counteract such adverse developments based on this information. Satellite remote sensing provides an ideal base, because it enables spatially and temporally precise analysis, which can also be easily integrated into spatial information systems and with other data, such as statistical information or geometries of cadastral surveying. Moreover, meaningful use potentials can be determined in order to determine, for example, areas with exploitable straw shares by the cultivation of grain or spatial potential through enhanced incurred hedge trimming at national care measures. In addition, relevant information for optimized positioning can be used for planning biogas plants.

Facing agriculture-related evaluations, the value of satellite remote sensing becomes more apparent particularly with the combined analysis of spectral information and phenological development curves (e.g., from vegetation indices). Esch et al. (2014) used high-resolution multi-seasonal recordings of Sensors LISS-3 (23.5 m spatial resolution) and AWiFS (56 m spatial resolution) of the satellite IRS-P6 to parcel-based classification of grassland and main crops of arable crops. The aim of this approach is to provide a methodology that allows for quick and flexible acquisition of cultivation patterns on predetermined areas of interest. The result of such analysis is shown in Figure 14.2.

14.2.3 DETERMINATION OF POPULATION STRUCTURES AND SITE CONDITIONS FOR CONSTRUCTION OF HEATING NETWORKS

The increased use of efficient energy technologies is indispensable in order to achieve climate change goals as well as the establishment of a sustainable energy supply. The expansion of local and district heating networks, especially in combination with plants for combined heat and power as well as the increased use of renewable district heating, is of central importance here. In order to evaluate the potential for establishing a wired heat supply, an analysis of settlement structural conditions and the

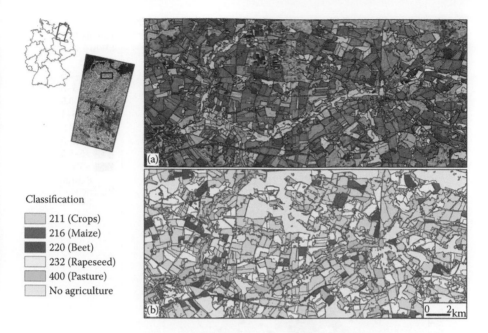

Classification

☐ 211 (Crops)
■ 216 (Maize)
■ 220 (Beet)
☐ 232 (Rapeseed)
▨ 400 (Pasture)
☐ No agriculture

FIGURE 14.2 Use of multiseasonal satellite imagery (a) to determine the agricultural cultivation pattern (b).

resulting potential for development is needed. An important element is the provision of current, nationwide heat power (energy) potential, which can be used to address the physical–structural location conditions for district heating. The heat power potential is calculated by the ratio of annual heat energy demand of the building and investment costs to provide the heat needed. The corresponding characteristic value represents the annual exploitable potential by the supply quantities of heat (kWh/a) per invested monetary unit (euros). The specific heat energy consumption value depends on the type and use of the building, the age of the building, and the climatic conditions. The estimation of costs for heat infrastructure is based on an analysis of the lengths and associated costs for main supply and service lines, which in turn can be modeled on an evaluation of the road network and the location of buildings. The heat demand is made up of the building volume and the specific heat demand value, while the investment costs include all expenses for distribution, connecting cables, house transfer stations, and savings for boilers.

Figure 14.3 shows the result of a Germany-wide analysis of heat power potential that has been identified through a combination of top-down and bottom-up approach as methods of digital image analysis and geographic information systems (DLR 2011). The top-down approach is used during the estimation of the heat demand on the basis of data on the housing stock and the climatic situation. The building stock is determined by Infas-geodata (INFAS 2015) of the building type (one/two-family houses, small and large apartment buildings) and building age (nine periods of construction) as well as information from the Federal Statistical Office for building use (residential buildings, nonresidential buildings). Climatic conditions

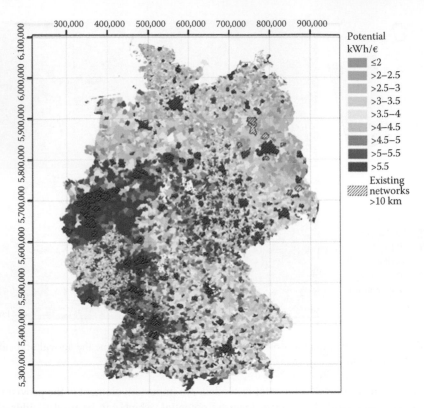

FIGURE 14.3 Heat power potential of communities in Germany in 2011, without considering the cost of a conventional heat supply.

can be displayed using the degree days provided by the German Weather Service (http://www.dwd.de). The calculation of the wireline investment costs is bottom-up involving data of the Authoritative Topographic–Cartographic Information System (accessible at http://www.atkis.de) on the situation of human settlements and the course of the road network (determining main supply and service lines) and makes use of information about the existing buildings (location, area) extracted from Digital Topographic Maps (DTK 25), which can be obtained at http://www .adv-online.de. Although data on the situation, floor space, and even the volume of buildings are available nationwide in the form of spatial data (e.g., house outlines or 3D building model LoD1), the high cost of ~100,000 to 350,000 euros currently prevents users to access them. The evaluation results show that the thermal power potential in addition to the local heat demand is highly dependent on the length of heat distribution. This is closely linked to the settlement structure. Therefore, core cities and their surrounding particularly exhibit high values, while rural areas have relatively low potential. For example, in the southern parts of Germany, relatively favorable structures are available, whereas the northeast has relatively unfavorable structures.

14.3 CONCLUSIONS

Satellite-based remote sensing offers versatile opportunities for the provision of geo-information for energy-related issues. Recent results and experience show that Earth observation can make important contributions in particular with regard to potential calculation, monitoring of current availability, and the monitoring of changes and trends.

Advantages over established approaches and data sets in this connection arise in particular with regard to timeliness, coverage, comparability, spatial detail, and update costs. Nevertheless, an operational provision of the necessary remote sensing–based geoinformation products in the form of reliable services is central for commercial production and ultimately for success in the planning and management sector. Cornerstones in this context ensure a high spatial, temporal, and qualitative continuity; transparent and reliable procurement costs and modalities; and the exploitation of synergies (e.g., with existing spatial data held by the cadastral survey). However, unclear responsibilities, requirements, and user requirements currently complicate targeted developments. Against this background, there is a particular need for the coordination and pooling of research and development activities between the different actors of planning/management, industry, and science. A networked and concerted approach enables the effective development and utilization of existing expertise and synergies. At the same time, a close connection to the GMES/Copernicus environment should be ensured in order to meet the requirements for data continuity and service delivery.

REFERENCES

DLR—Deutsches Zentrum für Luft- und Raumfahrt e. V. 2011. Potenzialanalyse zum Aufbau von Wärmenetzen unter Auswertung siedlungsstruktureller Merkmale. Endbericht (BMVBS/BBSR FKZ 3004775). www://elib.dlr.de/76816/1/Waermenetzpotenzial _DLR_Endbericht_final.pdf (accessed September 14, 2015).

DMK—Deutsches Maiskomitee e. V. 2013. Bedeutung des Maisanbaues in Deutschland. www .maiskomitee.de/web/public/Fakten.aspx/Statistik/Deutschland (accessed September 14, 2015).

Esch, T., Metz, A., Keil, M., Marconcini, M. 2014. Combined use of multi-seasonal high and medium resolution satellite imagery for parcel-related updating of cropland and grassland. *International Journal of Applied Earth Observations and Geoinformation*, 28C:230–237.

INFAS: Institut für angewandte Sozialwissenschaften GmbH. https://www.infas.de/ (accessed December 17, 2015).

Tum, M., Günther, K.P., McCallum, I., Kindermann, G., Schmid, E. 2013. Sustainable bioenergy potentials for Europe and the globe. *Geoinformatics & Geostatistics: An Overview*, S1.

Wißkirchen, K., Tum, M., Günther, K.P., Niklaus, M., Eisfelder, C., Knorr, W. 2013. Quantifying the carbon uptake by vegetation for Europe on a 1 km^2 resolution using a remote sensing driven vegetation model. *Geoscientific Model Development* 6:1623–1640.

15 Use of Nighttime Imaging Data to Assess Decadal Trends in Energy Use in China

Yanhua Xie and Qihao Weng

CONTENTS

15.1 INTRODUCTION

In general, energy refers to the ability to do work and can be in the form of electricity, thermal, chemical, mechanical, gravitational, nuclear, radiant, sound, and motion. The sources of energy can be fossil (e.g., petroleum, coal, natural gas, wood, etc.), renewable (wind, solar, geothermal, hydrogen, etc.), and fissile (e.g., uranium, thorium, etc.) (Bilgen 2014). Being one of the three crucial themes (i.e., energy consumption, urbanization, and carbon dioxide emission) in the 21st century, energy has caught much attention as it is essential for socioeconomic development for all countries (Al-mulali et al. 2012; Weng 2013). For the past decades, attention exerted on energy issues in developing countries (e.g., China and India) has been especially intense as these countries are currently experiencing fast urbanization. The leap of their economies has caused rapid growth of energy consumption, and this trend is expected to continue in the near future (Bilgen 2014). However, if the unsustainable energy consumption (e.g., overexploitation of nonrenewable resources and air pollution caused by fossil fuels) continues, environmental problems will definitely deteriorate (Adams and Shachmurove 2008; Ahmad et al. 2015; Bloch et al. 2015; Cherni and Kentish 2007). A better knowledge of the spatiotemporal patterns of energy consumption at multiple scales is thus a critical step for the sustainable development of energy as it provides policy makers with the guidelines for energy production, distribution, and management.

15.2 SOME BACKGROUND ON NTL-BASED ESTIMATION OF ENERGY CONSUMPTION

A major challenge for analyzing multiscale spatiotemporal patterns of energy consumption is the lack of detailed data in a temporally and spatially consistent manner, and this situation is more severe in developing countries (Parshall et al. 2009). Traditionally, the primary source of energy consumption data is census data with the interval of years and spatial resolution of administrative units (e.g., country, province), hardly satisfying the requirements of spatiotemporal analyses at multiple scales (especially for local scales such as urban level).

Remotely sensed nighttime light (NTL) imagery from the Operational Linescan System (OLS) of the Defense Meteorological Satellite Program (DMSP) has been proven effective in the estimation of electricity consumption (EC) in a spatially explicit and consistent manner, because it provides frequently repeated records of EC for outdoor lighting (Amaral et al. 2005; Cao et al. 2014; Chand et al. 2009; Elvidge et al. 2001; He et al. 2012, 2014; Letu et al. 2010; Lo 2002; Welch 1980). Meanwhile, the new generation of NTL image from the Visible Infrared Imaging Radiometer Suite onboard the Suomi National Polar-Orbiting Partnership satellite has shown an enhanced relationship between EC and NTL at the sub-country scale (Shi et al. 2014). NTL data from DMSP-OLS are especially attractive given its long time archive (1992 to the present).

The potential use of DMSP NTL images for characterizing EC pattern was first noted in the 1980s by Welch (1980). Since then, a number of studies have demonstrated the relationship between DMSP-OLS NTL imagery and EC at national, state/provincial, and county scales. For instance, Elvidge et al. (1997, 2001) reported the country-scale relationships between lit area and EC with 21 and 200 countries, respectively; Letu et al. (2010) also demonstrated a country-scale relationship between EC and saturation-corrected DMSP NTL intensity for 10 Asian countries; Amaral et al. (2005) modeled EC in the Brazilian Amazon from a lighted area at the municipal level; Lo (2002) established a logarithmic relationship between EC and lit area for 35 Chinese capital cities. In recent years, the examination of spatiotemporal dynamics of EC at different scales has been conducted since the availability of time-series DMSP-OLS NTL images. For instance, by using DMSP-OLS NTL data from 1993 to 2002, Chand et al. (2009) characterized spatiotemporal changes in EC patterns in the major cities and states of India; He et al. (2012, 2014) examined EC pattern at the county level in Mainland China using DMSP-OLS NTL data during 1995–2008 and 2000–2008, respectively; to detect the spatiotemporal pattern of EC at the sub-county level, Cao et al. (2014) proposed a top-down method to model pixel-based EC from 1994 to 2009 in China, using NTL, population density, and GDP as the independent variables.

Despite a number of research showing the capacity of NTL as an indicator of EC with varying degrees of success, most of them focused on global-, continental-, or national-level estimations. However, increasing municipal authorities and planners are addressing energy issues within the context of local climate and sustainability initiatives, requiring adequate supply of small-scale energy data (Parshall et al. 2009). Additionally, each study focused on the analyses at a single scale, ignoring the importance of multiscale analyses, which may offer potentials for better understanding the process over the interaction between socioeconomic developments and EC. Meanwhile, although studies have been undertaken to examine the relationship between NTL and EC, there is limited assessment from the current literature about whether NTL can be a good indicator of overall energy consumption (Wu et al. 2014; Xie and Weng 2016). According to our initial assessment, a significantly positive relationship exists between national NTL and overall energy consumption even though it includes both EC and energy consumed in other forms such as coal and natural gas, which cannot be directly detected by the DMSP-OLS sensor. In this chapter, we characterized the decadal trend (2000 to 2012) of EC at sub-country scales (i.e., provincial, prefectural, and urban scale) in Mainland China by using time-series DMSP-OLS NTL images. The same procedure could be followed for the spatiotemporal analyses of overall energy consumption. The results further provided a spatially explicit way to evaluate the relationship between China's urbanization and its EC in the first decade of the 21st century.

15.3 STUDY AREA AND DATA SETS

Mainland China was selected as the study area, where a rapid urbanization occurred since the economic reform in 1978, especially after the mid-1990s (Wang 2014). Specifically, urbanization in China accelerated for the first decade of the 21st century (Huang et al. 2015). Population size reached 1.34 billion at the end of 2010, and the urbanization rate of population increased from 36.22% to 51.27% between 2000 and 2011 (Wang 2014), with an annual increase of 3.21%. The rapid economic growth associated with urbanization steadily promoted the demand of energy, with the national consumption of electricity increased by 3.6 times, from 1361.78 billion kWh in 2000 to 4953.50 billion kWh in 2012.

To capture regional differences, the study area was divided into four regions (eastern: Beijing, Fujian, Guangdong, Hebei, Jiangsu, Shandong, Shanghai, Tianjin, and Zhejiang; northeastern: Heilongjiang, Jilin, and Liaoning; central: Anhui, Henan, Hubei, Hunan, Jiangxi, and Shanxi; and western: Chongqing, Gansu, Guangxi, Guizhou, Inner Mongolia, Ningxia, Qinghai, Shaanxi, Sichuan, Xinjiang, Xizang, and Yunnan). Table 15.1 shows that, within the overall EC across Mainland China, the amount consumed in eastern China increased steadily from 47.24% to 50.79% from 2000 to 2006, but this value reduced to 48.08% in 2012. For western China, the proportion remained stable until 2005 and started to increase from 21.85% in 2005 to 25.75% in 2012. The ratio of EC in northeastern China dropped dramatically, from 10.89% in 2000 to 6.79% in 2012. However, the ratio was around 19.50% for central China at all times.

TABLE 15.1
National EC and the Percentage Consumed in Each Region

	EC (Billion kWh)	Eastern (%)	Central (%)	Northeastern (%)	Western (%)
2000	1361.78	47.24	19.66	10.89	22.21
2001	1468.90	47.91	19.84	10.33	21.92
2002	1638.61	48.87	19.66	9.67	21.80
2003	1888.02	49.62	19.53	9.22	21.63
2004	2174.88	49.94	19.28	8.81	21.97
2005	2474.71	50.74	19.14	8.26	21.86
2006	2833.88	50.79	18.97	7.90	22.34
2007	3253.25	50.35	19.42	7.53	22.70
2008	3431.62	50.08	19.39	7.51	23.02
2009	3655.48	49.95	19.30	7.36	23.39
2010	4192.83	49.61	19.31	7.25	23.83
2011	4694.78	48.54	19.40	7.02	25.04
2012	4952.50	48.08	19.37	6.79	25.76

Source: *China Statistical Yearbook 2001–2013* (http://www.stats.gov.cn/tjsj/ndsj/). The original EC data
were reported by each province. Thus, they were reorganized by the authors into national and
regional statistics.

The data sets used included DMSP-OLS NTL imageries, MODIS 16-day enhanced
vegetation index (EVI) composites (MOD13A2), Gridded Population Density data
(GPWv3), statistical EC data, and administrative boundary (Table 15.2). All image-
based data sets were from 2000 to 2012 except for population density data for 2000,
2005, and 2010 because of their availability. The National Geophysical Data Center
(NGDC) provided publicly available DMSP-OLS NTL imageries (last access in
January 2015) with cloud-free composite, which contained persistent lights from

TABLE 15.2
Data Sets Used and Descriptions

Data Sets	Period	Data Source
DMSP-OLS NTL data	2000–2012	NOAA-NGDC (http://ngdc.noaa.gov/eog/dmsp.html)
MODIS 16-day EVI composites (MOD13A2)	2000–2012	USGS (https://lpdaac.usgs.gov/)
Gridded population density map (GPWv3)	2000, 2005, 2010	SEDAC (http://sedac.ciesin.columbia.edu)
Electricity consumption data (provincial level)	2000–2012	China Statistical Yearbook (http://www.stats.gov.cn/)
Electricity consumption data (district level)	2001, 2005, 2010	Urban Statistical Yearbook of China (http://www.stats.gov.cn/)
Administrative boundary	2015, SHP format	Global Administrative Areas (http://www.gadm.org/)

human activities. The digital number (DN) of pixels ranged from 0 to 63, with a spatial resolution of nearly 1 km (at the equator). MODIS vegetation index data (MOD13A2) were acquired from the United States Geological Survey (USGS), with the temporal resolution of 16 days and the spatial resolution of 1 km. EVI images were selected instead of normalized difference vegetation index images as they tend to saturate at suburban areas where dense vegetation and high NTL values may exist (Zhang et al. 2013). A Gridded Population Density map (GPWv3) was created by SEDAC (NASA Socioeconomic Data and Application Center), providing globally consistent and spatially explicit human population information and data. The grid cell resolution was ~5 km at the equator. Additionally, the statistical data of province-level EC and EC of urban districts (i.e., Shixiaqu in Chinese) at the prefecture-level were collected from China Statistical Yearbook and the Urban Statistical Yearbook of China, respectively. Furthermore, the administrative boundary file was acquired from Global Administrative Areas.

15.4 METHODOLOGY

The method included four major steps: preprocessing, estimating gridded EC and assessing the accuracy of estimations, mapping urban dynamics, and analyzing spatiotemporal pattern of EC at multiple scales (Figure 15.1).

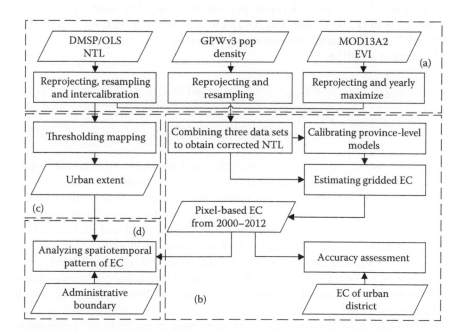

FIGURE 15.1 The flowchart of this study, with (a) preprocessing, (b) estimating pixel-based EC and accuracy evaluation, (c) mapping urban dynamics, and (d) analyzing spatiotemporal pattern of EC at multiple scales.

15.4.1 DATA PREPROCESSING

Image-based data sets were processed to the same projection (i.e., Lambert Azimuthal Equal Area) and spatial resolution (i.e., 1 km). Meanwhile, yearly maximum composite was computed for EVI:

$$EVI_{max} = max \ (EV_1, EV_2, \ldots EVI_{23}),\hspace{2cm}(15.1)$$

where EVI_{max} is the yearly maximum EVI of a total of 23 16-day composites for each year.

15.4.2 ESTIMATION OF PIXEL-BASED EC

In this chapter, a pixel-based EC was estimated instead of using administrative boundary as the unit so that EC at multiple scales could be estimated through aggregation. To obtain better gridded EC, DMSP-OLS NTL should be enhanced to eliminate the two notorious drawbacks (i.e., lacking onboard calibration system, and having a saturation effect and a blooming effect in urban cores and peri-urban areas, respectively). Thus, intercalibration was first applied to eliminate inconsistent DNs between NTL imageries. The method proposed by Elvidge et al. (2009) was adopted by assuming Mauritius, Puerto Rico, and Okinawa (in Japan) as the invariant regions (Wu et al. 2013). To eliminate the saturation effect of NTL in urban cores and to account for human activities in the unlit pixels, population density data were used (population density data of 2000, 2005, and 2010 were used to adjust NTL for 2000–2002, 2003–2007, and 2008–2012, respectively) followed by EVI adjustment (Meng et al. 2014; Zhang et al. 2013):

$$NTL_{cal} = w * NTL_{int} * \ln(PD + 1)\hspace{2cm}(15.2)$$

$$NTL_{adj} = (1 - EVI_{max}) * NTL_{cal},\hspace{2cm}(15.3)$$

where NTL_{int}, NTL_{cal}, and NTL_{adj} are intercalibrated, population density–calibrated, and population- and EVI-adjusted NTL DN value, respectively; PD refers to population density; and w is the weight for lighted and unlit pixel. w equaled to 1.0 for lighted pixels and 0.34 for unlit areas based on previous research on the gap between the detected electricity access in unlit and lighted area (Doll and Pachauri 2010; Meng et al. 2014). The adjustment using EVI was based on the assumption that vegetation cover and human activities were inversely correlated (Zhang et al. 2013). For the pixels with EVI < 0.1 or EVI > 0.9, NTL_{adj} was set to 0, based on the assumption that those regions were usually deserts, glaciers, water bodies, or dense forests with few human activities.

The province-level statistical EC data were used to calibrate the relationship between EC and NTL_{adj} owing to the difficulty of obtaining statistical data at sub-province levels. Additionally, to absorb the errors that might be introduced by intercalibration and to capture the yearly difference of the relationship between EC and NTL_{adj}, the model was calibrated for each year:

$$EC_{pro} = a * NTL_{pro} + b, \tag{15.4}$$

where EC_{pro} and NTL_{pro} refer to the EC and the sum of NTL_{adj} at the province level, respectively, and a and b are the coefficients. We then constructed pixel-based time-series EC from 2000 to 2012 based on the assumption that the calibrated provincial models were applicable to pixels:

$$EC_{yi} = a_y * NTL_{yi}, \tag{15.5}$$

where y refers to year index and i is the index of pixel. Further, normalization was conducted to regulate the total EC at the provincial level:

$$k_{yp} = EC_{yp} \Big/ \sum_p EC_{yi}, \tag{15.6}$$

where EC_{yp} is the statistical EC for province p in year y, $\sum_p EC_{yi}$ is the total estimated EC for each province, and k_{yp} refers to the normalization factor. Finally, gridded EC was estimated by

$$EC_{ypi} = k_{yp} * EC_{yi}. \tag{15.7}$$

A per-pixel evaluation was not possible because of the lack of high-resolution reference EC. Because of this, previous research on the estimation of gridded EC did not include the process of accuracy evaluation (Cao et al. 2014; Zhao et al. 2012). To validate the results of pixel-based EC in this chapter, we used the prefecture–city level statistical EC data as the reference. Specifically, EC by urban district (i.e., Shixiaqu in Chinese) at the prefectural level was used.

15.4.3 Analysis of Spatiotemporal Pattern of EC

To detect the spatiotemporal pattern of EC at different scales, pixel-based EC was aggregated to the provincial and prefectural scale. Additionally, to study the relationship between EC and urbanization, estimated gridded EC was aggregated to urban areas. Instead of using administrative units such as *Prefectural City* and *County* to define urban environment, we referred to built-up area as urban areas because (1) administrative units cover not only urban extent but also peri-urban and rural areas, and (2) the use of administrative boundaries to define urban areas is problematic as they are designed for political purposes, failing to capture urbanization dynamics and its impact on EC (Meng et al. 2014). Among the methods proposed to map regionally and globally consistent urban extents from DMSP-OLS

NTL data, including iterative unsupervised classification (Zhang and Seto 2011), SVM-based classification (Cao et al. 2009; Pandey et al. 2013), and threshold techniques (Liu et al. 2012; Zhou et al. 2014), we adopted the technique of threshold proposed by Liu et al. (2012) with the consideration of its simplicity and reasonable accuracy (Zhou et al. 2015) and the study area.

Global and local Moran's I indices were utilized to characterize the spatial pattern of EC at different scales. The global Moran's I index shows the nationwide spatial correlation of EC, while the local one mainly reflects the heterogeneity of EC between neighbor areas (Anselin 1995, 1996). The equation of global Moran's I is

$$I = \frac{N \sum_i \sum_j w_{ij}(x_i - \bar{x})(x_j - \bar{x})}{\left(\sum_i \sum_j w_{ij}\right) \sum_i (x_i - \bar{x})^2}, \tag{15.8}$$

where N is the sample size, w_{ij} is the matrix of spatial weight, x_i and x_j are EC for the ith and jth unit at the scale under consideration, and $\bar{x}$ refers to the average value of x. The value of global Moran's I index ranges from −1 to 1. Negative values indicate negative spatial autocorrelation and the inverse for positive values. Further, local Moran's I was applied to explore where high or low EC concentrations occurred. The equation of local Moran's I is

$$I_i = \frac{(x_i - \bar{x})}{s_i^2} \sum_{j, j \neq i} \left[w_{ij}(x_j - \bar{x}) \right], \tag{15.9}$$

where

$$s_i^2 = \frac{\sum_{j, j \neq i} (x_j - \bar{x})^2}{N - 1} - \bar{x}^2. \tag{15.10}$$

Local Moran's I describes four types of local spatial autocorrelation. The positive value of I_i indicates a high EC surrounded by similarly high values of EC or a low EC surrounded by low values of EC, while the negative value of I_i implies a low value surrounded by high values or a high value surrounded by low values.

Further, the index proposed by He et al. (2012) was used to examine the temporal pattern of EC. The formula of the index is

$$\text{SLOPE} = \frac{n \sum_{i=1}^n x_i \text{EC}_i - \sum_{i=1}^n x_i \sum_{i=1}^n \text{EC}_i}{n \sum_{i=1}^n x_i^2 - \left(\sum_{i=1}^n x_i\right)^2}, \tag{15.11}$$

where n is equal to 13, representing the number of years between 2000 and 2012, x_i refers to numbers 1 to 13 with increments of 1, and EC_i is the EC in the ith province, prefectural city, or urban area in year i. The negative value of SLOPE indicates a decreasing trend of EC and the inverse trend for positive SLOPE.

15.5 RESULTS

15.5.1 PIXEL-BASED EC FOR MAINLAND CHINA

Table 15.3 shows the calibrated yearly model with R^2 all above 0.84. The value of coefficient a for province-level models increased remarkably and steadily from 0.0215 in 2000 to 0.0536 in 2012, indicating the increasing trend of EC intensity per unit of NTL. Using the calibrated models in Table 15.3, we estimated pixel-based EC for Mainland China from 2000 to 2012. Figure 15.2 presents the estimated result from 2000 to 2012 with a temporal step of 4 years. It needs to note that the introduction of EVI greatly enhanced variation and reduced the blooming effect in the estimated EC maps. Visually, the overall pattern of EC in Figure 15.2a through d was almost identical, with some regions having remarkably high EC, such as the Beijing–Tianjin–Tangshan metropolitan region, the Yangtze River Delta, the Pearl River Delta, the Sichuan Basin, and most of the capital cities. It is also shown in Figure 15.2 that the urban cores tended to have a higher intensity of EC than their suburban and peri-urban regions owing to more intensive economic activities (e.g., service industry) in Chinese urban cores. The highest per-pixel EC in urban cores could be as high as 38.5, 39.8, 50.8, and 58.4 million kWh for 2000 to 2012, respectively. Nevertheless, much less electricity was consumed in rural regions and non-EC regions could be forests, deserts, and water bodies. Regionally, urban areas in eastern China used more electricity than other regions, while the majority of western China had less access to electricity during the past decade. Additionally, it is shown in Figure 15.2 that Shanghai has the most intense urban EC. The overall pattern of EC showed that urban areas had experienced the most increase in EC across China (e.g., rural regions).

TABLE 15.3
Yearly Regression Result for Provincial-Level Models (with $P = 0.0000$ for All Regressions)

Year	a	R^2	Year	a	R^2
2000	0.0215	0.8519	2007	0.0491	0.9092
2001	0.0236	0.8413	2008	0.0478	0.9168
2002	0.0264	0.8836	2009	0.0463	0.8492
2003	0.0301	0.9095	2010	0.0524	0.8744
2004	0.0335	0.9046	2011	0.0514	0.9081
2005	0.0387	0.8791	2012	0.0536	0.8933
2006	0.0434	0.9040			

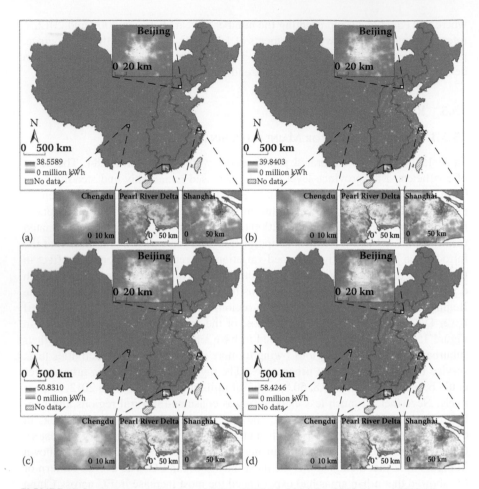

FIGURE 15.2   Estimated pixel-based EC in 2000 (a), 2004 (b), 2008 (c), and 2012 (d).

Figure 15.3 presents the comparison between the estimated and reference EC for 2001, 2005, and 2010. Although relatively strong correlations ($R^2 = 0.80$) were obtained by using all samples (Figure 15.3a), the slopes of the regressions demonstrated larger urban EC against EC of urban districts. Meanwhile, there were obvious outliers. These could be partly attributed to the different definition of urban district and urban area. For example, urban district does not include small urban areas within counties in China. Nevertheless, the gap between the estimated and statistical EC and the degree of overestimation were smaller for the capital cities, with R^2 of 0.94 (Figure 15.3b). The result in Figure 15.3 indicated that it was feasible to allocate China's grid-level EC using its province-level statistical data and DMSP-OLS NTL data, or at least for prefecture-level estimation of highly urbanized regions.

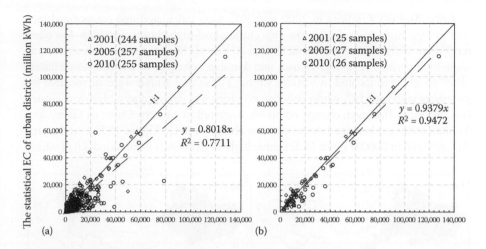

FIGURE 15.3 Scatterplot of estimated EC of urban area versus reference EC of urban district at the prefecture level, using samples across China (a), and samples of capital city of each province (b) in 2001, 2005, and 2010.

15.5.2 SPATIOTEMPORAL PATTERN OF EC AT MULTIPLE SCALES

Figure 15.4 shows significantly positive spatial autocorrelations of EC at multiple scales. However, the spatial autocorrelation showed different trends for different scales. Specifically, the global Moran's I of provincial EC remained stable from 2000 to 2012 around 0.18, while the value of urban EC increased steadily from 0.12 in 2000 to 0.27 in 2012 by an annual rate of increase of 0.0141 with an R^2 of 0.92, which means urban EC became more clustered during the past decade. For prefectural-level EC, the global Moran's I increased from 0.25 in 2000 to 0.41 in 2007 and remained around 0.40 until 2012.

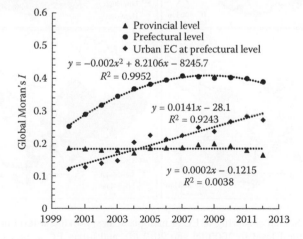

FIGURE 15.4 Global Moran's I of EC at multiple scales.

Local Moran's *I* indices in 2000 and 2012 were calculated and categorized into four groups: high–high cluster, low–low cluster, low–high cluster, and high–low cluster (Figure 15.5). For the provincial level in 2000, only Shandong showed a significant high–high cluster; Xinjiang, Gansu, and Yunnan belonged to the low–low cluster, all of which were located in western China; Fujian and Anhui were surrounded by provinces with high EC, while only one province was detected as a high–low cluster (i.e., Sichuan). However, change obviously occurred in western China as Gansu and Yunnan no longer belonged to the low–low cluster in 2012 (Figure 15.5b). At the prefectural level in 2000 (Figure 15.5c), three high–high clusters were found, all of which

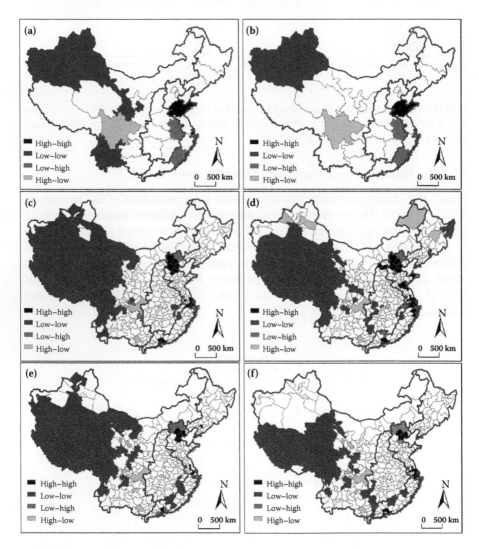

FIGURE 15.5 Spatial clustering of EC at multiple scales: provincial level in 2000 (a) and 2012 (b), prefectural level in 2000 (c) and 2012 (d), and urban EC at prefectural level in 2000 (e) and 2012 (f).

appeared in eastern China, specifically in the Beijing–Tianjin–Tangshan region, the Yangtze River Delta, and the Pearl River Delta; low–low clusters were identified in western China and in Anhui and Jiangxi provinces, especially the prefectural cities in Xinjiang, Xizang, and Qinghai; cities such as Chongqing and Nanning were surrounded by cities with low EC; six cities were surrounded by neighbors with high EC, including four prefectural cities around Chongqing (i.e., Tongren in Guizhou province, Shennongjia in Hubei province, and Guangan and Suining in Sichuan province), Chengde in Hebei province, and Xiamen in Fujian province. Two more high–high clusters were identified comparing Figure 15.5d and c, including Yantai, Linyi, and Zaozhuang in Shandong province and Lianyungang in Jiangsu province. Meanwhile, the size of low–low clusters shrank in western China but increased in central China, including some prefectural cities in Hubei and Hunan. Additionally, some high–low clusters appeared in northeastern, western, and central China, including Hulunbeier in Inner Mongolia, Harbin in Heilongjiang province, Ili Kazakh and Changji in Xinjiang, and Wuhan in Hubei province. Another significant change was that cities that belonged to the low–high cluster were all located in eastern China in 2012. For prefectural-level urban EC, the high–high cluster almost appeared in the same region but with smaller clusters as that in Figure 15.5c (Figure 15.5e). However, the high–high cluster in the Beijing–Tianjin–Tangshan region was larger than the other two clusters, which increased from 2000 to 2012, especially for Yangtze River Delta (Figure 15.5e and f). High–low clusters can be identified in western China, such as Chongqing, Chengdu, Lanzhou, Guiyang, and Nanning, all of which are the capital cities of each province. This indicated that urban EC in capital cities was significantly higher than that in cities around them in western China. This further implied the unbalanced development in western China. The low–low spatial clusters were mainly located in western China, yet the size of the cluster shrank; the low–high clusters were mainly located in eastern China, around high–high clusters. However, urban EC in central and northeastern China did not show statistically significant clusters.

Figure 15.6 shows the temporal variations of China's EC at different scales from 2000 to 2012. The calculated SLOPE index was categorized into five groups by using Natural Breaks (Jenks) in ArcMap 10.1: slow, relatively slow, moderate, relatively rapid, and rapid. It is revealed in Figure 15.6a and b that provinces that experienced rapid growth of EC were Shandong, Jiangsu, and Guangdong, all of which are located in eastern China; Inner Mongolia, Hebei, Henan, and Zhejiang demonstrated a relatively rapid growth of EC; most of the central provinces showed a moderate growth of EC except for Jiangxi. Particularly, a slow growth of EC was found at the provincial level for four municipal cities—Beijing, Shanghai, Tianjin, and Chongqing, mainly because of their smaller amount of EC than some provinces such as Zhejiang, Jiangsu, and Guangdong and their relatively highly urbanized status at the beginning of the 21st century (Figure 15.6b). Prefectural cities that experienced moderate to rapid growth of EC were either located in coastal regions or belonged to capital cities; the majority of western and northeastern cities and more than half of central cities detected a slow or relatively slow growth of EC (Figure 15.6c). The growth of prefectural-level urban EC demonstrated a similar pattern as prefectural-level EC (Figure 15.6d), highlighting capital cities and three metropolitan regions (i.e., the Beijing–Tianjin–Tangshan region, the Yangtze River Delta, and the Pearl

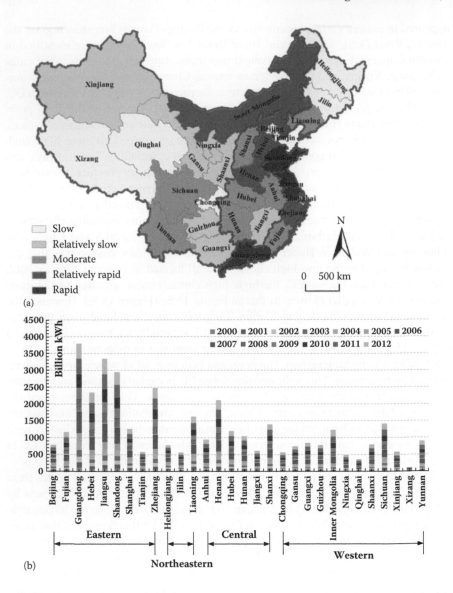

FIGURE 15.6 Temporal growth of EC at different scales: (a) provincial level, (b) provincial EC. *(Continued)*

River Delta). This further demonstrated the unbalanced development between four regions and capital cities and other prefectural cities.

15.5.3 National and Regional Contributions of Urban Areas to EC

Figure 15.7 shows a steady increase of urban extent during the past decade. The area increased from approximately 3.85×10^4 km^2 in 2000 to 8.83×10^4 km^2 in 2012, which occupied approximately 0.4% and 0.9% of the total area in Mainland China,

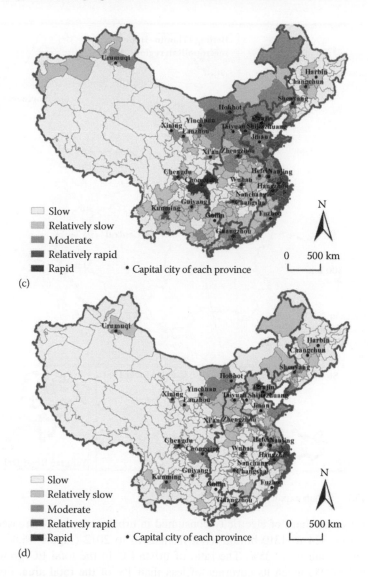

FIGURE 15.6 (CONTINUED) Temporal growth of EC at different scales: (c) prefectural level, and (d) urban EC at prefectural level for China from 2000 to 2012.

respectively. Regionally, urban extent in eastern China increased from 1.77×10^4 km² in 2000 to 3.70×10^4 km² in 2012, occupying approximately 2.0% and 4.2% of the total area in this region, respectively; in central China, urban area increased from 8.7×10^3 km² to 2.04×10^4 km² during the period, with an urbanization rate of 0.9% and 2.0%, respectively; the urbanization rate of northeastern China grew from approximately 0.6% to 1.4% (from 4.65×10^3 km² to 1.07×10^4 km², out of 7.92×10^5 km²); urban extent nearly tripled in the western part from 7.41×10^3 km² to 2.02×10^4 km² out of 6.65×10^6 km², occupying 0.1% and 0.3% of the total western region.

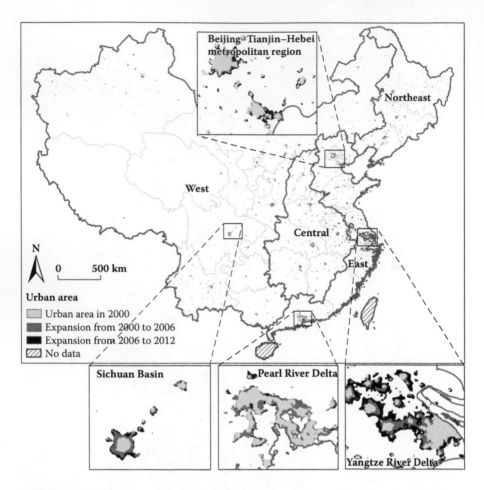

FIGURE 15.7 Urban expansion from 2000 to 2012.

Nationally, the share of electricity consumed in urban areas had increased continuously from 324 to 1310 billion kWh from 2010 to 2012 (Figure 15.8a), at an annual growth rate of 12.35%. The ratio of urban EC to the total EC grew from 23.80% to 26.45%, given its coverage of less than 1% of the total area. For total EC in urban areas across Mainland China, eastern China annually consumed more than 55%, but with a slightly decreasing trend of 0.11% per year (Figure 15.8a). The decreasing trend was also observed for northeastern China, but with a faster decreasing rate of 0.33% per year. The central and western regions experienced a growth of EC in urban areas during the period. However, EC in urban areas of western China increased with a faster rate than in the central region (0.32% vs. 0.12%). The results imply the accelerated urbanization in central and western China during the past decade and, hence, higher EC of these two regions.

The ratio of electricity consumed in urban areas showed an increasing trend for the four regions (Figure 15.8b), reflecting the growing urbanization rate across Mainland China. For electricity consumed in each region, eastern and northeastern China exhibited

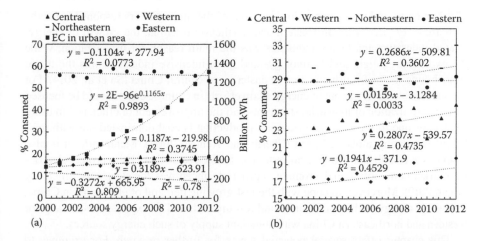

FIGURE 15.8 National and regional contribution of urban areas to EC in Mainland China from 2000 to 2012: (a) EC in Chinese urban areas and the percentage consumed in each region and (b) the percentage of EC in urban areas for each region.

the highest ratio of EC in urban areas, followed by central and western China. The ratio remained essentially constant for eastern China, but the increasing rate of urban EC for central, northeastern, and western China was 0.28%, 0.27%, and 0.19%, respectively.

15.6 DISCUSSION AND CONCLUSIONS

This chapter analyzed the decadal spatiotemporal pattern of EC across Mainland China at the provincial, prefectural, and urban scales by using time-series DMSP-OLS NTL images and auxiliary data sets. The results showed different spatio-temporal patterns of EC, both nationally and regionally. Specifically, the results demonstrated a moderate to rapid growth of EC for coastal regions at all scales, while a slow growth for the majority of western and northeastern cities and part of central China (except for the capital cities). The results of global Moran's I showed significantly positive spatial autocorrelations of EC with an increasing trend during the past decade, especially for EC at the prefectural level and urban EC. Moreover, the results of local Moran's I showed the phenomenon of high–high clusters in coastal regions, but high–low and low–low clusters in western China.

The disparity of economic development was a main factor for regional differences in EC, especially for the huge gap between western and eastern China. However, the results implied that the gaps between the western, central, and eastern regions were narrowing during the past decade. These changes may be attributed to the develop-ment policies applied to central and western China. For instance, in 2004, the Rise of the Central China Plan started to coordinate regional growth in central China (Tan 2015). To create regional economic development centers in western China, the Western Development Strategy was launched in 2000, calling for aid in infrastructure construc-tion, natural resource exploitation, and the establishment of market liberalization poli-cies from the central government (Tan 2015). Urbanization, including the intensification of urban cores and urban expansion, was one of the major factors responsible for the increase of EC in China. In western and northeastern China, the growth of energy-related industries and heavy industries was the main reason for urbanization, while the expansion of light and technology- and labor-intensive industries was the major contributing factor for urbanization in eastern and central China (Su et al. 2014). For the sustainable development of energy, main efforts in Chinese cities should be focused on the optimization of industrial structures and the improvement of efficiency in using electricity, due to the fact that electricity production is coal dominated but with inefficient generating units in China (Cherni and Kentish 2007). Meanwhile, it is also essential to establish a strong legal system and effective institutions to protect electricity market competition and promote appropriate incentives for electricity efficiency (Ngan 2010). Moreover, to cope with the fast-growing demand of electricity in China, renewable energy such as solar and wind resources should be exploited, especially for western and northeastern China with abundant supply of such energy sources.

This chapter offers several potential topics for further research. For example, to improve the accuracy of pixel-based EC estimation, exploring alternative methods to better eliminate the saturation and blooming effects in time-series DMSP/OLS NTL data is preferable. Although the application of EVI increased the variation of EC in urban regions in this study, the assumption that vegetation cover and urban surface were inversely correlated may weaken the contribution of urban areas to total EC given the higher DNs in urban areas than other regions. Except for developing methods to improve the quality of DMSP-OLS NTL data, incorporating other factors (e.g., GDP and topographic variables) into the model for NTL-based EC estimation may help better predict EC at multiple scales (Xie and Weng 2016). Additionally, a better method to extract urban dynamics, which reduces the overestimation of large urban areas and the underestimation of small ones, is required to better study the contribution of urban areas to total EC. To this extent, a cluster-based method proposed by Zhou et al. (2014), which estimated the optimal threshold for each cluster according to its size and the mean magnitude of NTL, can be more promising (Zhou et al. 2015). However, the incomparability of DMSP-OLS NTL DNs from different years and satellites may limit the usage of a calibrated model to other years and regions. Further studies on extending the cluster-based thresholding method to the temporal domain may provide some inspiring discoveries.

REFERENCES

Adams, F. G., and Y. Shachmurove. 2008. Modeling and forecasting energy consumption in China: Implications for Chinese energy demand and imports in 2020. *Energy Economics* 30 (3):1263–1278. doi: 10.1016/j.eneco.2007.02.010.

Ahmad, S., G. Baiocchi, and F. Creutzig. 2015. CO_2 emissions from direct energy use of urban households in India. *Environmental Science and Technology* 49 (19):11312–11320. doi: 10.1021/es505814g.

Al-mulali, U., C. N. B. C. Sab, and H. G. Fereidouni. 2012. Exploring the bi-directional long run relationship between urbanization, energy consumption, and carbon dioxide emission. *Energy* 46 (1):156–167.

Amaral, S., G. Câmara, A. M. V. Monteiro, J. A. Quintanilha, and C. D. Elvidge. 2005. Estimating population and energy consumption in Brazilian Amazonia using DMSP night-time satellite data. *Computers, Environment and Urban Systems* 29 (2):179–195.

Anselin, L. 1995. Local indicators of spatial association—LISA. *Geographical Analysis* 27 (2):93–115.

Anselin, L. 1996. The Moran scatterplot as an ESDA tool to assess local instability in spatial association. *Spatial Analytical Perspectives on GIS* 111:111–125.

Bilgen, S. 2014. Structure and environmental impact of global energy consumption. *Renewable and Sustainable Energy Reviews* 38:890–902. doi: 10.1016/j.rser.2014.07.004.

Bloch, H., S. Rafiq, and R. Salim. 2015. Economic growth with coal, oil and renewable energy consumption in China: Prospects for fuel substitution. *Economic Modelling* 44: 104–115.

Cao, X., J. Chen, H. Imura, and O. Higashi. 2009. A SVM-based method to extract urban areas from DMSP-OLS and SPOT VGT data. *Remote Sensing of Environment* 113 (10):2205–2209. doi: 10.1016/j.rse.2009.06.001.

Cao, X., J. Wang, J. Chen, and F. Shi. 2014. Spatialization of electricity consumption of China using saturation-corrected DMSP-OLS data. *International Journal of Applied Earth Observation and Geoinformation* 28:193–200. doi: 10.1016/j.jag.2013.12.004.

Chand, T. R. Kiran, K. V. S. Badarinath, C. D. Elvidge, and B. T. Tuttle. 2009. Spatial characterization of electrical power consumption patterns over India using temporal DMSP-OLS night-time satellite data. *International Journal of Remote Sensing* 30 (3):647–661. doi: 10.1080/01431160802345685.

Cherni, J. A., and J. Kentish. 2007. Renewable energy policy and electricity market reforms in China. *Energy Policy* 35 (7):3616–3629.

Doll, C. N. H., and S. Pachauri. 2010. Estimating rural populations without access to electricity in developing countries through night-time light satellite imagery. *Energy Policy* 38 (10):5661–5670.

Elvidge, C. D., K. E. Baugh, E. A. Kihn, H. W. Kroehl, E. R. Davis, and C. W. Davis. 1997. Relation between satellite observed visible-near infrared emissions, population, economic activity and electric power consumption. *International Journal of Remote Sensing* 18 (6):1373–1379.

Elvidge, C. D., M. L. Imhoff, K. E. Baugh, V. R. Hobson, I. Nelson, J. Safran, J. B. Dietz, and B. T. Tuttle. 2001. Night-time lights of the world: 1994–1995. *ISPRS Journal of Photogrammetry and Remote Sensing* 56 (2):81–99.

Elvidge, C. D., D. Ziskin, K. E. Baugh, B. T. Tuttle, T. Ghosh, D. W. Pack, E. H. Erwin, and M. Zhizhin. 2009. A fifteen year record of global natural gas flaring derived from satellite data. *Energies* 2 (3):595–622.

He, C., Q. Ma, T. Li, Y. Yang, and Z. Liu. 2012. Spatiotemporal dynamics of electric power consumption in Chinese Mainland from 1995 to 2008 modeled using DMSP/OLS stable nighttime lights data. *Journal of Geographical Sciences* 22 (1):125–136.

He, C., Q. Ma, Z. Liu, and Q. Zhang. 2014. Modeling the spatiotemporal dynamics of electric power consumption in Mainland China using saturation-corrected DMSP/OLS night-time stable light data. *International Journal of Digital Earth* 7 (12):1–22.

Huang, Q., C. He, B. Gao, Y. Yang, Z. Liu, Y. Zhao, and Y. Dou. 2015. Detecting the 20 year city-size dynamics in China with a rank clock approach and DMSP/OLS nighttime data. *Landscape and Urban Planning* 137:138–148.

Letu, H., M. Hara, H. Yagi, K. Naoki, G. Tana, F. Nishio, and O. Shuhei. 2010. Estimating energy consumption from night-time DMPS/OLS imagery after correcting for saturation effects. *International Journal of Remote Sensing* 31 (16):4443–4458.

Liu, Z., C. He, Q. Zhang, Q. Huang, and Y. Yang. 2012. Extracting the dynamics of urban expansion in China using DMSP-OLS nighttime light data from 1992 to 2008. *Landscape and Urban Planning* 106 (1):62–72.

Lo, C. P. 2002. Urban indicators of China from radiance-calibrated digital DMSP-OLS nighttime images. *Annals of the Association of American Geographers* 92 (2): 225–240.

Meng, L., W. Graus, E. Worrell, and B. Huang. 2014. Estimating CO_2 (carbon dioxide) emissions at urban scales by DMSP/OLS (Defense Meteorological Satellite Program's Operational Linescan System) nighttime light imagery: Methodological challenges and a case study for China. *Energy* 71:468–478. doi: 10.1016/j.energy.2014.04.103.

Ngan, H. W. 2010. Electricity regulation and electricity market reforms in China. *Energy Policy* 38 (5):2142–2148.

Pandey, B., P. K. Joshi, and K. C. Seto. 2013. Monitoring urbanization dynamics in India using DMSP/OLS night time lights and SPOT-VGT data. *International Journal of Applied Earth Observation and Geoinformation* 23:49–61.

Parshall, L., K. Gurney, S. A. Hammer, D. Mendoza, Y. Zhou, and S. Geethakumar. 2009. Modeling energy consumption and CO_2 emissions at the urban scale: Methodological challenges and insights from the United States. *Energy Policy* 38 (9):4765–4782.

Shi, K., B. Yu, Y. Huang, Y. Hu, B. Yin, Z. Chen, L. Chen, and J. Wu. 2014. Evaluating the ability of NPP-VIIRS nighttime light data to estimate the gross domestic product and the electric power consumption of China at multiple scales: A comparison with DMSP-OLS data. *Remote Sensing* 6 (2):1705–1724. doi: 10.3390/rs6021705.

Su, Y., X. Chen, Y. Li, J. Liao, Y. Ye, H. Zhang, N. Huang, and Y. Kuang. 2014. China's 19-year city-level carbon emissions of energy consumptions, driving forces and regionalized mitigation guidelines. *Renewable and Sustainable Energy Reviews* 35:231–243.

Tan, M. 2015. Urban growth and rural transition in China based on DMSP/OLS nighttime light data. *Sustainability* 7 (7):8768–8781. doi: 10.3390/su7078768.

Wang, Q. 2014. Effects of urbanisation on energy consumption in China. *Energy Policy* 65:332–339.

Welch, R. 1980. Monitoring urban population and energy utilization patterns from satellite data. *Remote Sensing of Environment* 9 (1):1–9.

Weng, Q. 2013. What is spacial about global urban remote sensing? In *Global Urban Monitoring and Assessment through Earth Observation*, edited by Qihao Weng, 1–12. Boca Raton, FL: CRC Press/Taylor & Francis.

Wu, J., S. He, J. Peng, W. Li, and X. Zhong. 2013. Intercalibration of DMSP-OLS night-time light data by the invariant region method. *International Journal of Remote Sensing* 34 (20):7356–7368. doi: 10.1080/01431161.2013.820365.

Wu, J., Y. Niu, J. Peng, Z. Wang, and X. Huang. 2014. Research on energy consumption dynamic among prefecture-level cities in China based on DMSP/OLS nighttime light. *Geographical Research* 33 (4):625–634 (In Chinese).

Xie, Y., and Q. Weng. 2016. World energy consumption pattern as revealed by DMSP-OLS nighttime light imagery. *GIScience & Remote Sensing* 53 (2):265–282. doi: 10.1080/15481603.2015.1124488.

Zhang, Q., C. Schaaf, and K. C. Seto. 2013. The vegetation adjusted NTL urban index: A new approach to reduce saturation and increase variation in nighttime luminosity. *Remote Sensing of Environment* 129:32–41.

Zhang, Q., and K. C. Seto. 2011. Mapping urbanization dynamics at regional and global scales using multi-temporal DMSP/OLS nighttime light data. *Remote Sensing of Environment* 115 (9):2320–2329.

Zhao, N., T. Ghosh, and E. L. Samson. 2012. Mapping spatio-temporal changes of Chinese electric power consumption using night-time imagery. *International Journal of Remote Sensing* 33 (20):6304–6320.

Zhou, Y., S. J. Smith, K. Zhao, M. Imhoff, A. Thomson, B. Bond-Lamberty, G. R. Asrar, X. Zhang, C. He, and C. D. Elvidge. 2015. A global map of urban extent from nightlights. *Environmental Research Letters* 10 (5):054011.

Zhou, Y., S. J. Smith, C. D. Elvidge, K. Zhao, A. Thomson, and M. Imhoff. 2014. A cluster-based method to map urban area from DMSP/OLS nightlights. *Remote Sensing of Environment* 147:173–185. doi: 10.1016/j.rse.2014.03.004.

16 Support of Wind Resource Modeling Using Earth Observation

A European Perspective on the Status and Future Options

Thomas Esch, Charlotte Bay Hasager,
Paul Elsner, Janik Deutscher, Manuela Hirschmugl,
Annekatrin Metz, and Achim Roth

CONTENTS

16.1 BACKGROUND

The fact that the power production of wind turbines is sensitive to the mean wind speed approaching the turbine means that small errors in wind speed modeling will translate to much larger uncertainties in the power output of wind energy installations.

The mean wind speed is in turn highly variable and sensitive to properties of the underlying land surface. Wind resource modeling is performed by using dedicated aerodynamic microscale models for limited areas as well as numerical weather prediction (NWP) for the meso- to continental scale. In general, two types of land surface description inputs are needed for this:

1. A digital elevation model (DEM) that describes the variation in terrain elevation
2. A set of surface parameters affecting the wind flow

For (1), the Shuttle Radar Topography Mission (SRTM) DEM is one of the most commonly used descriptions in both microscale and mesoscale models. However, the accuracy of SRTM data is too limited for built-up areas and areas with tall vegetation, where the building or forest height, rather than the terrain, is captured (e.g., Sun et al. 2008).

Regarding (2), the traditionally most central parameter for wind resource modeling is the land surface roughness (z_0), which determines the effect of the friction of the surface on the wind profile. Under ideal conditions, z_0 can be determined from wind measurements (e.g., Dellwik and Jensen 2005; Mölder and Lindroth 1999). Choudbury and Monteith (1988) as well as Raupach (1994, 1995) related z_0 and the displacement height parameter, which takes the height of vegetation into account, to physical properties such as height and density of the underlying vegetation. On the basis of such relationships, Tian et al. (2011) demonstrated that remote sensing techniques can be used to calculate z_0 for large areas. Some studies have also demonstrated the potential of Synthetic Aperture Radar (SAR) for mapping of z_0 (Bidaut et al. 2006). SAR sensors deliver a backscattering coefficient that is strongly affected by the composition of the ground cell and by its structure. However, these remote sensing–based approaches are rarely used in the wind resource community.

In current operational models, surface roughness and other surface parameters are often assigned for a certain land cover class via best-practice, well-established tabular values (Troen and Petersen 1989). Such classification is commonly based on either the Global Land Cover 2000 (GLC2000) database of the Joint Research Centre (JRC) (http://forobs.jrc.ec.europa.eu/products/glc2000/products), the fine-resolution European CORINE Land Cover (CLC) map (Hasager et al. 2003), the MODIS land cover product (Friedl et al. 2002), or the ESA GlobCover (Arino et al. 2008).

A simplified microscale approach is used, for example, in the commercial WAsP (Wind Atlas Analysis and Application Program) model, which is the wind energy industry standard PC software for bankable wind resource assessment and siting of wind turbines and wind farms. There are currently more than 4600 users in over 110 countries that use WAsP for all steps from analysis of wind and terrain effects to estimation of wind farm production. In WAsP, the roughness layer is usually not sufficiently accurate from CLC and look-up table alone and needs to be complemented by costly and time-consuming site visits.

Recent developments in microscale Reynolds' Averaged Navier Stokes (RANS) equation models as well as Large Eddy Simulation (LES) models allow for a direct simulation of the effects of vegetation as well as solid objects on the flow (Lopes et al. 2013; Sogachev and Panferov 2006), thereby short-cutting the need for land surface classification and uncertainties regarding the assignment of parameter values. For this advanced type of model, the dimensions of solid obstacles, as well as the height and density of tall vegetation, are necessary input parameters since they determine the development of the wind flow over smaller scale variations in the landscape at around 5 m resolution (Dellwik et al. 2013). The applicability of these models is severely limited by the lack of reliable land surface input data, and published studies are often limited to highly idealized cases (Dupont and Brunet 2008).

In NWP models used in wind resource applications, an accurate description of the elevation and surface roughness is important (Jiménez and Dudhia 2012; Santos-Alamillos et al. 2013) but not as critical as in aerodynamic models. To a certain extent, errors in the output from NWP models that are attributed to errors in surface properties can be corrected at the local scale by coupling them to aerodynamic models (Badger et al. 2014). However, an accurate depiction of the state of the land surface (i.e., its temperature and soil moisture content) is critical for an accurate simulation of the state of the planetary boundary layer (PBL) (Findel and Eltahir 2003; Hong et al. 2009; Refslund et al. 2013). It has also been established that an accurate PBL is critical for an accurate depiction of mesoscale circulations (Taylor et al. 2007) and the wind profile (Kelly and Gryning 2010), and ultimately these are important for an accurate wind resource assessment. On the other hand, the truthful modeling of the state of the land surface is highly dependent on the characteristic of the land surface itself and their variations in time, as well as the atmospheric forcing (Case et al. 2008, 2014; Chen et al. 2007; Hong et al. 2009).

16.2 EXTRACTION OF LAND SURFACE PARAMETERS IN SUPPORT OF WIND RESOURCE MODELING

Atmospheric models are employed for a wide range of operational services, including wind resource assessment. Despite the heterogeneity of such applications, all share very similar parameterization requirements. Standard inputs are information about surface roughness, land cover type, leaf area index (LAI), vegetation fraction, and phenological state. From a remote sensing point of view, these basic information requirements are split into the following methodological focuses:

1. Land cover– and land use–related geoinformation (2D)
2. Topography, surface morphology, and surface roughness (3D)
3. Multitemporal aspects describing process dynamics (4D)

16.2.1 LAND COVER AND LAND USE-RELATED GEOINFORMATION

Land has considerable heterogeneity because of the existence of different land cover types such as bare soil, water, urban land, trees, and snow, which vary over small areas. This surface variability not only determines the microclimate but also affects mesoscale atmospheric circulation (Weaver and Avissar 2001; Yang 2004). Accurate representation of the land surface is therefore important to precisely model the effect of the land surface. Remotely sensed derived global land cover products, such as Global Land Cover Characteristics (GLCC; Loveland et al. 2000), University of Maryland land cover classification (UMD; Hansen and Reed 2000), and GLC2000 (Bartholomé and Belward 2005), were implemented into various land surface schemes and climate models. In the new Global Wind Atlas released in 2015 (http://www.irena.org and http://globalwindatlas.com/), the GlobCover and MODIS land cover data are merged for improved characterization of the land surface roughness as shown in Figure 16.1.

Another example of using two different land cover maps is based on USGS and CLC land use data for a region in Denmark (see Figure 16.2a and b). The WRF model has been used to calculate the wind speed at different heights (10, 40, 60, 80, 100, and 160 m) for the period January 10, 2014 until September 12, 2014. The output frequency is every 10 min. The WRF model is run as in Floors et al. (2015) but

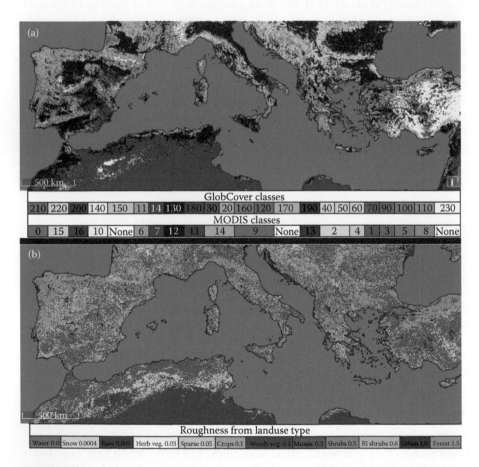

FIGURE 16.1 The Global Wind Atlas land cover classes in GlobCover and MODIS with corresponding color coding (a) and derived roughness map for the Mediterranean area (b). (From the Global Wind Atlas, supported by EUDP with explanation from Dr. Neil Davis. http://globalwindatlas.com/. With permission.)

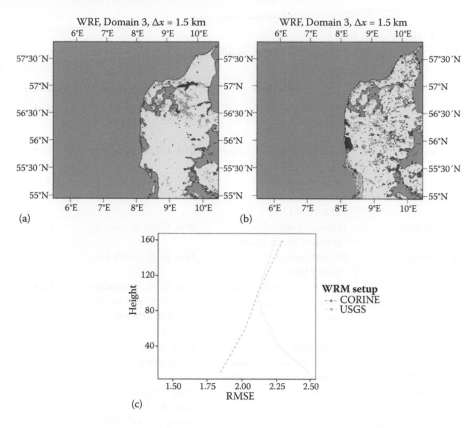

FIGURE 16.2 USGS (a) and CLC (b) land use data for a region in Denmark. The RMSE between the measured and modeled wind speed at the Høvsøre meteorological mast and WRF model using USGS and CLC land cover as input. The results are for the period January 10, 2014 until September 12, 2014 (c). (Courtesy of RUNE Project, supported by ForskEL with explanation from Dr. Rogier Floors.)

with ERA interim boundary conditions used and the horizontal resolution at 12, 4, and 1.333 km for domains 1, 2, and 3, respectively. The land use categories from the CLC data were reassigned to USGS categories using the method described in Pineda et al. (2004). The results on root mean square error (RMSE) between the measured and modeled wind speed at the tall meteorological tower at Høvsøre located in a flat coastal landscape near the North Sea (Peña et al. 2015) are shown in Figure 16.2c. At around 90 m above ground and upward, the results are comparable, while at lower levels, the higher-resolution CLC map yields the best results.

More detailed atmospheric/wind resource models could benefit from more detailed land cover information. A list of established land cover maps and products that would be available for Europe, for instance, is provided in Table 16.1.

Within Europe (EEA, 39 member countries), the CLC project is producing land use cartography in the form of detailed descriptions of land occupation and features, at an original scale of 1:100,000. It used 44 classes of the three-level CORINE

TABLE 16.1

List of Land Surface Parameter Maps Available for Europe

Name	Product Type	Resolution or MMU/Coverage	Provider
Corine CLC	Land cover with more than 30 types covering most of Europe	Minimum Mapping Unit (MMU) 25 ha (vector), EEA's 39 countries	EEA
GIO Forest	Tree cover density (0%–100%) Forest type (Coniferous/ deciduous/mixed)	20 m, EEA's 39 countries	EEA
GIO Imperviousness	Degree of impervious-ness (0%–100%)	20 m, EEA's 39 countries	EEA
GIO Grassland	Presence of permanent grassland (binary layer)	20 m, EEA's 39 countries	EEA
GIO Wetland	Wetland presence (binary layer)	20 m, EEA's 39 countries	EEA
GIO Water	Water presence (binary layer)	20 m, EEA's 39 countries	EEA
JRC Forest/Non-forest map	Classes: Forest, non-forest, clouds, no data	25 m, EU27 plus AL, BA, CH, HR, MN, MK, MR, LI, RS	JRC
GlobCover	22 land cover classes	300 m, global	ESA
GlobCover (bimonthly surface reflectance)	Average surface reflectance values over a 2-month period	300 m, global	ESA
Urban Atlas for Europe	Land use map	MMU 0.25 ha for urban classes, 1 ha for rural classes (vector), 305 cities throughout Europe	EEA
Geoland2 BioPar (Biophysical)	LAI, FCover, FAPAR	Variable	Geoland project
GLCC	Land cover 24 classes	~1 km, global	USGS
GLC2000	Land cover 22 classes	~1 km, global	JRC
EU-Hydro	Water bodies	MMU: 1 ha (vector), EU	EC
MODIS MCD12Q1	17 land cover classes	500 m, global	USGS
MODIS MCD15A2	LAI and FPAR every 8 days	1 km, global	USGS

nomenclature, for all European countries, with a definition of 25 ha. However, for the use of surface occupation cartography like the CLC in wind flow models, it is necessary to assign an equivalent characteristic roughness length to each different classification or land cover class. The CLC cartography has been adapted for its use in typical wind resource assessment studies, namely, for use with the WAsP model software. According to the particular terrain characteristics, the 44 different CLC classes were grouped into 14 roughness classes. The grouping was not always obvious. One has to take into account that the main goal of the CLC project is to classify the landscape and land cover to facilitate policymaking and environmental management at a European scale, and not the particular roughness of the surface with respect to its effect on atmospheric flow.

The application of CLC for dynamic roughness length simulation shows some important shortcomings. For example, CLC lacks some very important parameters: (a) the CLC class forest misses a reliable roughness parameter and an accurate forest borderline location and type description (hard cut or smooth transition); (b) the CLC class agriculture, specifically the permanent agricultural systems such as vineyards, olive groves, and fruit plantations, are very diverse and need a subdivision plus information on the direction of the planting rows; (c) the CLC class semi-urban is a very heterogeneous class in terms of surface roughness and lacks details such as building density and building height. These examples illustrate that the CLC data sets or low-resolution raster data sets have limited suitability for atmospheric/wind resource model parameterization. Inaccuracies and errors that exist at this level will invariably flow into the atmospheric/wind resource models and limit their performance.

16.2.2 Topography, Surface Morphology, and Surface Roughness

Relief information is freely available on a global scale but with relatively low resolution (3 to 1 arc sec, ~90 to 30 m). Table 16.2 provides the established products, their coverage, and sources. Most of these products, in particular the ones derived from spaceborne missions, refer to the height of the surface plus the coverage (top of the canopy). One can distinguish between a digital surface model (DSM), providing the top of a surface including the canopy, and the DEM, which refers to the land surface, typically with a height above sea level.

Surface morphology includes the detection of linear strips of woody vegetation such as hedgerows. It has been in the focus of several studies as they are important elements for landscape ecology as well as biodiversity. Besides their ecological functions as habitat, shelter, or corridor for specific species, these linear landscape elements serve as protection against soil erosion and as a biogeochemical barrier (Ducrot et al. 2012). Thus far, methods based on high-resolution and very high–resolution optical imagery have been developed to detect linear landscape features such as hedgerows (Aksoy et al. 2008, 2010; Ducrot et al. 2012; Vannier and

---

TABLE 16.2
List of Available Digital Elevation Data Available for Europe

Product	Type	Spatial Resolution	Provider
GTOPO30	Elevation	30 arc sec ~1 km	USGS
SRTM C-band	Elevation	Near-global (60°N to 56° S) ~1 arc sec, with the exception of Central Asia ~3 arc sec	NASA/NGA
SRTM X-Band	Elevation	Near-global (60°N to 56° S) ~1 arc sec, coverage not continuous: gaps in image strips	DLR
ASTER GDEM	Elevation	~30 m	NASA and METI
EU-DEM	Elevation	30 m	EC

Hubert-Moy 2008). To date, only Ducrot et al. (2012) have compared the results based on multitemporal optical data with multitemporal TSX data, achieving more than 80% overall accuracies for the optical analysis but only 75% for the SAR-based analysis. A detailed review on techniques for extracting linear features from Earth observation (EO) imagery is given by Quackenbush (2004). The landscape roughness is closely related to morphology.

16.2.3 MULTITEMPORAL ASPECTS DESCRIBING PROCESS DYNAMICS

Remotely sensed data collected by instruments such as AVHRR, Medium Resolution Imaging Spectrometer (MERIS), SPOT-Vegetation, and MODIS provide near-daily global observations of vegetation dynamics at 300 m to 8 km spatial resolution. Exploiting these dense time series, several remote sensing algorithms have been developed over the past decades. Most algorithms use as input time series of vegetation indices such as the normalized difference vegetation index (NDVI) or the enhanced vegetation index (EVI) that are derived from multispectral satellite data (e.g., Huete et al. 2002).

Multitemporal data can be used not only to investigate, for example, the phenological stage and changes of the land cover and surface but also for the improvement of existing land cover maps. On the basis of the filtered/noise-reduced time series, different algorithms have been applied to identify key phenological phases of vegetation. Algorithms based on user-defined thresholds depending on land cover types were analyzed, for example, by Schwartz et al. (2002) and by White and Nemani (2006). Algorithms based on significant and rapid increases in remotely sensed signals were developed, for example, by Moulin et al. (1997). Further algorithms determine phenological phases of vegetation by fitting of functions to the remotely sensed time series data (Beck et al. 2006; Zhang et al. 2003). The operationally generated MODIS Land Cover Dynamics product belongs to this group of algorithm, which identifies phenophase transition dates based on logistic functions fit to time series of the EVI (Huete et al. 2002). Taking into account dynamic processes can clearly improve the quality of the input data for wind modeling and thus also influence the final results.

16.3 PERSPECTIVES FOR IMPROVED EO-DERIVED PARAMETERS FOR WIND RESOURCE MODELING

The EO data sets currently in operational use by the wind resource modeling community in Europe do not reflect recent advances in EO. New and near-future missions and data sets offer enormous potential for improved data quality, resolution, and frequencies. Most notably, these are the ESA/Copernicus Sentinel missions and the TerraSAR-X and TanDEM-X missions for the German Aerospace Center (DLR). In this section, we will outline beyond state-of-the-art improvements and breakthroughs that can be expected from these missions in the context of atmospheric/ wind resource modeling parameters. The new sensors and missions provide complementary capabilities for beyond state-of-the-art improvement on the derivation of land surface parameters for atmospheric/wind resource modeling.

16.3.1 2D LAND COVER MAPPING

The land cover detail as represented in CLC or in low-resolution raster-based land cover maps is a generalization of the actual land cover, aggregating multiple land cover types into one object or pixels. Within these objects or pixels, a large heterogeneity in roughness length that is not reflected in the aggregated object is present. Significant improvements are already achievable by integrating existing data sets that have been developed very recently.

This includes the high-resolution land cover layers (HRL) data that provide additional details within the CLC objects. In this way, large agricultural areas (as represented in CLC) now include pixels representing small villages (HRL Imperviousness), small forest patches (HRL Forest), water and wetland areas (HRL Water and HRL Wetlands), and grassland areas (HRL Grasslands), all at a 20-m resolution (Figure 16.3). The heterogeneity within forests (species and density) can also be analyzed based on the HRL forest layer. The EEA Urban Atlas data can be integrated to create a more detailed view of the most densely populated areas. The Urban Atlas represents the land use/land cover within the functional urban areas of Europe with a mapping detail of 0.25 ha for the artificial areas and 1 ha for non-artificial areas. Compared to the low detail of the CLC data, especially in urban areas (which are very heterogeneous areas), the UA data offer an enormous upgrade in mapping detail (Figure 16.4).

The development of the European Sentinel satellite program offers substantial opportunities for further innovation. The objective of the Sentinel program is to meet the data requirements of the joint ESA/European Commission (EC) initiative Copernicus (previously known as Global Monitoring for Environment and Security) on an operational level. However, because of its advanced concepts and mission design, Sentinel has also significant potential to fulfill the data needs of the wind resource modeling community. The Sentinel-1, Sentinel-2, and Sentinel-3 missions

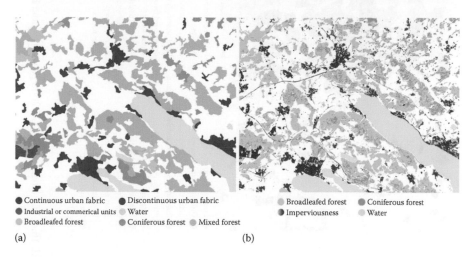

- ● Continuous urban fabric   ● Discontinuous urban fabric
- ● Industrial or commerical units   ● Water
- ● Broadleafed forest   ● Coniferous forest   ● Mixed forest

(a)

- ● Broadleafed forest   ● Coniferous forest
- ● Imperviousness   ● Water

(b)

FIGURE 16.3 Comparison of CLC product (a) and combined HRL layers: HRL Imperviousness, HRL Forest, and HRL Water (b).

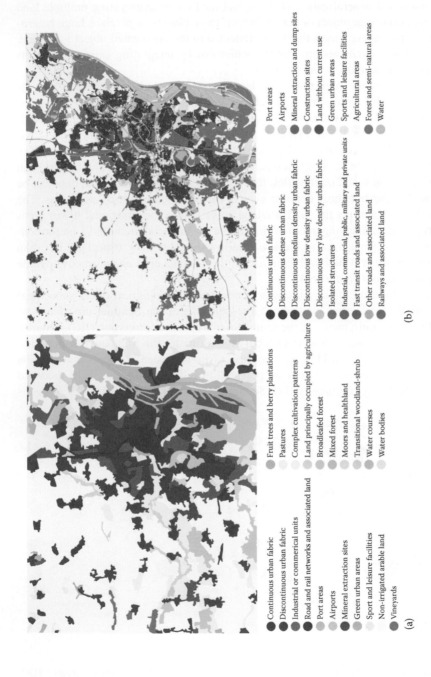

FIGURE 16.4   Comparison of CLC 2006 product (a) and European Urban Atlas layer of 2006 (b) for Strasbourg, France.

are of particular interest in the context of wind modeling parameterization, and we will briefly review each of the missions.

The concept of the Sentinel-1 mission is a constellation of two satellites orbiting 180° apart that carry a C-band SAR. The first of these satellites (Sentinel-1A) has been launched in April 2014, and the launch for Sentinel-1B is scheduled for 2016. While the current repeat cycle of Sentinel-1A is 12 days, the revisit time will increase to 6 days with both satellites in orbit. Sentinel-1 can be operated in several modes.

The Sentinel-2 mission is again a constellation of two identical satellites that have the key objective of collecting multispectral high-resolution data in the optical domain of the electromagnetic spectrum. The definition of the requirements of the Sentinel-2 program builds on the experience of previous multispectral campaigns such as the Landsat program and, in particular, the SPOT family (Satellite Pour l'Observation de la Terre) (Drusch et al. 2012; ESA 2014). Sentinel-2A was launched in June 2015 and Sentinel-2B is scheduled to be put in orbit in the second half of 2016. Their advantages compared to operational multispectral missions will be (a) higher spectral resolution with 13 bands in the visible, near-infrared, and short-wave infrared region; (b) high spatial resolution ranging from 10 to 60 m; (c) an unprecedented temporal resolution with a revisit time of <5 days; and (d) a dedicated ground segment that aims to rapidly process the acquired data and provide products to the user community in near-real time.

Sentinel-3 represents Europe's global land- and ocean-monitoring mission that will provide 2-day global coverage EO data. Sentinel-3A and 3-B are scheduled for launch early 2016 and 2017, respectively. One of Sentinel-3 mission's main objectives is to measure land surface color with high-end accuracy and reliability in support of ocean forecasting systems and for environmental and climate monitoring. The Sentinel-3 Ocean and Land Colour Instrument (OLCI) is based on heritage from Envisat's MERIS. With 21 bands, compared to the 15 on MERIS, it has a design optimized to minimize sun glint and a resolution of 300 m over all surfaces.

Wind resource modeling could benefit from the improved capabilities via better information on land use change, forest cover, photosynthetic activity, and phenology. Of all Sentinel missions, however, Sentinel-2 promises to be particularly useful for wind resource modeling activities, as its data allow the development of land cover map production systems that should be able to update the information monthly at a global scale. The temporal dimension will allow distinguishing classes whose spectral signatures are very similar during long periods of the year. The increased spatial resolution will enable one to work with smaller minimum mapping units. Additionally, the spectral bands are increased in number and decreased in width. With new image bands in the red-edge and specific bands for vegetation analysis and atmospheric analysis, Sentinel-2 images will show an increase in the separability of land cover classes, combined with an easier atmospheric correction.

It should therefore become feasible to update the land cover information with new data from a consistent source to move from mapping to monitoring. This monitoring will focus primarily on the vegetation evolution during a year/season by providing bio-geophysical parameters such as LAI, NDVI, green vegetation fraction, leaf chlorophyll content, and leaf cover. These parameters are a direct input for a more detailed land cover classification, allowing for diversification within otherwise aggregated areas such as agricultural fields and forested areas, thereby facilitating improved roughness estimation, taking into account the time of year. Besides being an input for land cover mapping, these parameters show a direct interrelation with atmospheric/wind resource models; for example, the green vegetation fraction is important for land surface heat flux calculation in coupled land–atmospheric models. Combining these parameter values with land cover data will provide a better estimation/understanding on the Earth's influences on the atmosphere.

16.3.2   3D Topography Characterization

One of the challenges to applying DEMs to modeling approaches—in particular on the microscale level—has been the absence of the availability of a standard DEM at a sufficient level of spatial detail. Low-resolution DEMs such as the STRM C-Band (30 m with the exception for Central Asia where the resolution is limited to 90 m) version or the one derived from stereo-processing of the Advanced Spaceborne Thermal Emission and Reflection Radiometer (ASTER) data at 30 m spacing are widely available free of charge, but offer limited use for detailed modeling. The reason is not only the relatively coarse spatial resolution but also the incomplete global coverage in case of SRTM or the varying local quality owing to the number of images per point in case of ASTER (Jacobsen 2013).

With the advent of scientifically and commercially available DEMs at higher resolutions such as the TerraSAR-X (TSX) and TanDEM-X (TDX) DEM at 12 m resolution, significant progress can be made in applying this data set to wind-related modeling. The outstanding characteristics of TDX are the provision of a real global and homogenous data set and the improved spatial and vertical resolution. The global coverage not only comes along with high-precision elevation information in areas above 60° latitude that were not covered by SRTM before. It also eases the transferability to other sites.

Figure 16.5 demonstrates the relevance of an increased spatial resolution. A coarse resolution levels out details and smoothens terrain features such as slopes and ridges. Smaller-sized obstacles can only be detected from high-resolution DEMs.

The corresponding elevation expresses the top of trees and other land surface elements rather than the ground level. Figure 16.6 shows the increased level of detail that can be achieved with the TDX DEM. A farmhouse, surrounded by hedges and a line of trees, is clearly visible, and the corresponding height values can directly be measured in the DEM. The arrow indicates the location of the farm in the DEM.

For the assessment of land cover and forestry parameters, DSMs generated by stereo imagery, SAR interferometry, or airborne laser scanning (ALS) can support atmospheric flow modeling both at the mesoscale and the microscale. EO-based DSMs provide information on important surface parameters such as edge morphology, forest borderline location and type, local obstacle properties, terrain slope, height of canopy, and canopy density.

At the mesoscale level, the land cover classifications (CLC, HRL layers) can be improved through the integration of height-derived parameters. This leads to substantial improvements in non-flat areas such as vineyards, orchards, and shrubby areas for which the calculation of the wind shear effect is inaccurate when solely

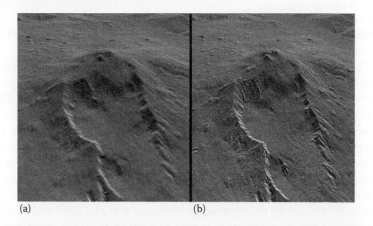

FIGURE 16.5 Digital elevation data of Mt. Etna in SRTM (90 m) (a) and new TanDEM-X (12 m) DEM (b).

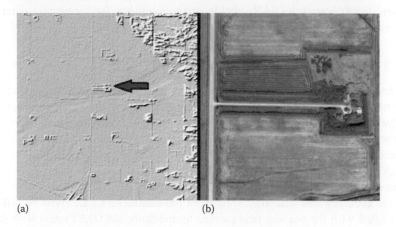

FIGURE 16.6 Color-shaded TanDEM-X DEM (a) for Winnipeg area, Canada, and zoom to farm house surrounded by hedges and trees via Google Earth (b).

based on the roughness length. Mesoscale models such as the widely used Weather Research and Forecasting (WRF) model can include the option of a canopy layer separated from the ground, such that the displacement height can be directly used as input to the model, but suitable values for the model parameters for the most common forest types need to be provided.

At the microscale level, the modeling community moves toward using flow models (RANS and LES models) in which the forest canopy is resolved. This creates a need for supplying the necessary surface parameters to the wind modelers. Boudreault et al. (2015) developed a method by which raw data from aerial Lidar scans of forested areas are transformed into gridded forest heights for a CFD model at a forested site. Furthermore, 3D approaches can be used to assess the attributes of the forest borders (e.g., hard cut vs. smooth transition from tall trees to non-forest) and 3D canopy structure/canopy density distribution.

The development of 3D approaches can be based on different sources according to availability, for example, optical (multi-)stereo data such as Pleiades, GeoEye and PRISM, radar from Sentinel-1, TSX/TDX, and airborne Lidar data.

Since the early years of SAR remote sensing, stereo-radargrammetric techniques have been applied to SAR image pairs (Toutin and Gray 2000). It has been demonstrated in previous studies that X-band radar data from COSMO-SkyMed and TerraSAR-X missions can provide useful information on forest cover and forest parameters (Deutscher et al. 2013). These sensors are able to collect images with a ground sampling distance down to 0.75 m in Spotlight mode at various look angles. In addition, they deliver imagery with very precise pointing accuracy so that remote regions where no reference data, that is, ground control points, are available can also be mapped and processed. One option to develop a DSM for forest assessment is based on the approaches by Raggam et al. (2010) and Perko et al. (2010) where 3D surface reconstruction was performed by stereo-radargrammetry in Europe. Perko et al. (2011) transferred the multi-image matching concept for digital surface modeling based on optical satellite images (Raggam 2006) to radar data and incorporated the SAR-specific image geometry for forest parameters in European test sites.

16.3.3 4D Surface Dynamics Mapping

The developing Copernicus Sentinel program also offers unprecedented data in terms of spectral, spatial, and, most significantly, temporal resolution. Revisit times of a few days will, for the first time, allow it to resolve highly dynamic phenological processes. This in turn will make it possible to parameterize atmospheric/wind resource models with near-real time and spatially explicit data of phenological stage and LAI at high spatial resolution.

One of Sentinel-3 mission's main objectives is to measure land surface color with high-end accuracy and reliability. Data from the Sentinel-3 OLCI could be utilized and coupled with the sea and land surface temperature (SST/LST) data with daily revisit times allowing for synergy products (e.g., phenological stage mapping).

Based on time series from future Sentinel-3 imagery and optionally other satellites (MODIS, Proba-V, MERIS), fully automatic, spatially explicit, wall-to-wall phenological models covering the whole EU could be developed by such an approach. The result of such phenological models would be maps that include entry dates of selected key phenological parameters. Separately for each vegetation type and biogeographic region, the temporal trajectories of vegetation index time series could then be analyzed on the basis of phenological field observation data. In addition to different algorithms/models, different features that are derived from the satellite data could also be analyzed, such as the cumulative vegetation index values. Following cross-validation, this would identify the best-suited algorithm/model and derive quality parameters. The best suited algorithms/models and feature combinations could then be applied over the whole monitoring period at a wall-to-wall basis.

The high revisit time of the fully operational Sentinel-1 and Sentinel-2 missions, combined with their large swath, will provide a high probability for a cloud-free coverage at key dates throughout the vegetation phase. This is of paramount importance since (i) it is mandatory for an accurate analysis of different vegetation types, and (ii) it enables the assessment of the intra-annual variability of the various land cover classes as well as the surface roughness parameters.

Sentinel-1 and Sentinel-2 time series data will enable new approaches for a detailed land cover classification that will further improve the currently available CLC classification scheme and can in particular be tailored to the data needs of increasingly sophisticated wind resource models. This will provide high thematic and spatial detail compared to the currently available CLC2006 database or the HRL layers. In particular, it will effectively discriminate between different vegetation types and enable the estimation of surface roughness and vegetation stand height. Moreover, the temporal variability of land cover classes and surface roughness can be analyzed and provided for wind resource modeling.

Exploitation of high-resolution DSMs such as TDX and stereo data from high-resolution satellites (Pléiades, Cartosat, TerraSAR, COSMO-SkyMed) will offer the opportunity to quantify the temporal dynamics of 3D surface parameters. This will allow the assessment of parameters such as forest growth, forest change detection (loggings, wind fall), changes in forest stock, and forest density. Quantification of, for example, forest growth can be used for computing the future aerodynamic flow at a potential wind turbine site.

16.4 CONCLUSION

The previous sections gave an overview on the potential of remote sensing data to support wind resource modeling especially through improved input parameterization regarding the state and characterization of the land surface. Three different topics in this relation have been discussed: land cover information (2D) from various remote sensing sources, DEM information (3D) at high spatial resolution, and the opportunities of monitoring changes in the state of the surface, which affects its roughness (4D). For all three topics, the currently used data sets and methods were briefly described and the related shortcomings were analyzed. Finally, the possibilities provided by new sensors (e.g., Sentinels, TerraSAR-X/TanDEM-X) and initiatives, such as Copernicus, are outlined. In conclusion, there is a wide amount of promising possibilities waiting to be tested for the specific needs in wind resource modeling. For 2D land cover information, existing high-resolution layers will need to be merged and integrated, possibly supported by specific classification of Sentinel-2 data. In terms of 3D data, now existing satellite-based DEMs at a resolution of 12 m can be used showing features such as hedges, forest borders, buildings, and so on in order to improve the modeling of surface roughness. Finally, the Sentinel-2 data are expected to bring additional benefit to the 4D component, as frequent update will allow monitoring the roughness situation over time, giving a more realistic scenario than just a snapshot assessment.

There is currently a substantial underexploitation of remote sensing data in the context of wind energy modeling. This chapter demonstrated that significant progress has been made in the past decade in terms of spaceborne data collection, represented by a range of new missions that collect global land cover and elevation information in unprecedented spatial, spectral, and temporal resolution. This is complemented by the increased availability of high-resolution ALS data. Yet, most operational wind resource models that are used in the wind energy sector still have to rely on data products that represent derivatives of data sets from sensors that have been put into orbit one or two decades ago.

There is hence an urgent need for a coherent effort that brings together both the remote sensing and wind modeling communities to develop bespoke downstream services for the wind energy sector, for example, within the framework of the European Copernicus program. Such an endeavor would not only constitute a major step forward in adding value to remote sensing data. It can also be expected that the new downstream data product will substantially improve wind energy planning and the subsequent commercial viability of wind energy projects.

REFERENCES

Aksoy, S., G. Akcay, G. Cinbis and T. Wassenaar (2008): Automatic mapping of linear woody vegetation features in agricultural landscapes. Proc. IGARSS, July 6–11, 2008, Boston, Massachusetts, pp. IV, 403–406.

Aksoy, S., G. Akcay and T. Wassenaar (2010): Automatic mapping of linear woody vegetation features in agricultural landscapes using very high resolution imagery. *IEEE Transactions on Geoscience and Remote Sensing*, 48(1), 511–522.

Arino, O., P. Bicheron, F. Achard, J. Latham, R. Witt and J. L. Webere (2008): GlobCover The most detailed portrait of Earth. *ESA Bulletin-European Space Agency*, 136, 24–31.

Badger, J., H. Frank, A. N. Hahmann and G. Giebel (2014): Wind climate estimation based on mesoscale and mircoscale modelling: Statistical–dynamical downscaling for wind energy applications. *Journal of Applied Meteorology and Climatology*, 53, 1901–1919.

Bartholomé, E., and A. S. Belward (2005): GLC2000: A new approach to global land cover mapping from Earth observation data. *International Journal of Remote Sensing*, 26(9), 1959–1977.

Beck, P., C. Atzberer, K. Høgda, B. Johansen and A. Skidmore (2006): Improved monitoring of vegetation dynamics at very high latitudes: A new method using MODIS NDVI. *Remote Sensing of Environment*, 100, 321–334.

Bidaut, S., T. Ranchin and L. Wald (2006): Mapping the aerodynamic roughness length using SAR images. ENSMP 300740, Paris, pp. 145–150.

Boudreault, L.-E., A. Bechmann, L. Taryainen, L. Klemedtsson, I. Shendryk and E. Dellwik (2015): A LiDAR method of canopy structure retrieval for wind modeling of heterogeneous forests. *Agricultural and Forest Meteorology*, 201, 86–97.

Case, J. L., W. L. Crosson, S. V. Kumar, W. M. Lapenta and C. D. Peters-Lidard (2008): Impacts of high-resolution land surface initialization on regional sensible weather forecasts from the WRF model. *Journal of Hydrometeorology*, 9(6), 1249–1266.

Case, J. L., F. J. LaFontaine, J. R. Bell, G. J. Jedlovec, S. V. Kumar and C. D. Peters-Lidard (2014): A real-time MODIS vegetation product for land surface and numerical weather prediction models. *IEEE Transactions on Geoscience and Remote Sensing*, 52(3), 1772–1786.

Chen, F., K. W. Manning, M. A. LeMone, S. B. Trier, J. G. Alfieri, R. Roberts, M. Tewari, D. Niyogi, T. W. Horst, S. P. Oncley, J. B. Basara and P. D. Blanken (2007): Description and evaluation of the characteristics of the NCAR high-resolution land data assimilation system. *Journal of Applied Meteorology and Climatology*, 46(6), 694–713. doi:10.1175/JAM2463.1.

Choudbury, B., and J. Monteith (1988): A four-layer model for the heat budget of homogeneous land surfaces. *Quarterly Journal of the Royal Meteorological Society*, 114, 373–398.

Dellwik, E., F. Bingöl and J. Mann (2013): Flow distortion at a dense forest edge. *Quarterly Journal of the Royal Meteorological Society*, 140 (679), 676–686; doi:10.1002/qj.2155.

Dellwik, E., and N. Jensen (2005): Flux–profile relationship over a fetch-limited beech forest. *Boundary-Layer Meteorology*, 115, 179–204.

Deutscher, J., R. Perko, K. Gutjahr, M. Hirschmugl and M. Schardt (2013): Mapping tropical rainforest canopy disturbances in 3D by COSMO-SkyMed Spotlight InSAR-stereo data to detect areas of forest degradation. *Remote Sensing*, 5, 648–663. doi:10.3390/rs5020648.

Drusch, M., U. Del Bello, S. Carlier, O. Colin, V. Fernandez, F. Gascon, B. Hoersch, C. Isola, P. Laberinti, P. Martimort, A. Meygret, F. Spoto, O. Sy, F. Marchese and P. Bargellini (2012): Sentinel-2: ESA's optical high-resolution mission for GMES operational services. *Remote Sensing of Environment*, 120, 25–36, ISSN 0034-4257.

Ducrot, D., A. Masse and A. Ncibi (2012): Hedgerow detection in HRS and VHRS images from different sources (optical, radar). In: Proc. IGARSS, July 22–27, 2012, Munich, Germany, pp. 6348–6351.

Dupont, S., and Y. Brunet (2008): Edge flow and canopy structure. *Boundary-Layer Meteorology*, 126, 51–71.

European Space Agency (ESA) (2014): The Copernicus Space Component: Sentinels Data Product List. Accessible at https://spacedata.copernicus.eu/documents/12833/14143/Sentinel_Products_List_Issue1_Rev1_Signed.pdf (July 2016).

Findel, K., and E. A. B. Eltahir (2003): Atmospheric controls on soil moisture—Boundary layer interactions. Part II: Feedbacks within the Continental United States. *Journal of Hydrometeorology*, 4, 570–583.

Floors, R. R., A. Peña and S.-E. Gryning (2015): The effect of baroclinicity on the wind in the planetary boundary layer. *Royal Meteorological Society. Quarterly Journal*, 141(687), 619–630. doi:10.1002/qj.2386.

Friedl, M. A., D. K. McIver, J. C. F. Hodges, X. Y. Zhang, D. Muchoney, A. H. Strahler, C. E. Woodcock, S. Gopal, A. Schneider, A. Cooper, A. Baccini, F. Gao and C. Schaaf (2002): Global land cover mapping from MODIS: Algorithms and early results. *Remote Sensing of Environment*, 83, 287–302.

Hansen, M. C., and B. Reed (2000): A comparison of the IGBP DISCover and University of Maryland 1 km global land cover products. *International Journal of Remote Sensing*, 21(6–7), 1365–1373.

Hasager, C. B., N. W. Nielsen, N. O. Jensen, E. Bøgh, J. H. Christensen, E. Dellwik and H. Søgaard (2003): Effective roughness calculated from satellite-derived land cover maps and hedge-information used in a weather forecasting model. *Boundary-Layer Meteorology*, 109, 227–254.

Hong, S., V. Lakshmi, E. E. Small, F. Chen, M. Tewari and K. W. Manning (2009): Effects of vegetation and soil moisture on the simulated land surface processes from the coupled WRF/Noah Model. *Journal of Geophysical Research*, 114(D18), D18118.

Huete, A., K. Didan, T. Miura, E. P. Rodriguez, X. Gao and L. G. Ferreira (2002): Overview of the radiometric and biophysical performance of the MODIS vegetation indices. *Remote Sensing of Environment*, 83(1–2), 195–213.

Jacobsen, K. (2013): DEM Generation from High Resolution Satellite Imagery, *PFG* 5(2013), 483–493.

Jiménez, P. A., and J. Dudhia (2012): Improving the representation of resolved and unresolved topographic effects on surface wind in the WRF model. *Journal of Applied Meteorology and Climatology*, 51(2), 300–316. doi:10.1175/JAMC-D-11-084.1.

Kelly, M. C., and S.-E. Gryning (2010): Long-term mean wind profiles based on similarity theory. *Boundary-Layer Meteorology*, 136(3), 377–390.

Lopes, A. S., J. Palma and V. J. Lopes (2013): Improving a two-equation turbulence model for canopy flows using large-eddy simulation. *Boundary-Layer Meteorology*, 149, 231–257.

Loveland, T. R., B. C. Reed, J. F. Brown, D. O. Ohlen, Z. Zhu, L. W. M. J. Yang and J. W. Merchant (2000): Development of a global land cover characteristics database and IGBP DISCover from 1 km AVHRR data. *International Journal of Remote Sensing*, 21(6–7), 1303–1330.

Mölder, M., and A. Lindroth (1999): Thermal roughness length of a boreal forest. *Agricultural and Forest Meteorology*, 98–99, 659–670.

Moulin, S., L. Kergoat, N. Viovy and G. Dedieu (1997): Global-scale assessment of vegetation phenology using NOAA/AVHRR satellite measurements. *Journal of Climate*, 10, 1154–1170.

Peña, A., R. R. Floors, A. Sathe, S.-E. Gryning, R. Wagner, M. Courtney, X. G. Larsén, A. N. Hahmann and C. B. Hasager (2015): Ten years of boundary-layer and wind-power meteorology at Høvsøre, Denmark. *Boundary-Layer Meteorology*. doi:10.1007/s10546 -015-0079-8.

Perko, R., H. Raggam, J. Deutscher, K. Gutjahr and M. Schardt (2011): Forest assessment using high resolution SAR data in x-band. *Remote Sensing*, 3, 792–815.

Perko, R., H. Raggam, K. Gutjahr and M. Schardt (2010): Deriving forest canopy height models using multi-beam TerraSAR-X imagery. In: Proceedings of 8th European Conference on Synthetic Aperture Radar, Aachen, Germany, June 7–10, 2010; pp. 568–571.

Pineda, N., O. Jorba, J. Jorge and J. M. Baldasano (2004): Using NOAA AVHRR and SPOT VGT data to estimate surface parameters: Application to a mesoscale meteorological model. *International Journal of Remote Sensing*, 25(1).

Quackenbush, L. J. (2004): A review of techniques for extracting linear features from imagery. *Photogrammetric Engineering and Remote Sensing*, 70(12), 1383–1392.

Raggam, H. (2006): Surface mapping using image triplets—Case studies and benefit assessment in comparison to stereo image processing. *Photogrammetric Engineering and Remote Sensing*, 72, 551–563.

Raggam, H., K. Gutjahr, R. Perko and M. Schardt (2010): Assessment of the stereo-radargrammetric mapping potential of TerraSAR-X multibeam spotlight data. *IEEE Transactions on Geoscience and Remote Sensing*, 48, 971–977.

Raupach, M. R. (1994): Simplified expressions for vegetation roughness length and zero plane displacement height as functions of canopy height and area index (research note). *Boundary-Layer Meteorology*, 71, 211–216.

Raupach, M. R. (1995): Vegetation–atmosphere interaction and surface conductance at leaf, canopy and regional scales. *Agricultural and Forest Meteorology*, 73(3), 151–179.

Refslund, J., E. Dellwik, A. N. Hahmann, M. J. Barlage and E. Boegh (2013): Development of satellite green vegetation fraction time series for use in mesoscale modeling: Application to the European heat wave 2006. *Theoretical and Applied Climatology*, 1–16.

Santos-Alamillos, F. J., D. Pozo-Vázquez, J. A. Ruiz-Arias, V. Lara-Fanego and J. Tovar-Pescador (2013): Analysis of WRF model wind estimate sensitivity to physics parameterization choice and terrain representation in Andalusia (Southern Spain). *Journal of Applied Meteorology and Climatology*, 52(7), 1592–1609. doi:10.1175/JAMC -D-12-0204.1.

Schwartz, M. D., B. C. Reed and M. A. White (2002): Assessing satellite-derived start-of-season measures in the conterminous USA. *International Journal of Climatology*, 22, 1793–1805.

Sogachev, A., and O. Panferov (2006): Modification of two-equation models to account for plant drag. *Boundary-Layer Meteorology*, 121, 229–266.

Sun, G., K. Ranson, D. Kimes, J. Blair and K. Kovacs (2008): Forest vertical structure from GLAS: An evaluation using LVIS and SRTM data. *Remote Sensing of Environment*, 112, 107–117.

Taylor, C. M., D. J. Parker and P. Harris (2007): An observational case study of mesoscale atmospheric circulations induced by soil moisture. *Geophysical Research Letters*, 34(15).

Tian, X., Z. Y. Li, C. van der Tol, Z. Su, X. Li, Q. S. He, Y. F. Bao, E. X. Chen and L. H. Li (2011): Estimating zero-plane displacement height and aerodynamic roughness length using synthesis of LiDAR and SPOT-5 data. *Remote Sensing of Environment*, 115, 2330–2341.

Toutin, T., and L. Gray (2000): State-of-the-art of elevation extraction from satellite SAR data. *ISPRS Journal of Photogrammetry and Remote Sensing*, 55, 13–33.

Troen, I., and E. L. Petersen (1989): *European Wind Atlas*. Riso National Laboratory, Roskilde, Denmark. 656 pp.

Vannier, C., and L. Hubert-Moy (2008): Detection of wooded hedgerows in high resolution satellite images using an object-oriented method. In: Proc. IGARSS, July 6–11, 2008, Boston, Massachusetts, pp. IV, 731–734.

Weaver, C. P., and R. Avissar (2001): Atmospheric disturbances caused by human modification of the landscape. *Bulletin of the American Meteorological Society*, 82, 269–281.

White, M. A., and R. R. Nemani (2006): Real-time monitoring and short-term forecasting of land surface phenology. *Remote Sensing of Environment*, 104, 43–49.

Yang, Z.-L. (2004): Modeling land surface processes in short-term weather and climate studies. In: *Observations, Theory, and Modeling of Atmospheric Variability* (ed. X. Zhu), World Scientific Series on Meteorology of East Asia, Vol. 3, World Scientific Publishing Corporation, Singapore, pp. 288–313.

Zhang, X. Y., M. A. Friedl, C. B. Schaaf, A. H. Strahler, J. C. F. Hodges, F. Gao, B. C. Reed and A. Huete (2003): Monitoring vegetation phenology using MODIS. *Remote Sensing of Environment*, 84(3), 471–475.

17 Assessing Solar Energy Potential and Building Energy Use in Indianapolis Using Geospatial Techniques

Yuanfan Zheng and Qihao Weng

CONTENTS

17.1 INTRODUCTION

By 2030, urbanized areas will expand to provide homes for 81% of the world's population, with the majority of the population increase coming from developing countries (Weng 2015). Continued urbanization will bring impacts to the environment, and there is a rapidly growing need for consideration of sustainable cities. Buildings are the major component in the urban environment, so accurate and timely spatial information about buildings (and associated attributes) in urban areas is needed as the basis to assist decision making in understanding, managing, and planning the continuously changing environment (Weng 2010). Regarding obtaining the information from buildings, the traditional field survey approach is costly in time and labor; on the other hand, the remote sensing technique can monitor the earth surface in a very high temporal and spatial resolution, which has a great potential to extract building information automatically from the urban area. Geographic Information Systems (GISs) have the strong ability to gather and analyze the information of urban buildings by using geospatial analysis technique. This chapter provides two case studies to present the most recent researches on buildings under the topic of sustainable city development, which includes the building solar potential assessment and correlation analysis of building annual energy use and building attributes, using the city of Indianapolis as an example.

17.2 BUILDING SOLAR ENERGY POTENTIAL ESTIMATION

Renewable energy systems (RESs), which include but are not limited to wind power, hydropower, solar energy, and geothermal energy, have become vital parts of future energy use resources because fossil fuels are declining but demands for energy keep growing. Solar energy has become a fast-growing energy source in the last few years. Compared to other renewable sources of energy, solar energy has the following advantages: it is inexhaustible, it can drastically reduce energy-related greenhouse gases to help limit climate change, it is relatively well spread over the globe (Philibert 2011), and it is not subject to geological, climatologic, and morphologic conditions. The largest demands of solar energy are coming from urban areas because more than half of the world's population lives in the urban environment (Weng 2015). Solar energy absorption equipment can be ground-mounted arrays close to where the energy is used or directly mounted on rooftops (SEIA 2014). There are three primary technologies by which solar energy is commonly harnessed: photovoltaic (PV), concentrating solar power, and heating and cooling systems (SEIA 2014). In the dense urban areas, the PV equipment are encouraged to be mounted on the rooftops. First, it is not realistic to allocate a large land to ground-mount them. Moreover, buildings are the largest energy consumers in the urban area today (Philibert 2011), and the PV modules mounted on the roofs can directly transfer the solar energy absorbed from sunlight into electricity and sold to the electric companies.

Not all buildings are suitable for installing PV modules on their roofs, such as historical buildings or buildings shaded by surrounding objects for most of the time. Therefore, there is a need to evaluate the solar potential for each building. A completed database including information on geographic location, height, volume, and ownership for each building is essentially important to estimate its solar potential. Building detection and reconstruction by using the remote sensed data and GIS technique has become a quite popular research topic for the last decade, especially after the LiDAR data became available because such data can reflect the absolute height of the ground feature and are not affected by solar shadows. Numerous approaches have been done for building boundaries detection and building reconstruction using LiDAR and other data sources (Haala et al. 1998; Henn et al. 2012; Huang et al. 2013; Lafarge et al. 2007; Miliaresis and Kokkas 2007; Sohn and Dowman 2007) in the last two decades.

The building detection studies separated buildings from other features based on a series of hypotheses, which are based on the spatial, spectral, and textural characteristics of buildings. Sohn and Dowman (2007) presented an approach for automatic extraction of building footprints in a combination of the LiDAR data with IKONOS imagery pan-sharpened multispectral bands. They achieved a detection accuracy of 90.1% and an overall accuracy of 80.5%. Miliaresis and Kokkas (2007) used a region-growing algorithm to extract building class from the LiDAR digital elevation models (DEMs) using the geomorphometric segmentation principles. The interpretation of the spatial distribution of clusters that is revealed by K-means classification method allows for identification of building class (Miliaresis and Kokkas 2007).

The three-dimensional (3D) building reconstruction studies became more and more popular since the 1990s. Haala et al. (1998) combined ground plans of buildings and LiDAR data to build the 3D building models for the application of virtual reality. Lafarge et al. (2007) reconstructed buildings from DEMs and associated rectangular building footprints based on a Bayesian approach. The Bayesian approach can automatically find the best-fit roof primitives in the predefined library to represent given blocks of LiDAR point clouds within the building footprints. Huang et al. (2013) presents a statistical approach to automatic 3D building roof reconstruction from LiDAR point clouds. The selection of roof primitives was driven by a variant of Markov Chain Monte Carlo technique with specific jump mechanism. Three-dimensional models were built for 21 buildings and 1 city block in Hannover, Germany. Henn et al. (2012) approached model selection among five roof prototypes using supervised machine learning methods, which achieved a classification accuracy of 95% for 6696 roofs.

Once a completed GIS-based building polygon database showing the clear boundaries is available, a geospatial technique can be used to calculate the annual solar energy yield for each building footprint. Many previous studies were focusing on the solar energy yield estimation based on application of GIS-based radiation models, LiDAR data, and extracted building footprints. Schuffert (2013) presented an automatic approach to extract suitable single roof planes from LiDAR data and assigned regional irradiation data to each extracted roof surface. A shadow analysis was also combined to obtain the realistic solar energy values since the shadowing effect might significantly affect the incoming solar radiation. Lukač et al. (2013) rated the solar potential of building roofs using the method that combines extracted urban topography from LiDAR data with the pyranometer measurement of global and diffuse solar irradiances. They also created multiresolution shadowing to complete the proposed method. Tooke et al. (2011) quantified the diurnal and seasonal impact of trees on solar radiation yield. Their results indicated that trees on average reduce 38% of the total solar radiation received by residential building rooftops. A point obstruction stacking approach was designed by Tooke et al. (2013) to model irradiance on the walls of multiple buildings by integrating contiguous data sets of surrounding urban form and topography with building footprints.

17.3   BUILDING ENERGY USE SIMULATION

Globally, buildings are responsible for more than 40% of energy demand and contribute more than 30% of CO_2 emissions (Tooke 2014). Building energy use is a major component of anthropogenic heat discharge. Zhou et al. (2011) estimated the relationship between remotely sensed anthropogenic heat discharge and building energy use, and the result suggests that there is an obvious consistency in terms of spatial distribution. The high anthropogenic heat discharge occurs in the dense residential and commercial areas, where the highest building energy use can be found.

In order to get a better understanding of the current urban energy balance and propose suggestions for sustainable city development, monitoring and analysis of the building energy consumption are vital. There are two major building energy use estimation approaches: top-down and bottom-up approaches (Heiple and Sailor 2008). The top-down approaches rely on extensive historical annual energy consumption data and are typically applied to a larger area with a relatively coarse scale (Heiple and Sailor 2008; Sailor and Lu 2004). The bottom-up approaches, on the other hand, simulate hourly consumption of different types of energies for each single building (Ward and Choudhary 2014). In order to examine the building energy use for the purpose of energy-efficient building design in future sustainable cities, the bottom-up approach should be chosen, since it can provided energy use information for each building. The building energy simulation basically requires the input of many attributes of buildings, such as geographic location, climate conditions, floor space, floor numbers, building prototypes, electricity systems, and occupancy status. Ward and Choudhary (2014) developed a methodology for the analysis of building energy retrofits for a diverse set of buildings. The methodology requires selection of appropriate building simulation tools based on the nature of the principal energy demand (Ward and Choudhary 2014). Wang et al. (2015) used a bottom-up approach to model residential heating energy consumption for China's hot summer–cold winter climatic region. They considered the occupant behaviors besides other building characteristics.

Remote sensing technique provides the data for large spatial and temporal scale monitoring. The GIS technique, on the other hand, can model individual building energy consumption and analyze the correlation between energy use and factors that might affect it. Results from various studies that combined remote sensing and GIS techniques and showed the climate, geographic location, building attributes, and occupant behavior can all contribute to the building energy consumption. Hemsath and Bandhosseini (2015) indicate that both the vertical and horizontal geometric proportion are equally sensitive to the material aspects related to building energy use, and the impact of the building geometry's effect on energy use depends on geographic locations. Premrov et al. (2015) pointed out that the total annual energy demand for heating and cooling has a stronger correlation with building shape factor if buildings are located in a cold climate with a lower solar potential. A statistical method applied by Aksoezen et al. (2015) showed that there is a strong interdependence between energy consumption, compactness, and building age.

17.4   CASE STUDIES

In this section, two case studies are presented. The first case study assesses the solar energy potential for commercial and residential buildings in an urban area using geospatial techniques. The second case study examines the relationship between building energy use and some building attributes.

17.4.1   STUDY AREA AND DATA

The city of Indianapolis, which is the capital of Indiana, was chosen for both case studies. The study area was 3.78 km², which covered the Central Business District of Indianapolis and some nearby residential areas. It is characterized by tall and complex commercial buildings and low apartment houses with different roof shapes.

Three data sources were used in this study: LiDAR data, building footprint polygons, and Marion County Assessor's parcel data. In the United States, LiDAR data are available nationwide. One of the most famous public data inventories is the United States Interagency Elevation Inventory (USIEI), which is a collaborative effort of the National Oceanic and Atmospheric Administration (NOAA) and the US Geological survey, and it is updated annually. Building footprints are available in almost all major cities in the United States, which are stored as GIS Shapefiles. The Marion County Assessor's parcel data are building Shapefile polygons with information on prototypes and ages that can be linked to the building footprint for building energy use simulation. The high-resolution orthophoto with 0.1 m spatial resolution was used as supplemental data.

LiDAR data digital surface model (DSM) and DEM were both rasterized from last return LiDAR point clouds, which were acquired from the Indiana Spatial Data Portal for 2009. They have the same spatial resolution of 0.91 m. The LiDAR normal height model (NHM), which represents absolute height information of objects above the ground surface, was calculated as follows:

$$NHM = DSM - DEM. \tag{17.1}$$

Building footprints data, which were provided by the city government of Indianapolis, were acquired in the year 2013. They were used to mask out the non-building features, such as tall trees in the LiDAR NHM. There are totally 411 building footprints in the study area. The building footprints less than 10 m² were removed from this study. Buildings having zero or negative height values were also removed since building footprints were created after the LiDAR data were obtained.

17.4.2   CASE STUDY I: ASSESSING SOLAR ENERGY POTENTIAL IN INDIANAPOLIS

Simulation of annual solar energy yield was first performed for building roofs that were reconstructed as a 3D city model from LiDAR DEM and building footprint. The shadowing effect from the surrounding objects was considered in the simulation since it will affect the total solar energy yield. Finally, the suitability of PV module installation was rated for all reconstructed roof segments based on their annual solar energy yield. The objectives of this case study were (1) to demonstrate a method to estimate the solar potential of building roofs in the densely urban area and (2) to create a complete 3D GIS-based city model that contains information of each building in the study area such as height, volume, total floor area, compactness, solar energy yield, and ownership for future use.

17.4.2.1 Building Reconstruction: 3D City Model

We reconstructed the commercial and residential buildings to a 3D city model. The 3D city model, which represents the geometry of urban environment, was used as input for the annual solar energy yield simulation in the following stage. Compared to the 2D building footprints, 3D city models with prototypical roofs are rarely available, but they are more valuable in applications of urban planning, environmental modeling, and particularly of solar radiation calculations or noise emission simulation (Henn et al. 2012) and urban structure analysis. LiDAR NHM data were used to calculate the height of each building, and the GIS-based building footprints were used to define the ground boundary of each building.

There are three steps in building reconstruction in this case study: step-edge detection, boundary refinement, and 3D building modeling. Not all buildings have a flat roof; many commercial buildings have substructures on their roofs. Direct use of those box-shaped models to represent buildings with complex and irregular roof shapes may cause loss of roof information and may lead to inaccurate reconstruction. Therefore, the step edge detection on building roofs has to be performed first. Those step edges may either be the boundaries of roof substructures or the transition with large variation in height between two parts in the rooftop (Figure 17.1). Ignoring step edges would lead to inaccurate reconstruction of building roofs.

Sobel's edge detection method (Sobel and Feldman 1968) was first used to find out the possible step edges on roofs. Figure 17.2 presents the result of applying the Sobel

FIGURE 17.1 Building with step edges on its roof.

FIGURE 17.2   Step edge detection on a building roof.

edge detector to the LiDAR NHM of a selected building. The step edge detection can split some building footprints into many small polygons, which causes an overseg-mentation problem. Therefore, refinement was carried out after the step edge–based segmentation. The purpose of refinement is to merge the small polygons into their adjacent larger polygons and simplify the boundaries of building footprints. An area threshold (30 m^2) was used to merge the cells with their adjacent larger cells. If there were multiple adjacent cells, the one sharing the longest boundary would be chosen. The *Bend Simplify* algorithm developed by Wang (1996) was used to simplify the boundaries into straight lines since it often produces results that are more faithful to the original and more aesthetically pleasing (Wang 1996). A classification for building footprints by their areas was performed before the simplification since foot-prints in different sizes require different simplification tolerance values. The same simplification tolerance may cause the small footprints to lose their essential shapes, while large footprints still keep the extraneous bends. A total of six groups of build-ing footprints were classified: (1) less than 400 m^2, (2) 400 to 1000 m^2, (3) 1000 to 2000 m^2, (4) 2000 to 4000 m^2, (5) 4000 to 10,000 m^2, and (6) greater than 10,000 m^2.

Mean height of pixels within the boundaries of each building segment was calcu-lated using LiDAR data. This research used box-shaped buildings, which extruded from the 2D building segments based on the mean height value to create the 3D city model.

17.4.2.2 Annual Solar Energy Yield Estimation

The annual solar energy yield for individual building roof segments is the most important indicator for their solar suitability rating in this study. Since the regional irradiation data used in Schuffert's (2013) study are not always available in every city, it would be good to propose a solar energy simulation method, which calculates the instant solar energy yield based on the angular motion of the sun.

The incoming solar radiation emitted by the sun has three major components when it reaches the earth's surface: direct radiation, diffuse radiation, and reflect radiation. Direct radiation is the largest component of total radiation, which travels on a straight line from the sun down to the surface of the Earth. Diffuse radiation is solar radiation that reaches the Earth's surface after being scattered from direct radiation. Reflected radiation is reflected from surface features, which constitutes only a small proportion of total radiation.

In this case study, a solar radiation map was generated by the Area Solar Radiation tool in the ArcGIS Spatial Analyst toolbox, using the LiDAR NHM and latitude of the study area as inputs. The simulation was based on the hemispherical viewshed algorithm developed by Rich et al. (1994) and further developed by Fu and Rich (2002). The total annual solar energy amount was calculated as the global radiation for each pixel in the solar radiation map. The global radiation was calculated as the sum of direct and diffuse radiation using the following equation:

$$I_{glo} = I_{dir} + I_{dif},\qquad(17.2)$$

where I_{glo} is the global radiation, I_{dir} is the direct radiation, and I_{dif} is the diffuse radiation. The reflect radiation was not considered in this study, because according to Lukač et al. (2013), the reflected radiation only contributed a little to the total amount of global radiation. The solar radiation tool in ArcGIS involves four steps for solar energy calculation:

1. Calculate an upward-looking hemispherical viewshed, such as a fisheye view, at each observation point location.
2. Calculate the direct radiation based on the overlay of the viewshed and direct sun map.
3. Calculate the diffuse radiation based on the overlay of the viewshed and diffuse sun map.
4. Repeat the same calculation process for every location in each simulation time.

The input latitude for this case study was defined as 39.77 for Indianapolis. The simulation time was set for every 14 days in the year 2010. Since the course of sun changes continuously over time in a few days, it is not necessary to calculate each single day (Schuffert 2013). The hourly interval within a day was set to 2 h. The sky size was set to 200, which is the highest resolution for the viewshed, sky map, and sun map grids. The slope and aspect input type was set to "from_DEM". Slope means steepness, aspect means the slope, and aspect grids are calculated from the input raster. The calculation direction was set to 32, which means the highest degree of complexity. Zenith division was set to 8, which means relative to zenith. Azimuth division was set to 8, which means relative to north. For a diffuse model type, *uniform sky* was the setting representing the incoming diffuse radiation, which is the same from all sky directions. Both the diffuse proportion and transmittivity proportion were set according to the *clear sky* condition for all times, which is the most probable weather condition in Indianapolis.

The instant direct irradiance (I_{dir}) for a given pixel can be calculated using the following equation:

$$I_{dir\theta,\alpha} = S_{Const} * \beta^{m(\theta)} * SunDur_{\theta,\alpha} * SunGap_{\theta,\alpha} * cos(AngIn_{\theta,\alpha}),\qquad(17.3)$$

where θ is the zenith angle, α is the azimuth angle, S_{Const} is the solar constant (1367 w/m^2), β is the atmospheric transmissivity, $m(\theta)$ is the relative optical path length, SunDur$_{\theta,\alpha}$ is the time duration represented by the sky sector, SunGap$_{\theta,\alpha}$ is the gap fraction, and AngIn$_{\theta,\alpha}$ is the angle of incidence between the centroid of the sky sector and the axis normal to the surface.

The instant diffuse irradiance (I_{dif}) for a given pixel can be calculated using the following equation:

$$I_{dif\theta,\alpha} = R_{glb} * P_{dif} * \text{Dur} * \text{Skygap}_{\theta,\alpha} * \text{Weight}_{\theta,\alpha} * \cos(\text{AngIn}_{\theta,\alpha}), \qquad (17.4)$$

where R_{glb} is the global normal irradiance, P_{dif} is the proportion of the global normal radiation flux that is diffused, Dur is the time interval used for analysis, SkyGap$_{\theta,\alpha}$ is the proportion of visible sky for the sky sector, Weight$_{\theta,\alpha}$ is the proportion of diffuse irradiance originating in a given sky sector relative to all sectors, and $\cos(\text{AngIn}_{\theta,\alpha})$ is the angle of incidence between the centroid of the sky sector and the intercepting surface.

The instant solar energy yield in all time nods was calculated and summed up to obtain the daily solar energy yield map. All daily solar energy yield maps were summed up to obtain the annual solar energy yield map.

Roof can be shadowed by its surrounding objects such as other buildings and trees that are taller. In order to obtain the realistic estimation of annual solar energy yield, a shadowing effect should be considered because it can affect the absorption of direct radiation. The shadow balance for a pixel at a given time was calculated using the following equation:

$$S = 255.0 * ((\cos(\text{Zenith\_rad}) * \cos(\text{Slope\_rad})) + (\sin(\text{Zenith\_rad})$$
$$* \sin(\text{Slope\_rad}) * \cos(\text{Azimuth\_rad} - \text{Aspect\_rad}))), \qquad (17.5)$$

where Zenith_rad and Azimuth_rad are the zenith and azimuth angles of the illumination source at a given location and given time, which can be obtained in the NOAA solar calculator website (http://www.noaa.gov/). The zenith angle can be obtained by subtracting the altitude angle from 90°. Slope_rad is the slope value at the given pixel, and Aspect_rad is that of the aspect value. In order to match the simulation time matched in the annual solar energy yield simulation, a subset of days in each 14 days was chosen. In each chosen day, the sun position was calculated for six representative time nods: 8 a.m., 10 a.m., 12 p.m., 2 p.m., 4 p.m., and 6 p.m. The instant shadow balance maps in six time nods were summed up to obtain the daily shadow balanced map, and all daily shadow balanced maps were summed up to obtain the annual shadow balance map pixels within the study area were classified into three categories based on their annual shadow balance values: always-shaded zone, half-shaded zone, and unshaded zone. This classification result was applied to annual direct irradiance value in each pixel to obtain the realistic annual solar energy yield simulation result using the following equation:

$$I_{glo} = I_{dir} * \text{SC} + I_{dif}, \qquad (17.6)$$

where SC is the shadowing coefficient. SC was set to 0.1 for the always-shaded zone, because pixels in this zone can hardly receive direct irradiance owing to the shadowing effect. SC was set to 0.5 for the half-shaded zone, which means that pixels in this zone receive partial amount of direct solar irradiance. SC was set to 1 for the unshaded zone, which means pixels in this zone receive full amount of direct solar irradiance. The annual solar energy yield for a specific reconstructed roof segment was assigned using the mean value of annual global solar irradiance values of all pixels within it.

17.4.2.3 Results
Figure 17.3 shows the annual solar energy yield in downtown Indianapolis; the warmer colors represent larger values of annual global solar radiation absorption, while cooler colors represent smaller values of annual global solar radiation absorption. The solar potential index for each roof segment can be calculated using the following equation:

$$SPi = (SEi)/(SEmax),$$ (17.7)

where SPi is the solar potential index, SEi is the annual solar energy yield for the given roof segment, and SEmax is the roof segment with the highest annual solar energy yield in the study area. All the roof segments were assigned to five categories according to their solar potential index, as presented in Figure 17.4.

Figure 17.4 shows the result of a building roof's PV suitability with a view of the 3D city model after assigning the solar potential index into the roof segments' attribute table. Purple and blue represent the roof segments that are unsuitable or with low suitability for PV module installation, green represents the roof segments with medium level of suitability, and yellow and red represent the roof segments with high

Annual global
solar irradiance
(Unit: watts hours
per square meter)

■ High: 2707375

■ Low: 847

FIGURE 17.3   Annual global solar irradiance in Indianapolis.

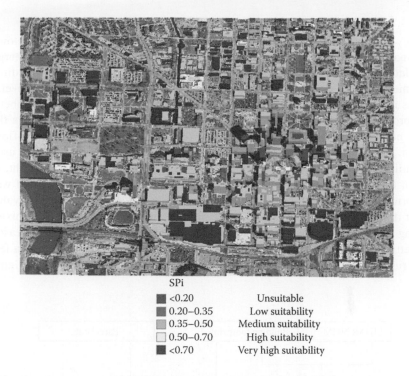

SPi

■ <0.20	Unsuitable
■ 0.20–0.35	Low suitability
▨ 0.35–0.50	Medium suitability
☐ 0.50–0.70	High suitability
■ <0.70	Very high suitability

FIGURE 17.4   Assessment of building PV suitability in downtown Indianapolis with a 3D city model.

or very high suitability. With the 3D city model, it can be determined that the common characteristics of roof segments with lower PV suitability (represented by blue and purple colors) are located in the downtown area, surrounded by higher objects, and have a smaller floor space area. Shading by the higher surrounding objects for most of the time is probably the reason. Since they have a small floor space area, a large portion of them are shaded. Roof segments with higher PV suitability (represented by red and yellow colors) usually have greater heights than the surrounding objects or are located at a distance from the surrounding objects, making sure that they are not shaded for most of the time.

17.4.3   CASE STUDY II: CORRELATION ANALYSIS BETWEEN BUILDING ENERGY USE AND BUILDING ATTRIBUTES IN INDIANAPOLIS

Buildings are a major component in urban areas, and building energy use accounts for a large proportion of anthropogenic heat. Understanding the characteristic of energy use for different types of buildings is vital for current sustainable city research. The objectives of this case study were to analyze the correlation between building energy use and their attributes in different levels, to find out the most significant contributor among these attributes, and to propose an energy-saving suggestion based on the results.

17.4.3.1 Building Energy Consumption Simulation in Indianapolis

Energy consumption from commercial and residential buildings in downtown Indianapolis was simulated using the eQUEST software, which is an energy simulation tool designed to model the annual energy cost for a single building. The methodology was illustrated in Figure 17.5. The simulation requires several inputs related to the buildings such as building shape, floor area, floor numbers, and prototype. LiDAR NHM and building footprint were used to determine the floor numbers through dividing the height of the building by the floor-to-floor height. In order to obtain a more accurate estimation, eave height should be used for buildings with non-flat roofs, rather than using the top height, mean height, or height for major parts of the roofs. A classification of flat and non-flat roof was performed, based on a slope threshold. The percentage of pixels with a slope value lower than 5° in each cell was calculated, and the cells with at least 40% of pixels having a slope value lower than 5° would be recognized as flat roofs. The reason for setting the threshold to 5° rather than 0° is that many flat roofs contain a few substructures on their top. For building roofs classified as non-flat roofs, inner

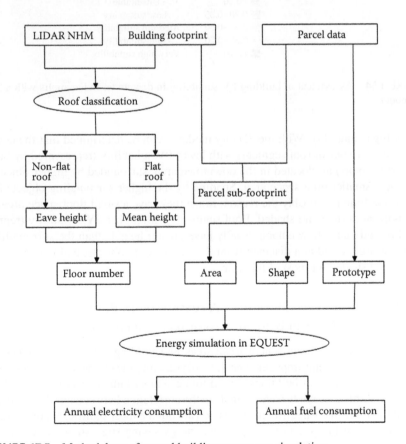

FIGURE 17.5 Methodology of annual building energy use simulation.

buffer zones were created around the boundary of building footprints to extract the eave area and calculate the eave height.

For individual flat roofs, the mean height of the LiDAR NHM pixels within it was used to estimate the floor numbers. The same height value was assigned to parcel sub-footprints within the same building footprint. The floor-to-floor height was set differently for buildings with different prototypes, according to a standard defined by the US Department of Energy.

Building footprint polygon and parcel polygon were overlaid to create a new polygon layer. Polygons in the new layer maintain the original boundary of building footprints as their outside boundary as well as contain the parcel boundaries within the building footprint. One building footprint may have multiple parcel sub-footprints, which can be explained by the fact that some commercial buildings are shared by different owners for different purposes of use.

A total of 10 building prototypes, which include 8 for commercial and 2 for residential buildings, were determined according to the parcel data for each parcel building sub-footprint. Each single parcel sub-footprint obtained a unique setting of parameters for simulation according to their own characteristics. Assigning building prototype in the first step of simulation would automatically affect the default setting of several major parameters such as weekday and weekend schedule, internal load (e.g., light and equipment use), activity area allocation, and cooling and heating equipment characteristics. Specific meteorological information will be retrieved and will contribute to the simulation based on the input of *location setting* and *year for simulation*. With the combination of prototypes, related parameters, area, shape, number of floors, age, location, and retrieved meteorological data, the annual electricity and annual fuel consumption were simulated for each parcel sub-footprint. Indianapolis and the year 2014 were chosen for the *location setting* and *year of simulation*, respectively. The energy consumption for each building was calculated by summing up energy consumption for all parcels within it. Figure 17.6 shows the amount of annual fuel and electricity consumption for each commercial and residential building in downtown Indianapolis.

17.4.3.2   Correlation Analysis

This section provides a correlation analysis to examine the relationship of energy use and building attributes including ground area, total floor area, height, surface area, compactness, aspect ratio, and orientation. The ground area represents the area of individual GIS building footprint. The total floor area represents the total floor space of all stories in the building, which was calculated by multiplying the ground floor area by floor numbers. The floor numbers were calculated by dividing the building height by the floor-to-floor height of the building, which is generally 7.5 ft in the case of residential buildings and 10 ft for commercial buildings (Chun 2010). Surface area (SA) represents the outer exposed building surface, including the roof surface and walls, can be calculated by the following equation:

$$SA = A + P * h, \tag{17.8}$$

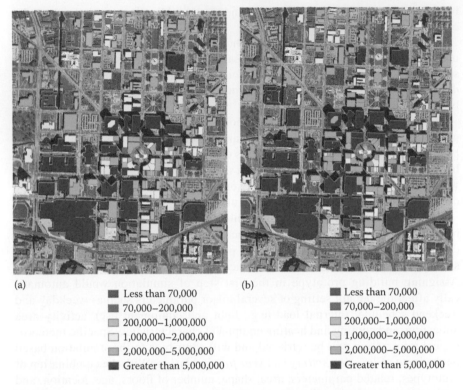

(a)

Less than 70,000
70,000–200,000
200,000–1,000,000
1,000,000–2,000,000
2,000,000–5,000,000
Greater than 5,000,000

(b)

Less than 70,000
70,000–200,000
200,000–1,000,000
1,000,000–2,000,000
2,000,000–5,000,000
Greater than 5,000,000

FIGURE 17.6 Amount of annual fuel consumption (a) and electricity (b) consumption (kilowatt hour) for each commercial and residential building in downtown Indianapolis.

where A denotes ground area, P is the perimeter of building footprint, and h is the height of the building. Compactness is the ratio of building surface to volume, and is a proxy for thermal energy emission from a building's exposed surface (Chun 2010). The aspect ratio is the proportional relationship between building width and height. Orientation, which is the main angle of building, ranges from $-90°$ to $90°$. The values $-90°$, $0°$, and $90°$ denote that the longer side of the building points to the west, north, and east, respectively. Figure 17.7 presents the average amount of annual electricity and fuel consumption for all commercial and residential buildings in the study area. Since a large contrast between the amounts of annual energy consumption in commercial and residential buildings can be observed, correlation analysis will be performed to these two major types of buildings separately.

The relationships between building energy consumption and their area are presented in Figures 17.8 and 17.9 as scatterplots. It can be found out that building energy use has a positive correlation with building ground area and total floor area. In order to avoid the impact from building area, the relationship of the other attributes and energy use per square foot is examined. Figures 17.10 and 17.11 present the relationship between annual energy consumption per square foot and building height, surface area, compactness, aspect ratio, and orientation in scatterplots.

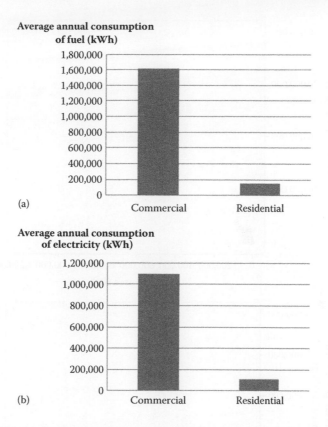

FIGURE 17.7 Average amount (kilowatt hour) of annual consumption of fuel (a) and electricity (b) for commercial and residential buildings in downtown Indianapolis.

Table 17.1 defines the correlation between energy use and all attributes using Pearson product–moment correlation coefficients.

A very strong correlation can be observed between building energy use and total floor area for both commercial and residential buildings with a correlation from 0.889 to 0.977. Relatively lower correlations are observed for the relationship between energy use and ground area, which indicates that energy use might have a positive correlation with building height. As Table 17.1 suggests, for commercial buildings, height has stronger positive correlations of 0.647 and 0.638 with fuel and electricity consumption per square foot, respectively, compared to residential buildings. Although it appears that height has a weak correlation with residential building energy use, a higher positive correlation between height and ground area explains the fact that residential building energy use has a higher correlation with total floor area than with ground area. It can also be found out that area has a stronger impact on energy use than height. It is also worth noting that energy use per square foot has a negative correlation with compactness for commercial buildings, indicating that commercial buildings with complicated shapes consume less energy than others with simple shapes and similar sizes in Indianapolis. No significant correlations can be observed between the energy use of buildings and their attributes, including surface area, aspect ratio, and orientation.

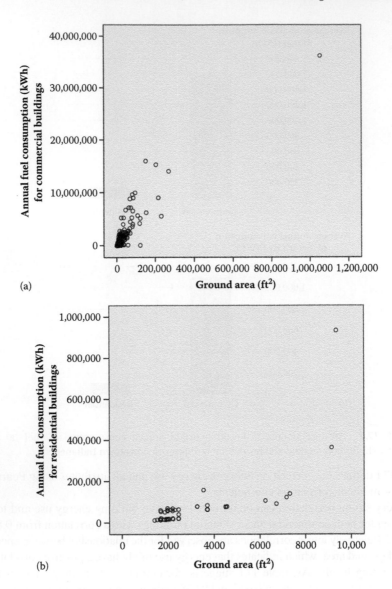

(a)

(b)

FIGURE 17.8 Relationship between annual fuel consumption and ground area (a and b) for commercial and residential buildings. *(Continued)*

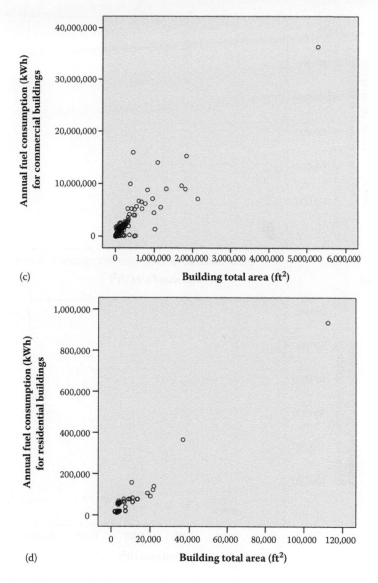

(c)

(d)

FIGURE 17.8 (CONTINUED) Relationship between annual fuel consumption and building total area (c and d) for commercial and residential buildings.

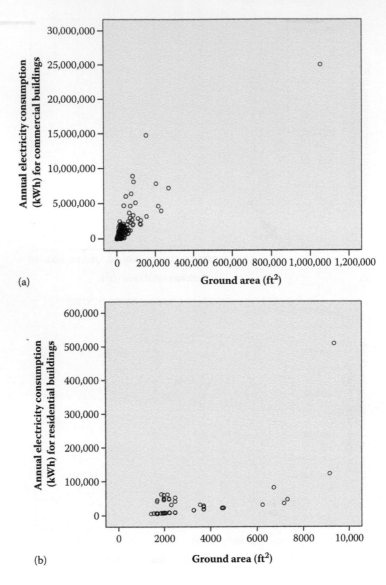

(a)

(b)

FIGURE 17.9 Relationship between annual electricity consumption and ground area (a and b) for commercial and residential buildings.   (*Continued*)

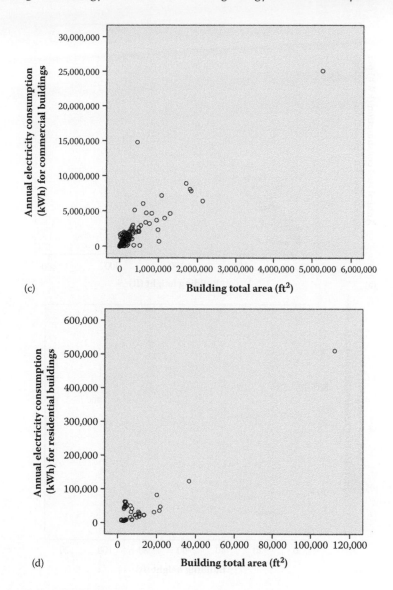

(c)

(d)

FIGURE 17.9 (CONTINUED)   Relationship between annual electricity consumption and building total area (c and d) for commercial and residential buildings.

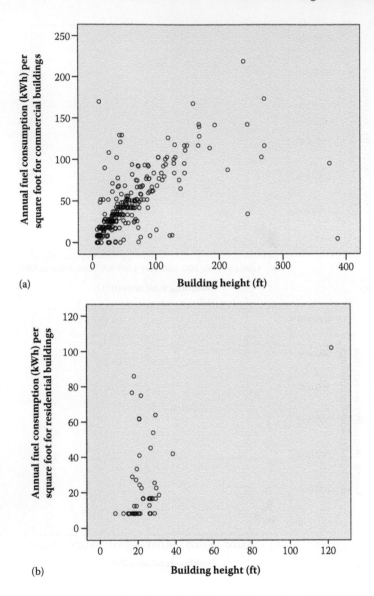

FIGURE 17.10   Relationship between annual fuel consumption per square foot and building height (a and b) for commercial and residential buildings.   (*Continued*)

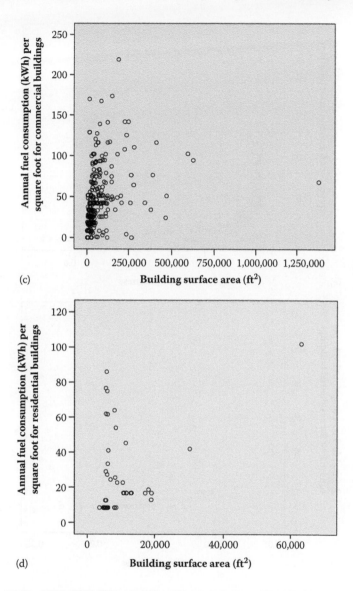

(c)

(d)

FIGURE 17.10 (CONTINUED) Relationship between annual fuel consumption per square foot and building surface area (c and d) for commercial and residential buildings.

(Continued)

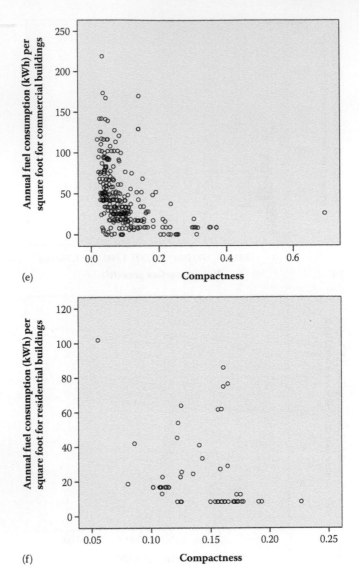

FIGURE 17.10 (CONTINUED)   Relationship between annual fuel consumption per square foot and compactness (e and f) for commercial and residential buildings.   *(Continued)*

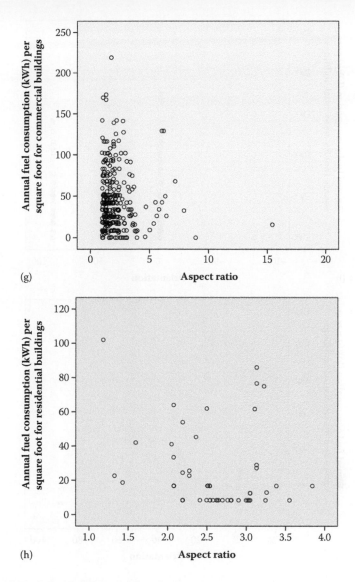

(g)

(h)

FIGURE 17.10 (CONTINUED)   Relationship between annual fuel consumption per square foot and aspect ratio (g and h) for commercial and residential buildings.   (*Continued*)

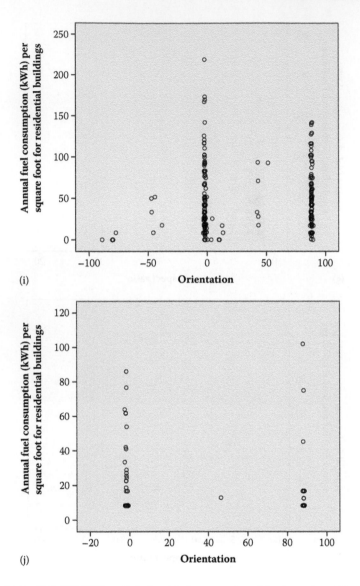

(i)

(j)

FIGURE 17.10 (CONTINUED)   Relationship between annual fuel consumption per square foot and orientation (i and j) for commercial and residential buildings.

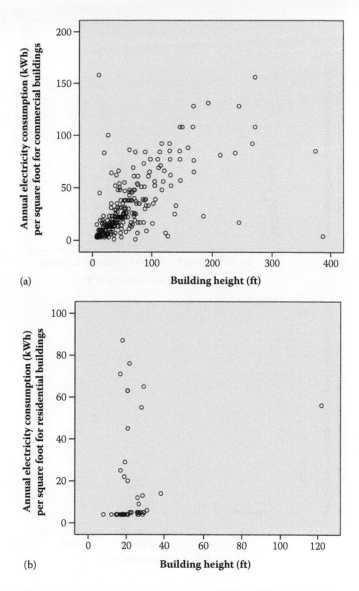

(a)

(b)

FIGURE 17.11   Relationship between annual electricity consumption per square foot and building height (a and b) for commercial and residential buildings.   *(Continued)*

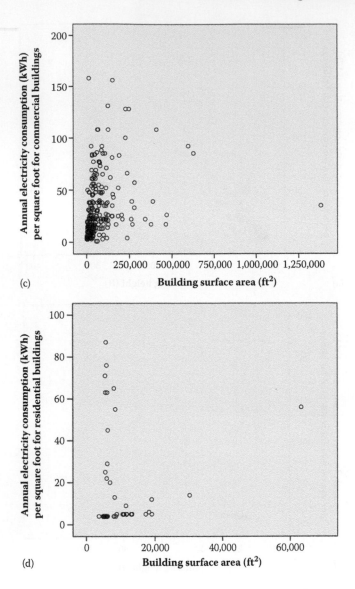

FIGURE 17.11 (CONTINUED) Relationship between annual electricity consumption per square foot and building surface area (c and d) for commercial and residential buildings.

(Continued)

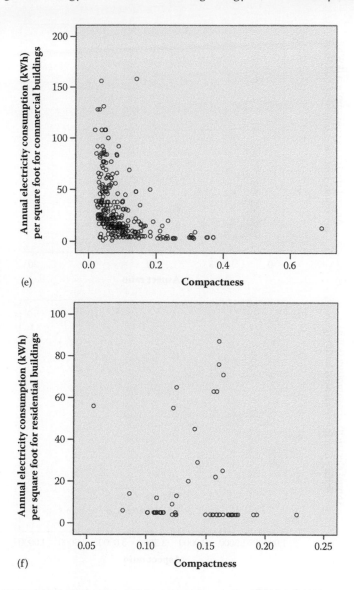

FIGURE 17.11 (CONTINUED) Relationship between annual electricity consumption per square foot and compactness (e and f) for commercial and residential buildings. *(Continued)*

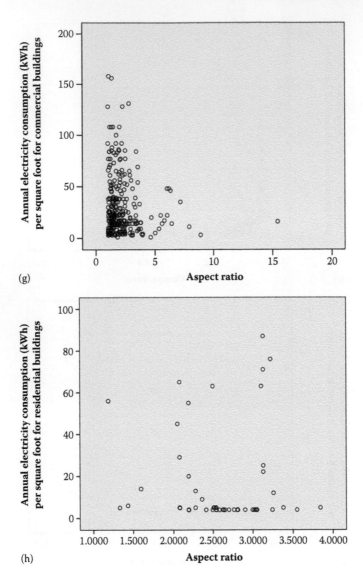

FIGURE 17.11 (CONTINUED) Relationship between annual electricity consumption per square foot and aspect ratio (g and h) for commercial and residential buildings. (*Continued*)

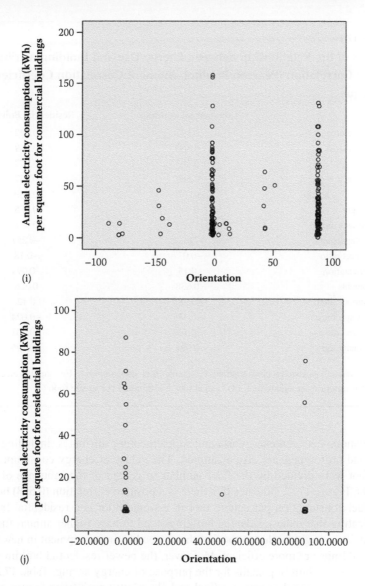

FIGURE 17.11 (CONTINUED) Relationship between annual electricity consumption per square foot and orientation (i and j) for commercial and residential buildings.

TABLE 17.1

Analysis of the Relationship between Energy Use and Building Attributes Using a Correlation (Pearson Product–Moment Correlation Coefficients) Technique

	Commercial Buildings	Residential Buildings
F vs. ground area	0.893	0.722
E vs. ground area	0.860	0.589
F vs. total area	0.903	0.977
E vs. total area	0.889	0.937
Ground area vs. height	0.076	0.659
FU vs. height	0.647	0.495
FU vs. surface area	0.3	0.411
FU vs. compactness	−0.412	−0.251
FU vs. aspect ratio	−0.077	−0.18
FU vs. orientation	0.15	−0.094
EU vs. height	0.638	0.226
EU vs. surface area	0.28	0.12
EU vs. compactness	−0.388	−0.024
EU vs. aspect ratio	−0.107	−0.002
EU vs. orientation	0.093	−0.196

Note: E, annual electricity consumption; F, annual fuel consumption; FE, annual electricity consumption per square foot; FU, annual fuel consumption per square foot.

Correlations between energy use and supplementary attributes, including building age and prototypes, are also examined. The values of energy consumption per square foot were divided by the floor number to get rid of the influence of building height. Figure 17.12 indicates that there is a positive correlation (0.558) between annual fuel consumption per square foot in a single floor and residential building age, indicating that older residential buildings tend to have higher annual fuel consumption. This might be due to the fact that space heating equipment in newer residential buildings are more efficient. Moreover, the newer residential buildings have a better design of building shells for the purpose of energy saving. Table 17.2 presents the average annual energy use (fuel and electricity) per square foot in a single floor for eight types of commercial buildings. It is worth noting that compared to other subtypes, office buildings have significantly higher annual fuel and electricity consumption per unit floor area. The possible reason is that office buildings have a higher density of people and more equipment than other buildings such as computers, printers, and space cooling and heating facilities. The other reason is that in US cities, office buildings in the downtown area tend to keep most of the internal lights on for the whole night.

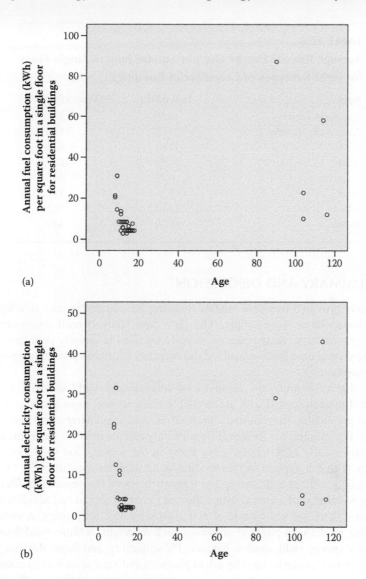

(a)

(b)

FIGURE 17.12 Relationship between annual fuel consumption per square foot in single floor and building age (Pearson's $r = 0.558$) (a) and between annual electricity consumption per square foot in single floor and building age (Pearson's $r = 0.331$) (b).

TABLE 17.2

Average Annual Energy Use per Square Foot in Single Floor for Eight Subtypes of Commercial Buildings

Types	Fuel (kWh)	Electricity (kWh)
Community center	19.56	15.52
Conference/convention center	26.70	18.84
Hotel	17.92	9.31
Office	51.51	48.29
Religious	9.23	4.75
Retail	9.76	5.61
School	11.34	5.25
Theater	6.49	3.97

17.5   SUMMARY AND DISCUSSION

This chapter provided two case studies focusing on sustainable city development, using Indianapolis as an example. The first case study overall demonstrated a method to estimate the solar potential of building roofs in densely populated urban areas. The second case study examined the correlation between building energy use and their attributes.

In the first case study, the accuracy of solar energy yield simulation can be improved if meteorological data are available since several input parameters, such as diffuse proportion, transmittivity proportion, and shadowing effect, are related to instant sky conditions. Seasonal-based analysis can also be performed since Indianapolis usually experiences heavy snow in the winter. One limitation of this case study is that it applied the same shadowing analysis algorithm for buildings and vegetation without considering the transmittivity of the vegetation. Moreover, regarding building wall material data, reflected radiation should be considered in the simulation if there are walls made of materials with high reflectivity. A completed 3D building model with physical attributes such as height, volume, total floor area, annual solar energy yield, shadow balance, PV suitability, and ownership has potential applications that can be used by urban planners and local solar energy companies since it will provide a multistage decision-making scheme regarding PV suitability. With the ownership parcel data, historical preserved buildings and government buildings can be removed from the list in the first stage, which can also save a lot of calculation time. Buildings with low annual solar energy yield can be removed in the next stage. In the final stage, urban planners can make predictions for each chosen roof surface (e.g., how many kilowatt hours of solar energy can be absorbed theoretically per square meter annually). Local solar energy companies and local governments will be able to determine the profits incurred for each particular building, each neighborhood and district, and the entire city if the cost of PV installation is considered.

In the second case study, the significant contributors to building energy use were determined to be area and age for residential buildings and area, height, compactness, and prototype for commercial buildings. For the purpose of saving energy, attention should be given to the management of office buildings since such structures consume more energy than others. Many office buildings can immediately save energy using some short-term solutions, such as turning off some of the lights, computers, office equipment, and space heaters during the night. Long-term solutions should also be provided, such as replacing the current lights with energy-saving lights to reduce internal loads. In order to save energy in residential buildings, the refurbishment of older buildings should be advocated. For example, the traditional space heating facilities in such houses can be replaced by energy-saving ones. Future studies on building energy use should consider more potential factors such as the income level and the number of residents.

REFERENCES

Aksoezen, M., Daniel, M., Hassler, U., and Kohler, N. 2015. Building age as an indicator for energy consumption. *Energy and Buildings* 87: 74–86.

Chun, B. S. 2010. Three-Dimensional City Determinants of the Urban Heat Island: A Statistical Approach. PhD diss., Ohio State University.

Fu, P., and Rich, P. 2002. A geometric solar radiation model with applications in agriculture and forestry. *Computers and Electronics in Agriculture* 37: 25–35.

Haala, N., Brenner, C., and Anders, K. H. 1998. 3D urban GIS from laser altimeter and 2D map data. *International Archives of Photogrammetry and Remote Sensing, Remote Sensing and Spatial Information Science* 32: 339–346.

Heiple, S., and Sailor, D. J. 2008. Using building energy simulation and geospatial modeling techniques to determine high resolution building sector energy consumption profiles. *Energy and Buildings* 40: 1426–1436.

Hemsath, T. L., and Bandhosseini, K. A. 2015. Sensitivity analysis evaluating basic building geometry's effect on energy use. *Renewable Energy* 76: 526–538.

Henn, A., Groger, G., Stroh, V., and Plumer, L. 2012. Model driven reconstruction of roofs from sparse LiDAR point clouds. *ISPRS Journal of Photogrammetry and Remote Sensing* 76: 17–29.

Huang, H., Brenner, C., and Sester, M. 2013. A generative statistical approach to automatic 3D building roof reconstruction from laser scanning data. *ISPRS Journal of Photogrammetry and Remote Sensing* 79: 29–34.

Lafarge, L., Descombes, X., Zerubia, J., and Deseilligny, M. 2007. Structural approach for building reconstruction from a single DSM. *IEEE Transactions on Pattern Analysis and Machine Intelligence* 6: 135–147.

Lukač, N., Zlaus, D., Seme, S., Zalik, B., and Stumberger, G. 2013. Rating of roofs surfaces regarding their solar potential and suitability for PV systems, based on LiDAR data. *Applied Energy* 102: 803–812.

Miliaresis, G., and Kokkas, N. 2007. Segmentation and object-based classification for the extraction of the building class from LIDAR DEMs. *Computers & Geosciences* 33: 1076–1087.

National Oceanic and Atmospheric Administration, http://www.noaa.gov/ (accessed September 20, 2015).

Philibert, C. 2011. *Renewable Energy Technologies Solar Energy Perspectives*. OECD/IEA.

Premrov, M., Leskovar, V. Z., and Mihalic, K. 2015. Influence of the building shape on the energy performance of timber-glass buildings in different climatic conditions. *Energy* 1–11.

Rich, P., Dubayah, R., Hetrick, W., and Saving, S. 1994. Using Viewshed models to calculated intercepted solar radiation: Applications in ecology. *American Society for Photogrammetry and Remote Sensing, Technical Paper* 524–529.

Sailor, D., and Lu, L. 2004. A top-down methodology for developing diurnal and seasonal anthropogenic heating profiles for urban areas. *Atmospheric Environment* 17: 2737–2748.

Schuffert, S. 2013. An automatic data driven approach to derive photovoltaic-suitable roof surfaces from ALS data. *Urban Remote Sensing Event* 267–270.

SEIA. Solar energy technologies solutions for today's energy needs, http://www.seia.org/sites/default/files/Solar%20Energy%20Technologies%20Overview%2011-13-2014.pdf.

Sobel, I. and Feldman, G. 1968. A 3 × 3 isotropic gradient operator for image processing. Presented at a talk at the Stanford Artificial Project.

Sohn, G., and Dowman, I. 2007. Data fusion of high-resolution satellite imagery and LiDAR data for automatic building extraction. *ISPRS Journal of Photogrammetry & Remote Sensing* 62: 43–63.

Tooke, T. R. 2014. Building energy modeling and mapping using airborne LiDAR. PhD diss., The University of British Columbia.

Tooke, T. R., Coops, N. C., and Christen, A. 2013. A point obstruction stacking (POSt) approach to wall irradiance modeling across urban environments. *Building and Environment* 60: 234–242.

Tooke, T. R., Coops, N. C., Voogt, J. A., and Meitner, M. J. 2011. Tree structure influences on rooftop-received solar radiation. *Landscape and Urban Planning* 102: 73–81.

U.S. Department of Energy, Building Energy Codes Program, https://www.energycodes.gov/development/commercial/90.1_models (accessed October 15, 2015).

Wang, Z. 1996. Manual versus Automated Line Generalization. *ESRI* White Paper Series* 94–106

Wang, Z., Zhao, Z., Lin, B., Zhu, Y., and Ouyang, Q. 2015. Residential heating energy consumption modeling through a bottom-up approach for China's hot summer–cold winter climatic region. *Energy and Buildings* 109: 65–74.

Ward, R., and Choudhard, R. 2014. A bottom-up energy analysis across a diverse urban building portfolio: Retrofits for the buildings at the Royal Botanic Gardens, Kew, UK. *Building and Environments* 74: 132–148.

Weng, Q. 2010. *Remote Sensing and GIS Integration Theories, Methods, and Applications.* McGraw-Hill.

Weng, Q. 2015. *Remote Sensing for Urbanization in Tropical and Subtropical Regions—Why and What Matters.* CRC Press.

Zhou, Y., Weng, Q., Gurney, K., Shuai, Y., and Hu, X. 2011. Estimation of the relationship between remotely sensed anthropogenic heat discharge and building energy use. *ISPRS Journal of Photogrammetry and Remote Sensing* 67: 65–72.

Index

Note: Page numbers ending in "f" refer to figures. Page numbers ending in "t" refer to tables.

Printed and bound by CPI Group (UK) Ltd, Croydon, CR0 4YY

01/11/2024

01782619-0012